Vaccine Cell Substrates 2004

Developments in Biologicals

Vol. 123

This series «Developments in Biologicals» begins with Vol. 102 and is the continuation of IABs symposia series «Progress in Immunobiological Standardization», Vols 1-5, «Immunobiological Standardization», Vols 1-22 and «Developments in Biological Standardization», Vols 23-101.

Organized and published by the International Association for Biologicals (IABs)

KARGER

Basel · Freiburg · Paris · London · New York ·
Bangalore · Bangkok · Singapore · Tokyo · Sydney

Vaccine Cell Substrates 2004

National Institutes of Health (NIH)
Doubletree Hotel, Rockville, MD, USA, June 29-July 1, 2004

Volume editors

Rebecca Sheets
NIAID/NIH, Bethesda, MD, USA

John Petricciani
International Association for Biologicals (IABs)
Geneva, Switzerland

Proceedings of a Symposium co-sponsored by the
National Institutes of Health
National Institutes of Allergy and Infectious Diseases and
International Association for Biologicals

35 figures and 51 tables, 2006

KARGER

Basel · Freiburg · Paris · London · New York ·
Bangalore · Bangkok · Singapore · Tokyo · Sydney

• • • • • • • • • • • • • • • • •

Developments in Biologicals

Library of Congress Cataloging-in-Publication Data

Vaccine cell substrates 2004 : National Institutes of Health (NIH), Doubletree Hotel, Rockville, Md., USA, June 29-July 1, 2004 / volume editors, Rebecca Sheets, John Petricciani.
p. ; cm. -- (Developments in biologicals ; v. 123)
"Proceedings of a conference co-sponsored by the National Institutes of Health, National Institutes of Allergy and Infectious Diseases and International Association for Biologicals."
Includes bibliographical references.
ISBN 3-8055-7999-3 (soft cover : alk. paper)
1. Vaccines--Congresses. 2. Human cell culture--Congresses. 3. Biological products--Congresses. 4. Cancer cells--Congresses. I. Sheets, Rebecca. II. Petricciani, John C. III. International Association for Biologicals. IV. National Institute of Allergy and Infectious Diseases (U.S.) V. Series.
[DNLM: 1. Viral Vaccines--Congresses. 2. Cell Culture Techniques--Congresses. 3. Cell Transformation, Neoplastic--Congresses. 4. Cells, Cultured--Congresses. 5. Drug Contamination--prevention & control--Congresses. W1 DE997VKF v.122 2005 / WC 500 V116 2006]
RS189.D45 vol. 123
[QR189]
615'.372--dc22

2005036046

Bibliographic Indices. This publication is listed in bibliographic services, including Current Contents® and Index Medicus.

Drug dosage. The authors and the publisher have exerted every effort to ensure that drug selection and dosage set forth in this text are in accord with current recommendations and practice at the time of publication. However, in view of ongoing research, changes in government regulations, and the constant flow of information relating to drug therapy and drug reactions, the reader is urged to check the package insert for each drug for any change in indications and dosage and for added warnings and precautions. This is particularly important when the recommended agent is a new and/or infrequently employed drug.

Worldwide distribution by S. Karger AG, Basel (Switzerland).
Printed in Switzerland by Médecine et Hygiène, Genève.
ISBN 3-8055-7999-3

Scientific Steering Committee

Rebecca Sheets	NIAID/NIH – Conference Co-chair
John Petricciani	IABs – Conference Co-chair
Robert Johnson	DMID/NIAID/NIH
Anthony Lubiniecki	Glaxo SmithKline
David Onions	BioReliance
James Robertson	NIBSC
Peter Wright	Vanderbilt University

Organization of the Meeting

Introduction	
Session I	Potential Oncogenicity of Cellular DNA
Session II	Potential Oncogenicity of Viruses Latent in Cells
Session III	Routine Adventitious Agent Tests – Can in Vivo Methods be Replaced?
Session IV	Level of Assurance Provided by Guidelines, Especially ICH Q5A and Q5D
Bonus Session	Experience with Novel Cell Lines
Session V	Bovine (and Porcine) Viruses
Session VI	Bovine Spongiform Encephalopathy

Contents

Petricciani J, Sheets R (eds): Vaccine Cell Substrates 2004.
Dev Biol (Basel). Basel, Karger, 2006, vol 123, p 1.

Introduction

Petricciani J, Sheets R (eds): Vaccine Cell Substrates 2004.
Dev Biol (Basel). Basel, Karger, 2006, vol 123, pp 3-7.

Introduction to Vaccine Cell Substrates 2004

R.L. Sheets

Commander, U.S. Public Health Service, NIH/NIAID, Bethesda, MD, USA

Cell substrates are the cells used in the manufacture of viral vaccines. Essentially all viral vaccines (except synthetic peptide vaccine candidates) are manufactured in some type of cell substrate. Vaccine cell substrates must, thus, be characterized, by the manufacturer, to ensure their suitability for use in the production of their product.

Why are cell substrates an issue requiring a three-day conference (and many in the past)? The risk of a production substrate to the safety and quality of biological products was most famously, and perhaps originally, identified in a case known as "a horse named Jim" – Jim was the name of a horse (an animal substrate – precursors to animal cell culture substrates) used to obtain diphtheria anti-toxin (a serum product). Unfortunately, Jim became infected with tetanus and thus, so was the last lot of diphtheria anti-toxin made from his serum. At least 13 children receiving this anti-toxin died of tetanus as a result of this contamination. This infamous tragedy led to the U.S. Congress approving the Pure Food and Drug Act of 1906, forming the precursor to today's Food and Drug Administration [1].

When considering the substrate used for the manufacture of vaccines, the primary consideration must be given to the clinical indications for vaccines – which are generally given to healthy people, most often infants and children, to prevent developing a disease in the future upon the possible exposure to the pathogen the vaccine is intended to prevent. Thus, vaccines are expected to be quite safe and not to contain contaminants that would risk the health of otherwise healthy individuals.

As viral vaccines began to be manufactured in cell cultures in the 1950s, another well-documented event occurred. It was discovered that some primary monkey kidney cell cultures were contaminated with a previously unidentified simian virus, SV40. This virus is endemic in the species (rhesus or cynomolgus) that were used to harvest kidney cells for production and thus does not result in obvious signs of viral infection, notably cytopathic effect (CPE). However, when passaged on to cells of cercopithecus origin, CPE was observed and thus a test for SV40 was established to test future lots of vaccine for possible contamination [2,3]. In fact, many tests developed in the 1950s and early 1960s to detect the contamination of cell substrates for poliovirus vaccines, measles, mumps, and rubella vaccines, smallpox vaccines and influenza vaccines are still in use today.

In addition, in the 1950s and early 1960s, scientists knew that human leukaemias occurred but had not yet identified human leukaemia viruses, even though animal leukaemia and sarcoma viruses had been identified. Thus, early cell substrate policy (the history of which will be further outlined in the next article by John Petricciani) developed limiting vaccine cell substrate choices to cells that were "normal". What "normal" meant was not clearly elucidated, but the concern was to prevent the contamination of vaccines from agents that could cause cancer or other latent diseases. Over time, "normal" has come to mean healthy primary cells and tissues, secondary cultures derived from these primary tissues, and diploid cell strains, which senesce over passage in culture. In the past two decades, this policy has evolved further. It has become recognized that primary cells, which were preferred in the 1960s, are problematic from the perspective of more common, non-oncogenic, adventitious contaminants and thus they are no longer considered an optimal cell substrate although they continue to be used safely to manufacture licensed vaccines. The definition of "normal" has evolved to mean non-tumorigenic. And immortalized but non-tumorigenic continuous cell lines (CCLs) are acceptable for vaccine production now. In fact, even tumorigenic CCLs are being explored for the production of well-purified investigational vaccines and are used to produce an inactivated split-virion influenza vaccine licensed in Europe. It is these latter continuous cell lines that formed the primary focus for this conference.

What has driven this evolution in policy are advances in science. There is a much better elaboration of the molecular pathology of cancer now than in the early 1960s. It is understood that cancer is a multi-step process requiring many factors, not simply a single event [4]. There have been advances in the state-of-the-art for testing for adventitious viruses. There are better methods for detecting specific oncogenic viruses, although there are still no validated broadly general (i.e. detects most or all families) screening methods for them. We now have many years of clinical experience with biological products of a variety of types manufactured safely in various cell substrates. Importantly, we have urgent public health needs to develop vaccines for viruses that do not necessarily propagate in previously acceptable cell substrates. For example, many HIV vaccine candidates are under development that are based on novel viral vector strategies and some of these will not propagate unless a cell line is engineered to complement the growth of the vector or to allow the vector to package into a non-replicating particle. Often, this engineering results in a cell line that is tumorigenic when inoculated into immunosuppressed rodents, even in the cases when the parental cell line was not derived from a tumor.

Cell substrates used to produce U.S. licensed vaccines include the following:

(i) simple (i.e. unicellular organisms) substrates, such as baker's or brewer's yeast, which is used to produce Hepatitis B vaccines;

(ii) primary cells or animals to produce vaccines such as measles, mumps, and influenza;

(iii) diploid cell strains, such as:

(a) MRC-5, used to produce varicella, polio, and Hepatitis A;

(b) WI-38, used to produce rubella, and

(c) FRhL-2, used to produce a rabies vaccine (and the licensed rhesus rotavirus vaccine which is no longer marketed); and

(iv) the only continuous cell line used to produce currently U.S. licensed vaccines is Vero cells, used to produce purified, inactivated poliovirus vaccines [5].

In addition to these cell substrates, additional substrates are used to produce investigational vaccine candidates or proposed candidates. These include other simple substrates, such as other species of yeast or E. coli, insect cell lines, other types of primary cells or animals, and other uses of continuous cell lines. Vero cells are being used to produce minimally purified live, attenuated viral vaccine candidates. Transformed cell lines, such as PerC.6 or 293 cells (and their derivatives), which are tumorigenic in immunosuppressed rodents, are being used to produce purified replication-incompetent viral vectors [6,7]. Even tumour-derived cell lines, such as HeLa cells, have been engineered to produce novel viral vectors proposed for investigational use [8].

The potential sources of contaminants to a viral vaccine include not only the cell substrate used in production, but also the viral seed(s), the cell substrate(s) used in the passage history of the viral seed(s), animal-derived or human-derived raw materials used in culture (e.g. in the culture medium) during the passage history or in production or in down-stream processing (e.g. albumin added as a virus stabilizer or antibodies used for affinity purification), personnel, facilities, and other production materials (e.g. flasks, bioreactors, etc.). The types of contamination a novel cell substrate may introduce include adventitious agents, tumorigenic potential (e.g. from an oncogenic element), or residual cellular DNA (which may result in infectivity from a latent provirus or episomal viral genome or in tumorigenic potential from an activated oncogene). In addition, concern has been expressed that viral elements in cellular DNA may recombine with the vaccine virus genome to form a novel recombinant virus. Whether this is likely to occur between virus families is only theoretical; however, it is possible that if a retroviral element were expressed, it could recombine with a defective retroviral vector being manufactured to produce an infectious recombinant retrovirus [9,10].

The following have previously been identified as potential sources of tumorigenic potential that could derive from a cell substrate. The cells themselves would be the most likely source of tumorigenicity; however, even the most minimally purified vaccines are generally filtered to remove potential bacterial contamination (although this is not always feasible) and this process can be validated to demonstrate the clearance of substrate cells. Cellular proteins (oncoproteins, growth factors, etc. but excluding PrP^c which may be self-replicating, but which is not tumorigenic) would not replicate and would be unable to exert any adverse effect for an extended period as they would be quickly degraded in the recipient. Cellular DNA as discussed above has long been the focus of concern. Finally, and most likely to be a realistic source of concern, are endogenous or adventitious oncogenic viruses [11].

Each manufacturer characterizes the cell substrates used in production in their facility. The characterization of cell substrates [12-14] includes assessment of the following: the history of isolation and banking, growth characteristics, karyology, identity testing (e.g. iso-enzyme analysis), tumorigenicity, and freedom from adventitious agents. Tumorigenicity assessments include tumour formation (progressively growing nodules and/or metastases) in immunosuppressed rodents and/or colony formation in soft agar. Tumorigenicity need not be performed on cells of rodent origin (as most are tumorigenic and so they are simply presumed to be so). Other cells are expected to pass the tumorigenicity test (i.e. not form progressively growing nodules in immunosuppressed rodents) as the current guidance documents are written.

Adventitious agent tests include: bacterial and fungal sterility (according to 21 CFR 610.12), mycoplasma (both cultivatable and non-cultivatable for all cells and spiroplasma for insect cells), mycobacteria (animals and/or culture), viruses (broadly general in vitro and in vivo methods) including acute (lytic, haemadsorbing, haemagglutinating) or latent (e.g., retroviruses or other oncogenic viruses), specific tests for retroviruses, other specific viruses (particularly by polymerase chain reaction [PCR]), and bovine or porcine agents as specified in 9 CFR 113. In addition, bovine-derived materials should be sourced from countries other than those listed by the USDA where BSE is known to exist or which represent an undue risk for BSE because their import requirements are less restrictive than would be acceptable for import into the U.S. (9 CFR 94.18).

The purpose of the conference was to inform the field of existing data and progress on addressing specific scientific issues related to cell substrates; to discuss the continued use of existing tests and the appropriateness of new ones at this time; and to develop consensus recommendations for the field to address these issues, either recommending implementation or identifying research gaps that preclude decision-making. The scientific topics covered in the conference included oncogenicity of cellular components, both latent viruses and cellular DNA, viral adventitious agent test methods, level of assurance provided by current tests, bovine (and porcine) viruses in (primarily) bovine-derived raw materials (particularly foetal calf serum), and bovine spongiform encephalopathy agents as potential contaminants of cell substrates. In addition, a bonus session included presentations on testing performed to ensure the safety and quality of specific novel vaccine cell substrates used for investigational vaccine production (or in one case, a European licensed vaccine). To meet these aims, a panel discussion was held on each topic area to address specific questions, and these are presented in subsequent sections of these Proceedings.

References

1 White Junod S: Biologics Centennial: 100 Years of Biologics Regulation. Update, bimonthly publication of Food and Drug Law Institute, 2002.

2 Sweet BH, Hilleman MR: The Vacuolating Virus, S.V.$_{40}$. Proc Soc Exp Biol Med 1960; 105: 420-427.

3 Katz SL: Efficacy, Potential and Hazards of Vaccines. New Engl J Med 1964; 270: 884-889.

4 Vogelstein B, Kinzler KW: The multi-step nature of cancer. Trends in Genetics 1993; 9: 138-141.

5 Package inserts for licensed vaccines indicate the cell substrate in which they are manufactured.

6 Nichols WW, Lardenoije R, Ledwith BJ, Brouwer K, Manam S, Vogels R, Kaslow D, Zuidgeest D, Bett AJ, Chen L, van der Kaaden M, Galloway SM, Hill RB, Machotka SV, Anderson CA, Lewis J, Martinez D, Lebron J, Russo C, Valerio D, Bout A: Propagation of Adenoviral Vectors: Use of PER. C6 Cells, in Curiel DT, Douglas JT (eds): Adenoviral Vectors for Gene Therapy. New York, Academic Press, 2002, pp 129-166.

7 Butman BT et al: Comprehensive Characterization of the 293-ORF6 Cell Line, in Petricciani J, Sheets R (eds): Vaccine Cell Substrates 2004. Dev Biol (Basel). Basel, Karger, 2006, vol 123, pp 227-235.

8 The International AIDS Vaccine Initiative Mid-Year Progress Report: January-June 2001, 2001, p 9.

9 Chong H, Vile RG: Replication-competent retrovirus produced by a "spit-function" third generation amphotropic packaging cell line. Gene Therapy 1996; 3: 624-629.

10 Weiss RA, Mason WS, Vogt PK: Genetic Recombinants and heterozygotes derived from endogenous and exogenous avian RNA tumor viruses. Virology 1973; 52 (2): 535-552.

11 Lubiniecki AS: Continuous Cell Substrate Considerations. Bioprocess Technology 1990; 495-513.

12 CBER's Points to Consider on the Characterization of Cell Lines Used to Produce Biologicals. CBER Guidance Documents, 1993.

13 Q5D: Derivation and Characterisation of Cell Substrates Used for Production of Biotechnological/Biological Products. International Conference on Harmonisation, 1998.

14 Q5A: Viral Safety Evaluation of Biotechnology Products Derived from Cell Lines of Human or Animal Origin. International Conference on Harmonisation, 1998.

Dr. R.L. Sheets, NIH/NIAID, 6700B Rockledge Dr., Rm. 5145, Bethesda, MD 20892, U.S.A.
E-mail: rsheets@nih.gov

Petricciani J, Sheets R (eds): Vaccine Cell Substrates 2004.
Dev Biol (Basel). Basel, Karger, 2006, vol 123, p 9.

SESSION I

Potential Oncogenicity of Cellular DNA

Petricciani J, Sheets R (eds): Vaccine Cell Substrates 2004.
Dev Biol (Basel). Basel, Karger, 2006, vol 123, pp 11-21.

Cell Substrates: Where do we Stand after 50 Years of Discussion?

J. Petricciani

CancerVax, Inc., Carlsbad, CA, USA

Abstract: The issue of which cells are acceptable as substrates for the production of biological products, and especially vaccines, has been with us in one form or another since the development of cell cultures in the 1950s.
The major cell substrate events that occurred over the past 50 years are reviewed briefly, including the outcomes of the seven most significant meetings. Cell DNA has been discussed for 30 years with incomplete resolution of some aspects of the issue. The currently available information on the potential oncogenicity of cellular DNA derived from continuous cell lines for which there is no evidence of either viral integration or extrachromosomal viral elements is reviewed. Issues associated with the continuing regulatory uncertainty about cell DNA are discussed.
A perspective is provided on the answers to four questions: (i) what do we know now about the DNA issue? (ii) what can we conclude from what we know now? (iii) what more information, if any, is needed to provide updated guidance? and how do we reach a consensus and updated guidance? And finally, a way forward is presented.

INTRODUCTION

The issue of which cells are acceptable as substrates for the production of biological products, and especially vaccines, has been with us in one form or another since the development of cell cultures in the 1950s. The fundamental issues are safety and acceptability, and safety has concentrated on the potential of a cell to transmit an infectious agent or other elements that could cause disease.

The first major decision about cell substrate acceptability occurred exactly 50 years ago in 1954, when a U.S. committee discussed which cells would be an acceptable substrate to use for the production of an experimental live adenovirus vaccine [1]. They had two fundamental choices: (i) HeLa, a human cancer cell line; or (ii) primary monkey kidney cells. Because they were worried about risks associated with a cell substrate derived from a human tumour, they decided to recommend primary monkey kidney cell cultures. That decision essentially ruled out as unsafe the use of cells other than those derived from normal tissues of animals in good health, and it set the tone for attitudes towards cell cultures in vaccine production for the next 30 years. And even now, after 50 years, it is obvious that we are still living in the shadow of that decision since a number of issues to be discussed at this conference derive from the 1954 decision. While it is an interesting intellectual exercise to speculate on where we would be today if the decision had gone the other way, it is purely hypothetical, so I will spare you that digression. Table 1 highlights the major cell substrate events along that 50-year pathway.

Table 1 : Major cell substrate events 1954-2004.

Time	Cell type	Product
1954	HeLa	Live adenovirus vaccine
1960s	Human diploid cells	Viral vaccines
1980s	• Tumorigenic cells – Animal (CHO) – Human (Namalwa) • Non-tumorigenic continuous cell lines – VERO	 – rDNA therapeutics – Interferon – Viral vaccines
1990s and 2000s	Human and animal tumour cell lines	Experimental vaccines

Although I was not at the 1954 meeting, I sometimes feel I was because of past discussions with people closer to it than I, and because of the number of times I have written and spoken on the issues surrounding cell substrate acceptability. During the past 50 years, there must have been more than 100 meetings to discuss various issues, and the specifics are well documented in the published literature. The most relevant of those meetings are listed in Table 2.

Table 2 : Major cell substrate meetings : 1954 – 2004.

Year	Meeting	Major outcome	Ref.
1954	Armed Forces Epidemiology Board	1° monkey kidney	1
1967	NIH	Consider human diploid cells, develop monkey diploid cells	2
1978	NIH	Consider alternate cell substrates (Namalwa)	3
1984	NIH/FDA	DNA, viruses, transforming proteins 10 pg DNA/dose	4
1986	WHO Study Group	DNA, viruses, transforming proteins 100 pg DNA/dose	5
1996	WHO ECBS	10 ng DNA/dose	6
1999	FDA, NIH, WHO, IABs	DNA risk issues unresolved	7

Why should one worry about DNA in the first place? From a theoretical point of view, cellular DNA in products might contain cancer cell genes, in the form of an activated oncogene, viral genes, or both. If that were the case, then cellular DNA in products might result in an oncogenic event or pathology of another type. It is also important to note that, based on theoretical considerations, residual cellular DNA from any cell substrate might give rise to: (i) an infectious event, or (ii) an oncogenic event. An oncogenic event could arise through insertional mutagenesis, resulting in activation, inactivation, up-regulation, or down-regulation of critical genes.

The 1984 conference was the first in-depth discussion of the three elements of risk with the objective of arriving at interim guidance [4]. In that regard, it is instructive to take a look at the DNA recommendation which states: "Procedures for production of biologicals must demonstrate that no cellular or other unwanted DNA molecules will be in the final product at a level which would have a biological activity. That is, activities which could induce changes of normal cellular processes. Until more information on the determination of the biological activities of DNA becomes available, a level of unwanted DNA in the pg range per dose appears acceptable. There were discussions about specific quantities of DNA that might be acceptable, and then, as an example, the currently accepted level of ten picograms of DNA per dose of polio virus produced in VERO cells was given as an example of the sort of thing that would be a good acceptable range." The 10 pg of DNA was provided as an example for guidance with the expectation that it would be revised when more in-depth consideration of the available data could be undertaken and new data could be considered.

A re-examination of the DNA risk issue took place in 1986 by a WHO Study Group who concluded that continuous cell lines are acceptable in principle, and that the primary concern should be viral safety with an emphasis on the elimination of potential viruses that could be pathogenic for humans [5]. DNA was of clearly lesser concern, and the group concluded that 100 pg/dose would provide an acceptable margin of safety when data are provided to validate the ability of the manufacturing process to achieve that level.

The last major meeting on cell substrates was held five years ago (1999), and cell DNA was among the issues discussed [7]. Following several presentations and a panel discussion, there was no consensus on the level of risk that should be attached to cell DNA as an impurity in a product. At that time, I stated to the group that if we failed to come to any conclusions, but asked for more data, we would be back together again in five years going over the same issues with essentially the same data.

The reason I said that is because I have been around long enough to know that the nature of the beast is to do nothing once the heat is off. One of my roles today is to turn up the heat to the point where, after 30 years of discussing DNA, we can begin to move to a resolution of at least some aspects of the issue and identify a clear pathway forward on those areas where resolution is not possible at this conference. For that reason, we have divided the DNA issue into two parts, the first of which is dealt with in Session 1, and is intended to review the potential oncogenicity of cellular DNA derived from continuous cell lines for which there is no evidence of either viral integration or extrachromosomal viral elements. The other part of the cellular DNA issue (those cases where a whole viral genome or viral elements are present) is covered in Session 2. Obviously, there is overlap between the two sessions since, for example, DNA uptake and processing is common to both situations. However, there are differences between the two types of cell DNA and our aim is to separate them so that recommendations can be made on each type rather than on cell DNA in general.

So here we are, just as I predicted, five years later with essentially the same issues on the table and precious little progress. In fact, very little has changed over the past 20 years on the DNA issue that might help resolve any lingering concerns, with the exception of the fact that purification technology has vastly improved. At the very least, it is an embarrassment for the regulatory process to have allowed this to drag on for so many years when the evidence has pointed to a clearly minimal risk. If we do not make progress at this meeting it will be a disgrace.

Impact of Risk Uncertainty on Product Improvement and Product Development

One of the outcomes of the continuing uncertainty about cell DNA is that it has acted as an impediment to product improvements as well as to research and development. Ultimately, the people who may benefit from products that might emerge from such efforts are the ones who suffer. Let me give two examples.

Rabies is a global disease with the heaviest burden in the developing world. Contemporary rabies vaccines for human use in industrialized nations are manufactured in cell culture systems. However, the vast majority of human exposures to rabies occur in developing countries. In India alone, over 500,000 people are treated for exposure to the virus every year. World wide, more than one million people receive rabies vaccines annually. To satisfy local needs, many developing countries have been producing their own vaccines derived from animal neural tissue, usually sheep brain, which often results in an immune response to neural-tissue antigens along with significant morbidity and mortality.

Because rabies is a disease of animals as well as humans, veterinary vaccines were developed to control animal reservoirs and have been used globally for many years. Most of those vaccines are produced in cell-culture systems such as BHK-21 and Vero, some of which are produced locally in developing countries. Why have cell-culture rabies vaccines for human use lagged so far behind those for animals in the developing world? The answer lies largely in the attitudes of many national regulatory authorities towards what is acceptable for human administration. These regulatory positions are entrenched in cell substrate issues, and in the industrialized countries' lingering reservations about using continuous cell lines for vaccines. Those concerns have been linked to the use of cells such as BHK-21 and Vero to produce rabies vaccines for humans.

While developed countries can indulge in the luxury of debating the technical and medical issues of benefit versus theoretical risk in relation to the use of various cell substrates, the tragedy is that those issues are essentially irrelevant to rabies vaccines because the manufacturing process includes a stringent inactivating step that has been shown not only to inactivate the rabies virus but also to destroy the biological activity of any contaminating cellular DNA in the vaccine. However, concern has been expressed that the DNA content of such vaccines will have to be reduced to an "acceptable" level. The other tragedy is that because developing countries look to the World Health Organization (WHO) and to the regulatory authorities of developed countries for guidance on the acceptability of specific products and manufacturing techniques, the uncertainty and lingering concerns over such things as cell DNA have had a spill-over effect of inhibiting the replacement of sheep brain rabies vaccines with those derived from cell culture.

The absurdity of that situation should be obvious to anyone in this room if faced with the need for a rabies vaccine in a developing country and had been given a choice between a veterinary rabies vaccine produced locally on BHK-21 or a sheep brain vaccine for humans. The choice is clear. This does not mean that developing countries should be relegated to using vaccines of lower quality or higher risk than the human vaccines in the developed world. What it does mean is that low-cost rabies vaccines derived from cells such as BHK-21 and VERO should be equally acceptable for human use in both the developed and developing world. A clear signal from a group such as this that a rabies vaccine produced in a continuous cell line could be acceptable for human use might result in significant improvements. Ultimately that would benefit very many people in both industrialized and developing countries.

Uncertainty and lack of uniform views on DNA risk also have a negative impact on new product development. For example, some regulatory agencies take the point of view that there is more risk associated with some cell substrates than others, while other agencies focus on the question of how well-characterized the cell substrate is, and how the manufacturing process addresses any elements of risk that might be associated with the product derived from that substrate. One of the potential consequences of encouraging the use of cell substrates that are perceived to be "lower risk" is the less efficient production of a new vaccine. That in turn has a direct impact on the potential for a commercially viable product. The issue of relative risk has been discussed on many occasions in the past, including the 1986 WHO Study Group. Their position was that the cell substrate itself was not the main issue, but rather that the focus of attention should be on the adequacy of the manufacturing process to address elements of risk that are presented by a given cell substrate. A uniform regulatory position on relative risk associated with the DNA from various cell types would be of significant step in removing uncertainty and inconsistency on the acceptability of various cell substrates so that new vaccine research and development can proceed more efficiently.

A Way Forward

To make progress on the cell DNA issue, it seems to me that there are really only a few basic questions that need to be answered:

1. What do we know now about the issue?
2. What can we conclude from what we know now?
3. What more information, if any, is needed to provide updated guidance?
4. How do we get to a consensus and updated guidance?

While these questions can be applied generally to cell substrate issues, I will concentrate on DNA and provide a perspective to each of the four questions.

WHAT DO WE KNOW NOW ABOUT DNA RISK?

Actually we already know a considerable amount, which is not surprising since we have been working on this for at least 30 years. Tables 3, 4, and 5 summarize the major pieces of information that we have in hand. The overall message is that there are no data in the literature to demonstrate that cell DNA has tumorigenic potential.

DNA from two out of 26 human tumour specimens tested could transform 3T3 cells. Complicating the interpretation of those data, however, was the finding that normal mouse and human DNA also scored positive on occasions, suggesting that 3T3 cells are already on the pathway to becoming transformed [8-10]. Nevertheless, there was some evidence that DNA from human cancer cells could cause changes in 3T3 mouse cells in vitro.

Table 3: Cell DNA can cause changes in 3T3 cells.

- 2/26 human tumour DNAs scored positive
- Normal mouse and normal human DNA scored positive
- 20 µg of high MW DNA (30 x 10^6) required
- Uptake facilitator required

Table 4: Human tumour DNA tested in vivo.

Study description	Ref.
a) Hybridoma – 250 µg (mice and rats)	11
b) HeLa – 100 µg (ATS newborn rat)	12
c) T-24 – 10 µg (ATS newborn rat)	12
d) T-24 – 1 mg (i.m., i.c., & i.v. in Rh monkeys)	13

Table 5: Human experience with tumour cells and tumour cell DNA.

Study description	Ref.
a) Adenovirus vaccine produced in HeLa cells	14
b) Human tumour cell inoculations	15, 16, 17
c) Blood	
– transfusion/leukaemia	18
– DNA in units of blood for transfusion	19
d) Melacine™ in melanoma (melanoma cell lysate)	20
e) Canvaxin™ in melanoma (irradiated melanoma cells)	21
f) Onyvax-P™ in prostate cancer (irradiated prostate cells)	22

Several human tumour cell DNAs have been tested in animal models for their ability to induce tumour formation. The results were uniformly negative and include in vivo studies of hybridoma DNA in which up to 250 µg did not induce tumours in mice [11]. Also, 100 µg of HeLa cell DNA, and 10 µg of T-24 cell DNA, which contains an activated *ras* oncogene, were negative in the anti-thymocyte treated newborn rat assay [12].

T-24 cell DNA in the form of chromatin was used in a non-human primate study in which approximately 1 mg of cell DNA was inoculated intramuscularly, intravenously, and intracerebrally into immunosuppressed Rhesus monkeys [13]. Follow-up examinations over an eight-year period failed to show signs and symptoms of tumour formation. Unfortunately, the monkeys were reassigned to an HIV programme at that point which precluded the types of follow-up studies that could be very helpful at this point.

An experimental live adenovirus vaccine in 1956 turned out in retrospect to have been produced in HeLa cells [14]. The vaccine, which had only minimal purification, almost certainly would have contained HeLa cell DNA impurities. Only six people were inoculated with that experimental vaccine, but a 25-year follow up showed no history of cancer in those six people.

During the 1960s a number of studies were reported in which human were inoculated with tumour cells. It is clear from human studies that individuals with an intact immune system are capable of rejecting heterologous human tumour cells even at inocula of 10^6 to 10^7 cells [15-17]. 10^4 cells of even very aggressive human tumour cells (Hep 3) failed to produce tumour nodules in humans, including patients with non-resectable disease. On the other hand, non-human primates immunosuppressed with antithymocyte globulin [23,24], and humans who are immunosuppressed because of their underlying malignancy, are able to support tumour nodule growth after the inoculation of 10^5 to 10^6 cells.

An epidemiological study conducted in the mid-1970s examined the recipients of blood from donors who subsequently developed neoplasms of the lymphatic or haematopoietic tissues [18]. The results showed that no recipient of the pre-cancerous cells had developed leukaemia or lymphoma after an average of seven years of follow-up. Since it was recently shown that blood for transfusion contains 75 to 450 µg of DNA in the plasma of a unit [19], it is reasonable to conclude that substantial amounts of human DNA from pre-malignant and malignant white blood cells, along with the cells themselves, were transfused into those patients. The fact that neoplastic disease did not develop in any of the transfusion recipients argues strongly that substantial amounts of DNA from pre-neoplastic cells, as well as the cells themselves, do not carry a measurable risk, at least during the first seven years after exposure.

More recently, three therapeutic cancer vaccines that contain substantial amounts of tumour cell DNA have been administered to patients. In the first case, Melacine™, which is a lysate of two melanoma cell lines, was administered to > 1,000 patients in Phase 2 and Phase 3 studies, and is now an approved product in Canada [20]. The other two vaccines are composed of three tumour cell lines that have been irradiated to render them replication-incompetent, but the cells remain viable. Canvaxin™ has been administered to approximately 3,000 late stage melanoma patients in Phase 2 and Phase 3 studies [21], and Onyvax-P™ has been administered to > 50 prostate cancer patients in Phase 1 and Phase 2 studies [22]. Those human exposures to tumour cell DNA may provide important information on risk.

Pre-clinical and clinical studies in the fields of DNA vaccines and gene therapy should provide quantitative information on the issue of how difficult it is to get exogenous DNA into human cells in vivo where it can be expressed. In addition, valuable information on the potential for insertional mutagenesis should contribute to the discussion of the risk of cellular DNA. These points will be covered by Dr. Cichutek later in this session. Data have recently been published on the uptake and processing of DNA that enters the gastro-intestinal tract in the form of food, and this will be mentioned by Dr. Strain.

Finally, recent data have been published that show foetal DNA is present in the maternal circulation at a level of from three to 300 genomes per ml of maternal plasma [25]. Most of the foetal DNA was < 313 base pairs in length in the plasma of the 23 pregnant women tested. This provides an example of several months of human exposure to foreign DNA on a very broad scale since most, if not all, pregnant women would be expected to be exposed to foetal DNA.

Table 6: Human experience with non-tumour DNA.

Study description
• DNA vaccines
• DNA in gene therapy
• DNA from ingested food
• Maternal and foetal DNA

What Can we Conclude from What we Know Now?

The clear message from the data that are available now is that if there is a risk associated with cell DNA, it has virtually been eliminated and must be considered theoretical. Perhaps the most important point from all this information is that the negative results are consistent among the various studies, whether they are in vivo animal models or human experience. And that, of course, is consistent with the theoretical calculations listed in Table 7 that have been made by several authors [22-27]. The low level of risk should not come as a surprise since, as Dr. Strain will discuss more fully, there are formidable barriers to the uptake and expression and/or integration of foreign DNA into mammalian cells.

Table 7: Summary of risk estimates of the potential for DNA to cause an oncogenic or infectious event.

Source for estimate of risk	Ref.	Type of risk	Estimated level of risk
1986 WHO Study Group	5	Oncogenic	1 in 2×10^{10}
1987 Petricciani and Regan	26	Oncogenic	1 in 1×10^{10}
1990 Temin	27	Oncogenic	1 in 1×10^{12}
1995 Kurth	28	Oncogenic	1 in 1×10^{12}
1997 Dortant	11	Oncogenic	1 in 5×10^{8}
1999 Krause and Lewis	29	Infectious	1 in 4×10^{9}

What More Information, if Any, is Needed to Provide Updated Guidance?

It has been a hallmark of most of the previous discussions on the cell DNA issue to conclude that there is not enough information to make firm decisions or to provide updated guidance. In the background, however, there has always been the additional issue of the psychological barrier associated with the topic. It is all too easy to say that there is not enough information. However, as I have already pointed out, that position is not serving the public well. When we talk about more information, I think it is important to divide it into two parts: (i) essential for decision-making and (ii) nice-to-have. The nice-to-have category is virtually endless and it keeps people occupied for long periods of time. I believe this is part of the problem. What we need to concentrate on are the essential additional pieces of information that are needed to reach some conclusions and provide updated guidance. This is not to say that the nice-to-have category is unimportant. For example, it would certainly be helpful to have DNA dose-response data. But that is not really essential at this point in the context of all the other information that is available. I would submit that there is no additional information that is essential at this point to be able to move forward on DNA from cells without evidence of an integrated virus or viral genes. Others, however, would argue that we need more data.

How do we Get to a Consensus and Updated Guidance?

I think it is clear from the past 30 years that developing a consensus that will lead towards updated guidance requires a mechanism that builds on the outcomes of a conference such as this and takes it to the next level. In that regard, I would like to propose that the conclusions and recommendations of this conference on the DNA issue be considered by WHO. A working group could be established to identify specific studies that should be undertaken to provide data that would form the basis for developing a true international consensus and updated regulatory positions. Doing so will be a very welcome step, and could begin the process to remove the inconsistencies that currently exist among regulatory agencies and WHO.

Summary

After 50 years of experience with cell substrates for the production of vaccines and other biological products, we have learned several important lessons. First, conventional wisdom is not always correct. That was certainly the case when primary cells were established as the preferred substrate for vaccine production. The theoretical concerns expressed about an oncogenic agent or element in human diploid cells did not translate into actual risk, and diploid cells eventually became the new substrate of choice. No study has yet demonstrated that tumour cell DNA can cause tumours in animal models, and all the currently available human data suggest that if there is a risk from cell DNA in biological products, it is extremely small.

It is possible to apply the well-established principles of rigorous cell substrate and product characterization to identify and deal with potential risk factors such as DNA using current technology. In some cases, special studies specific for the cell and vaccine may need to be undertaken. It is also important to note that in assessing the level of potential risk, the underlying assumptions should be examined critically to be sure that they are credible and realistic.

REFERENCES

1 Hilleman MR: Cells, vaccines, and the pursuit of precedent. NCI Monogr 1968; 29: 463-468.

2 NCI Monogr 1968; 29.

3 Petricciani JC, Hopps HE, Chapple PJ (eds): Cell Substrates. New York, Plenum, 1979.

4 In vitro cellular and developmental biology. Monograph No.6: Abnormal Cells, New Products, and Risk, in Hopps HE, Petricciani JC (eds): Tissue Culture Association, Gaithersberg, MD, 1985.

5 Report of the meting of the study group on biologicals (1987). WHO Technical Report Series, No. 747.

6 Requirements for cell substrates used for production of biologicals (1996). WHO Technical Report Series 1998, No. 878, 19-56.

7 Brown F, Lewis AM, Peden K, Krause P (eds): Evolving Scientific and Regulatory Perspectives on Cell Substrates for Vaccine Development. Dev Biol. Basel, Karger, 2001, vol 106.

8 Krontiris TG, Cooper GM: Transforming activity of human tumor DNAS. Proc Nat Acad Sci (USA) 1981; 78: 1181-1184.

9 Shih C et al: Transforming genes of carcinomas and neuroblastomas introduced into mouse fibroblasts. Nature 1981; 290: 261-264.

10 Shih C et al: Passage of phenotypes of chemically transformed cells via transfection of DNA and chromatin. Proc Nat Acad Sci (USA) 1979; 76: 5714-5718.

11 Dortant PM, Claassen JTM, van Krey CF, van Steenis V, Wester PW: Risk assessment on the carcinogenic potential of hybridoma cell DNA: implications for residual contaminating cellular DNA in biological products. Biologicals 1997; 25: 381-390.

12 Levinson AD, Svedersky LP, Palladino MA Jr: Tumorigenic potential of DNA derived from mammalian cell lines, in Hopps HE, Petricciani JC (eds): In vitro cellular and development biology. Monograph No. 6: Abnormal Cells, New Products and Risk. Tissue Culture Association, Gaithersburg, MD, 1985, pp 168-169.

13 Wierenga DE, Cogan J, Petricciani JC: Administration of tumor cell chromatin to immunosuppressed and nonimmunosuppressed nonhuman primates. Biologicals 1995; 23: 221-224.

14 Hilleman MR: Cell line saga – an argument in favor of production of biologics in cancer cells. Adv Exp Biol Med 1978; 118: 47-58.

15 Moore AE: Evidence of oncogenic properties of heteroploid cell lines – Discussion. NCI Monograph 1968; 29: 291-294.

16 Brunschwig A et al: Host resistance to cancer – clinical experiments by homotransplantants, autotransplants and admixture of autologous leucocytes. Ann Surg 1965; 162: 416-425.

17 Southam CM, Moore AE, Rhoads CP: Homotransplantation of human cell lines. Science 1957; 125: 158-160.

18 Greenwald P et al: Morbidity and mortality among recipients of blood from preleukemic and prelymphomatous donors. Cancer 1976; 38: 324-382.

19 Duxbury M, Jezuit M, Letwin B, Wright J: DNA in plasma of human blood for transfusion. Biologicals 1995; 23: 229-231.

20 Corixa Corp. Press release, 15 February 2001.

21 http://www.cancervax.com/vaccine/immune_data.htm#ongoing.

22 Onyvax Ltd. Press release, 17 June 2004.

23 Petricciani JC, Krichstein RL, Hines JE, Wallace RE, Marin DP: Tumorigenicity assays in nonhuman primates treated with antithymocyte globuline. J Natl Cancer Inst 1973; 51: 191-196.

24 Petricciani JC, Krichstein RL, Wallace RE, Marin DP: Assays for cell tumorigenicity in subhuman primates treated with antilymphocte globuline. J Natl Cancer Inst 1972; 48: 705-173.

25 Chang KCE et al: Size distributions of maternal and fetal DNA in Maternal. Plasma Clin Chem 2004; 50: 88-92.

26 Petricciani JP, Regan PJ: Risk of neoplastic transformation from cellular DNA: Calculations using the oncogene model, in Petricciani JC, Hennessen W (eds): Cells – Products – Safety. Dev Biol Stand. Basel, Karger, 1987, vol 68, pp 43-49.

27 Temin H: Overview of biological effects of addition of DNA molecules to cells. J Med Virol 1990; 31: 13-17.
28 Kurth R: Risk potential of the chromosomal insertion of foreign DNA. Ann NY Acad Sci 1995; 772: 140-151.
29 Krause PR, Lewis AM Jr: Safety of viral DNA in biological products. Biologicals 1998; 26: 317-320.

Dr. J. Petricciani, CancerVax, Inc., 2110 Rutherford Dr., Carlsbad, CA 92008, USA.
E-mail: jpiabs@aol.com

Petricciani J, Sheets R (eds): Vaccine Cell Substrates 2004.
Dev Biol (Basel). Basel, Karger, 2006, vol 123, pp 23-28.

The Uptake and Fate of Exogenous Cellular DNA in Mammalian Cells

A.J. Strain

School of Biosciences, University of Birmingham, Edgbaston, Birmingham, U.K.

Abstract: Mammalian cells take up exogenous DNA very inefficiently. However, in the absence of viral vectors, DNA can be transfected into cells by co-precipitation with calcium phosphate and usually also with carrier DNA or by lipofection or electroporation. Such DNA can be expressed efficiently by cells. Alternatively, direct injection can also result in uptake and expression of transgenes. Without carefully designed means to target DNA specifically to integrate into the host genome, the vast majority of internalised DNA remains extra-chromosomal and is degraded. The likely fate of DNA which in low levels may contaminate vaccines derived from mammalian cell lines, will also be destruction. There is a theoretical risk of DNA integration events with random sequences of donor-derived DNA but the probability of that leading to serious adverse effects to the host is extremely small.

Introduction

Because of the increasing use of mammalian cells for the development and production of biologicals, including recombinant proteins and vaccines, concern has been raised about the potential risks which may be associated with the use of mammalian cell lines. Although we are routinely exposed to a variety of external sources of DNA including food, infectious agents, foetal DNA in maternal blood and even plant DNA in pollen, none of these poses a particular risk. Thus it is only with the advent of gene therapy and the production of biologicals such as vaccines and recombinant proteins that we need to consider the risks associated with exogenous DNA which may exist and the harm it could cause.

In particular, the issue of residual contaminating DNA arising from cell lines used to produce biologicals is under close scrutiny. The choice of cells is important since the cells most readily amenable to multiple and continuous use, may be perceived as having the greatest risk. Thus, immortalised cells and malignant transformed cell lines will be associated with, for example, activated oncogenes. Therefore the risk that DNA from such a cell which may constitute one (or a series of) activated oncogenes being taken up by and integrating into the genome of cells of a recipient must be assessed.

This does of course present investigators with a paradox. Protocols must be developed which minimise this risk; in other words which keep the level of contaminating DNA to an absolute minimum, and which selects procedures or means of administration most likely to help cells destroy exogenous DNA. Conversely, for gene therapy to be effective, protocols must ensure maximum uptake and stability of DNA.

The questions to be answered include: (i) what is the likely nature of the residual DNA? (ii) what is the uptake efficiency by cells of this exogenous source and (iii) what happens to it if and when it does enter the cell? (iv) what is the efficiency of recombination events, and how does this affect the complex multi-component process of tumorigenesis? This latter question will be the subject of other sections of this monograph.

Uptake of DNA by Mammalian Cells

Overall, cells take up DNA very inefficiently. In the 1970s and 1980s, considerable effort was expended to determine the fate of DNA following transfection and treatment with a variety of facilitators. These included principle methods of co-precipitation with calcium phosphate, complexing with macromolecules including DEAE-dextran, or use of other reagents such as DMSO [1-3]. Without the facilitator, very little DNA was internalised. Additionally, it was quickly established that efficiencies were substantially enhanced by combining the target DNA with high molecular weight but sheared carrier DNA such as calf thymus, [4]. More recently lipofection has developed as the favoured method as well as electroporation [5,6].

All these treatments and routes of administration highlighted and confirmed the presumed route of entry, namely endocytosis and targeting to the lysosomal compartment. A number of treatments which were known to influence lysosomal function by raising the pH thereby altering the optimal conditions for nuclease activity were at the time shown to enhance efficiency further [7-9].

Much of this early work was carried out with viral DNA and often SV40 DNA using permissive cell lines and giving an infectivity read-out as end point [10]. By definition this would have been supercoiled as closed circular DNA and would be stripped of protein components. While providing worthwhile information regarding the efficiency of uptake, insight into the route of uptake and cellular handling, it was widely accepted that the information so derived would not reflect the position encountered when residual DNA from the nuclei of mammalian cells was the principal element.

In the event in practical terms, contaminant DNA deriving from cellular sources would be none of these. Indeed, it would almost certainly be complexed with proteins and might even retain the properties of chromatin-like structure with the likely effect that it would be more readily targeted to the lysosomes and destroyed by them. There are very few studies of the fate of DNA uptake when administered as protein/chromatin complexes. In one such study, however, human tumour-derived DNA/chromatin was inoculated into non-human primates by direct injection intramuscularly, intravenously and intracerebrally [11]. After eight years of monitoring, none of the animals showed any evidence of neoplastic disease.

There have now been many experimental gene therapy studies where DNA is taken up by cells and DNA can be monitored by expression studies [reviewed in 12]. That the plasmid DNA does enter the nucleus is assumed since it is highly unlikely that plasmid DNA containing promoters could be expressed anywhere else. However, very little is known about the nuclear uptake process. For gene therapy purposes it would be very helpful to understand this. The uncertainties further underline the inefficiency of the survival of DNA and the low risk it poses in the circumstances where DNA is contaminant and every effort is taken to reduce its presence and to favour its destruction (in contrast to gene therapies).

Cellular Nucleases

Surprisingly little is known about the nucleases which may degrade DNA other than those in the lysosomes. Lysosomal nucleases act as an effective barrier to DNA taken up by cells and have been reported to prevent efficient delivery to the nucleus [13,14]. Deoxyribonuclease II has recently been shown to be the major and effective lysosomal barrier to transfection of cells [15] and therefore equally effective at degrading contaminant DNA co-incidentally taken up by mammalial cells. Although there is a large number of nucleic acid processing enzymes in the nucleus of cells, including those involved in replication, recombination and repair [16], it is not clear whether a "protective role" can be ascribed to these enzymes.

If such enzymes are found to help destroy "foreign" exogenous DNA which enters cells, then this would have the effect of further reducing the risk associated with the uptake of contaminant DNA. Those striving to enhance the efficiency of gene therapy approaches, however, would have to consider how to overcome this potential natural barrier to stable DNA expression and/or integration.

Receptor-Mediated Endocytosis and Endosome Escape

DNA which enters cells under the experimental conditions described above, that is with the use of facilitators, and under the conditions which might prevail during exposure to contaminant DNA following vaccine administration, enters cells and is dealt with by the endocytic pathway. This acts as a highly efficient degradative pathway. We know that multiple pathways maintain functionally distinct endosomes and there is a sophisticated system of proteins which direct the endosomes to the appropriate site [17]. Endocytic compartments can contribute to processes such as recycling, storage, regulated secretion and membrane repair [18]. However, the view that the endocytotic pathway leads inevitably to lysosomal destruction is no longer held since the phenomenon of endosome escape is now quite well understood, at least in regard to viral uptake [19]. Understanding the molecular mechanisms of endosome escape will facilitate the development of future generations of non-viral vectors with better efficiencies of endosomal escape of non-viral vector DNA for successful gene therapy approaches. There is evidence that viruses can subvert these regulatory events [19], which lead to lysosomal destruction, and gene therapists are interested in exploiting these. At present, in the absence of viral machinery, the non-viral vectorologists have a problem in overcoming the natural barriers cells present, which re-inforces the low risk faced from chance contaminant DNA in preparations.

Nature of Residual DNA

The risks posed by residual DNA are subject to a number of important (and difficult to estimate) influences. One of the most critical is the size of the DNA fragments. Most will be sheared below that associated with a normal functional gene. However, there may be small sequence fragments which could incorporate promoters and if inserted may have a significant effect. The likelihood of a chance integration at a site which would have a serious consequence is small. It would be relatively straightforward to investigate the size of DNA fragments found in the biological samples in question. However, although the risk is difficult to estimate, it is probably very small indeed.

Oral Administration

Recent carefully conducted studies on the fate of DNA entering cells have included two reports of orally administered foreign DNA in cells of the gastrointestinal tract [20,21]. Using a variety of molecular techniques to monitor the uptake and progression of DNA, persistence in cells was not observed beyond the first day. Complexing DNA with high fibre diet, protamine or lipofectin did not extend the persistence [21]. In another study DNA fragments were found in the plasma but were short-lived, being detectable only for a few hours [20]. It can be argued that these studies may not have direct relevance to the risk assessment exercise which is pertinent because the specific role of these cells (following uptake from the lumen) is to digest and process. Nevertheless, they provide further evidence that mammalian cells are well equipped to handle "foreign" DNA.

Direct Injection and Fate of DNA

The route of administration for gene or DNA vaccine therapy is of course a major factor in the assessment of stability versus risk from adverse effects. Intravascular delivery is probably the favoured route for gene therapy, whereas direct (intra-muscular) injection is the preferred route for vaccination.

Injection of polyoma virus DNA complexed to calcium phosphate precipitates was described some years ago. This was found to be non-stable and degraded within four hours when injected directly into the liver or spleen [22]. While it is clear that naked DNA injected into muscle tissue of experimental animals does appear to remain detectable for some time, the rate of integration has been found to be extremely low. Ledwith et al [23] showed that although substantial amounts of free plasmid DNA remained six weeks after intramuscular injection, the frequency of integration was found to be at least three orders of magnitude below the spontaneous mutation rate. Thus, although as indicated earlier, DNA injected directly into muscle can be detected extra-chromosomally, the evidence that integration occurs with other than a minute frequency is weak [12].

To enhance the efficiency for in vivo gene therapy and to induce a higher frequency of integration following intramuscular integration, investigators have studied the plasmid sequence (by virtue of recombination signals), mode and route of administration, the nature of the transgene protein and use of adjuvants. Despite their best efforts, these and other studies have revealed little evidence for plasmid integration [24]. Thus, as far as plasmid DNA is concerned, plasmid integration in vivo is a theoretical concern [25], and the same is likely to be true for residual contaminant DNA currently under scrutiny. Furthermore, even resorting to complex protocols including "in vivo" electroporation treatment requiring direct exposure of the muscle, the frequency of integration was higher [24], although it was still very small.

There have been several attempts to assess and estimate the risk associated with contaminant DNA and its transforming potential. Inevitably there are many factors and variables which have to be taken into account. Temin worked carefully through the sequence of events following DNA entry into a cell and decided that the probability of a molecule surviving nucleases and entering a cell was lower than one in 10^{-10} pre DNA molecule [26]. Thus, in what was described as a "worst-case scenario", a maximum cumulative probability of having harmful effects is less than 10^{-16} to 10^{-19} per DNA molecule from a cell without activated proto-oncogenes [26]. Although it is difficult to put definitive figures on the risk, estimates have been confirmed by others [24,25,27-29].

In conclusion, none of the methodologies used to facilitate DNA 'infectivity' of cells, with the exception of viral vectors, is likely to be encountered under physiological conditions in persons receiving vaccines or other products containing cellular DNA. Therefore the risk of residual cellular DNA uptake and persistent expression must be considered to be exceedingly low.

References

1 Scangos G, Ruddle FH: Mechanisms and applications of DNA-mediated gene transfer in mammalian cells. Gene 1981; 14: 1-10.

2 Spandidos DA, Wilkie NM: Expression of exogenous DNA in mammalian cells, in Hames BD, Higgins SJ (eds): Transcription and Translation a Practical Approach. Oxford, IRL Press, 1984, pp 1-48.

3 Kawai S, Nishizawa M: New procedure for DNA transfection with polycation and dimethylsulfoxide. Mol Cell Biol 1984; 4: 1172-1174.

4 Strain AJ, Wallace AH, Wyllie AH: Enhancement of DNA-mediated gene transfer by high Mr carrier DNA in synchronised CV-1 cells. Biochem J 1985; 225: 529-533.

5 Friend DS, Papahadjopoulos D, Debs RJ: Endocytosis and intracellular processing accompanying transfection mediated by cationic liposomes. Biochim Biphys Acta 1996; 1278:41-50.

6 Zuhorn IS, Kalicharan R, Hoeskstra D: Lipoplex mediated transfection of mammalian cells occurs through the cholesterol-dependent clathrin-mediated pathway of endocytosis. J Biol Chem 2002; 277: 18021-18028.

7 Ohkuma S, Poole B: Fluorescence probe measurements of the intralysoosmal pH in living cells and the perturbation of pH by various agents. Proc Natl Acad Sci 1978; 73: 3327-3331.

8 Luthman H, Magnusson G: High efficiency polyoma transfection of chloroquine treated cells. Nuc Acids Res 1983; 11: 1295-1308.

9 Strain AJ: Inhibitors of ADP-ribosyl transferase enhance the transformation of NIH3t3 cells following transfection with SV40 DNA. Exptl Cell Research 1985; 159: 531-535.

10 Strain AJ, Wyllie AH: The uptake and stability of SV-40 DNA after calclium phosphate transfection of CV-1 cells. Biochem J 1984; 218: 475-482.

11 Wierenga DE, Cogan J, Petricciani JC: Administration of tumor cell chromatin to immunosuppressed and non-immunosuppressed nonhuman primates. Biologicals 1995; 23: 221-224.

12 Herweijer H, Wolff JA: Progress and prospects: naked DNA gene transfer and therapy. Gene Ther 2003; 10: 453-458.

13 Lechardeur D, Sohn KJ, Haardt M, Joshi PB, Monck M, Graham RW, Beatty B, Squire J, O'Brodovich H, Lukacs GL: Metabolic instability of plasmid DNA in the cytosol: a potential barrier to gene transfer. Gene Ther 1999; 6: 482-497.

14 Pollard H, Toumaniantz G, Amos JL, Avet-Loiseau H, Guihard G, Behr JP, Escande D: Ca2 + sensitive cytosolic nucleases prevent efficient delivery to the nucleus of injected plasmids. J Gene Med 3,153-164.

15 Howell DP, Krieser RJ, Eastman A, Barry MA: Deoxyribonuclease II is a lysosomal barrier to transfection. Mol Therapy 2003; 8: 957-963.

16 Mistra NC: Biological function of nucleases. In: Nucleases, Wiley, London, 2002, pp 255-287.

17 Katzman DJ, Odorizzi G, Emr SD: Receptor down regulation and multi-vesicular body sorting. Nat Rev Mol Cell Biol 2002; 3: 893-905.

18 Reddy A, Caler EV, Andrews NW: Plasma membrane repair is mediated by Ca^{2+}regulated exocytosis of lysosomes. Cell 2001; 106: 157-169.

19 Pornillos H, Garrus JE, Sundquist WI: Mechanisms of enveloped RNA virus budding. Trends Cell Biol 2002; 12:56-582.

20 Forsman A, Ushamekis D, Bindra A, Yun Z, Blomberg J: Uptake of amplifiable fragments of retrotransposon DNA from the human alimentary tract. Mol Gen Genomics 2003; 270: 362-368.

21 Palka-Santini M, Schwarz-Herzke B, Hosel M, Renz D, Auerochs S, Brondke H, Doerfler W: The gastro-intestinal tracts as a portasl of entry for foreign macromolecules: fate of DNA and proteins. Mol Gen Genomics 2003; 270: 201-215.

22 Dubensky TW, Campbell BA, Villarreal LP: Direct transfection of viral and plasmid DNA into the liver or spleen of mice. Proc Natl Acad Sci U S A 1984; 81: 7529-7533.

23 Ledwith BJ, Manam S, Troilo PJ, Barnum AB, Pauley CJ, Griffiths TG, Harper LB, Schock HB, Zhang H, Faris JE, Way PA, Beare CM, Bagdon WJ, Nichols WW: Plasmid DNA vaccines: assay for integration into host genomic DNA, in Brown F, Cichutek K, Robertson J (eds): Development and Clinical Progress of DNA Vaccines. Dev Biol. Basel, Karger, 2000, vol 104, pp 33-43.

24 Manam S, Ledwith BJ, Barnum AB, Troilo PJ, Pauley CJ, Harper LB, Griffiths TG 2nd, Niu Z, Denisova L, Follmer TT, Pacchione SJ, Wang Z, Beare CM, Bagdon WJ, Nichols WW: Plasmid DNA vaccines: tissue distribution and effects of DNA sequence, adjuvants and delivery method on integration into host DNA. Intervirology 2000; 43: 273-281.

25 Wang Z, Troilo PJ, Wang X, Griffiths TG, Pacchione SJ, Barnum AB, Harper LJ, Pauley CJ, Niu Z, Denisova L, Follmer TT, Rizzuto G, Ciliberto G, Fattori E, Monica NL, Manam S, Ledwith BJ: Detection of integration of plasmid DNA into host genomic DNA following intramuscular injection and electroporation. Gene Therapy 2004; 11: 711-721.

26 Temin HM: Overview of biological effects of addition of DNA molecules to cells, in Brown F, Griffiths E, Horaud F, Petricciani JC (eds): Safety of Biological Products Prepared from Mammalian Cell Culture. Dev Biol Stand. Basel, Karger, 1998, vol 93, pp 37-44.

27 Kurth R: Risk potential of the chromosomal insertion of foreign DNA: in DNA vaccines. Ann N Y Acad Sci 1995; 772: 140-151.

28 Meulien P: DNA Issues, in Brown F, Griffiths E, Horaud F, Petricciani JC (eds): Safety of Biological Products Prepared from Mammalian Cell Culture. Dev Biol Stand. Basel, Karger, 1998, vol 93, pp 57-60.

29 Ledwith BJ, Manam S, Troilo PJ, Barnum AB, Pauley CJ, Griffiths TG, Harper LB, Beare CM, Bagdon WJ, Nichols WW: Plasmid DNA vaccines. Investigation of integration into host cellular DNA following intramuscular injection into mice. Intervirology 2000; 43: 258-272.

Prof. A.J. Strain, School of Biosciences, University of Birmingham, Edgbaston, Birmingham B15 2TT, U.K.
E-mail: a.j.strain@bham.ac.uk

Petricciani J, Sheets R (eds): Vaccine Cell Substrates 2004.
Dev Biol (Basel). Basel, Karger, 2006, vol 123, pp 29-34.

Lessons Learned from Gene Therapy Concerning and the Use of Integrating Vectors and the Possible Risk of Insertional Oncogenesis

K. Cichutek

Paul-Ehrlich-Institut, Langen, Germany

Gene Transfer Products Include Plasmid DNA and Vectors Used as Investigational Preventive Vaccines Against Infectious Diseases

A number of clinical trials are being performed using vectored vaccines and nucleic acid vaccines intended to protect against infectious diseases. Nucleic acid vaccines used in vivo are typically bacterial plasmid DNAs prepared form *E. coli* cells encompassing an expression vector which harbours a CMV promoter driving the expression of a viral, parasitic or bacterial antigen gene and a bovine growth hormone gene-derived poly-adenylation site. Newer nucleic acid vaccines no longer contain antibiotic resistance genes because such genes are not recommended to be used in humans. Preventive vectored vaccines are viral vectors used in vivo, many of them inoculated s.c. or i.d. They mainly include replication-incompetent adeno-associated viral (AAV) vectors, adenoviral vectors and poxviral vectors encompassing an expression vector encoding one or several antigens derived from a human pathogen. The expression vector mediates the expression of antigens in the transduced cells, thus mediating the induction of an immune response.

The potential of a vector to integrate chromosomally varies with the variety of its properties. It depends on the amount of vector or plasmid DNA used, the route of inoculation and other circumstances. Plasmid DNA inoculated intra-muscularly via needle or gene gun is normally deemed non-integrating [1], whereas integration has been observed to a certain extent following electroporation of plasmid DNA into mouse muscle [2]. AAV vectors do not integrate to detectable levels following i.m. inoculation, but they do integrate in liver cells following i.v. inoculation [3,4].

Adenoviral vectors may integrate at a low level under certain circumstances, although adenovirus itself mainly keeps its genome in an episomal state.

From the databases publicly available [5] it can be ascertained that about 1 % of the about 6,000 humans enrolled in gene transfer clinical studies were healthy volunteers, most of whom probably received preventive vaccines. In addition, in about 2,000 of the subjects treated with gene therapy products adenoviral, poxviral or AAV vectors have been used, some of them probably also for preventive uses. Preventive vaccination thus is a small but considerable part of the human application of gene therapy products (Fig. 1). According to the European legal definition, gene therapy products include medicinal products for therapeutic, diagnostic and preventive use.

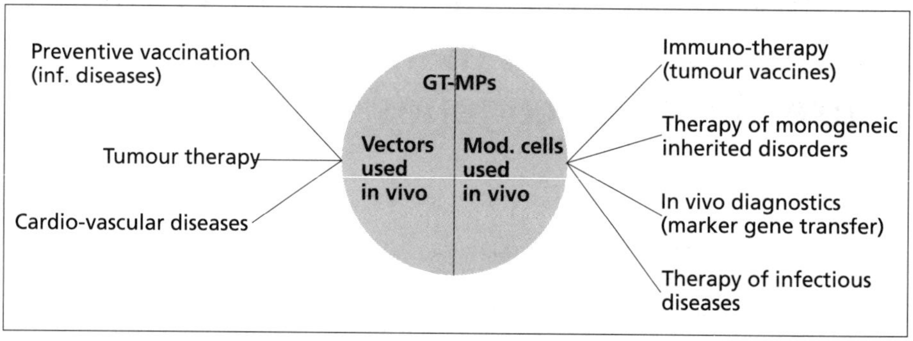

Fig .1: **Gene therapy medicinal products (GT-MPs) include preventive nucleic acid and vectored vaccines.** They are intended to be used in a variety of diseases as listed above.

Conversion of a Theoretical Risk to a Practical Risk: Leukaemias Induced in SCID-X1 Patients Following Transfer of MLV Vector-Modified Bone Marrow Stem Cells

In a gene therapy clinical trial performed in Paris on patients suffering from the inherited monogeneic immunodeficiency disease SCID-X1 (X1 type of severe combined immunodeficiency disease), three out of about 10 treated patients developed a T-cell leukaemia [6]. These two patients were among nine of the ten patients who were successfully treated. One additional patient did not benefit from the treatment. All patients enrolled in this trial had no other causative therapy options due to the non-availability of a haplo-identical donor for conventional bone marrow transplantation. The γc-chain gene known to be defective in SCID-X1 patients was transferred to alleviate the immunodeficiency by mediating the generation of functional immune cells in the treated patients.

Subsequent analysis of the molecular events possibly mediating leukaemia induction has provided evidence that all three leukaemias were indeed caused by chromosomal integration of the Murine leukaemia virus (MLV) vector used to transfer the functional γc-chain gene to the bone marrow stem cells that were subsequently re-infused [7]. More specifically, the MLV vector has been found to be integrated in the known human proto-oncogene in the leukaemic cells. Integration occurred 5´ of the LMO2 gene in the leukaemic cells of one patient and between exon 1 and 2 in the leukaemic cells of the other patient. Analysis of the crucial integration sites in the third patient is in progress. It has been demonstrated that the LMO2 allele suffering from the vector integration was activated by an enhancer mechanism involving the 3´ Long Terminal Repeat (LTR) of the MLV vector (Fig. 2). The proto-oncogene activation probably resulted in over-expression and deregulation of its expression. This is the first case of tumour induction in humans clearly linked to the chromosomal integration of a non-replicating vector. Chromosomal integration has been previously known as a theoretical risk of all integrating vectors and is mentioned as such in the relevant guidelines and named as a risk in the patient information leaflets of clinical trials involving the use of vectors to which this risk pertains. Due to this clinical observation integration has now been converted to a practical risk for MLV vectors.

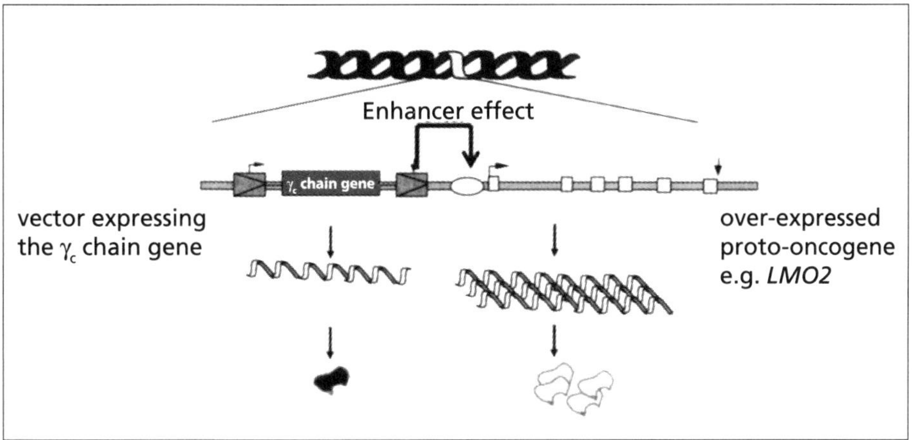

Fig. 2: **Enhancer mechanism-mediated activation of the human proto-oncogene LMO2 leads to transcriptional deregulation and over-expression.** Above is a model of the chromosomal integration of the MLV vector 5´ of LMO2 observed in the leukaemic cells of patient 5.

However, about 300 other patients who have been treated with MLV vector-modified human cells have so far not shown any signs of tumour development suspected to be related to vector use or integration [8]. It is currently believed that specific risk factors attributable to the SCID-X1 patient population and the specific gene therapy approach used have contributed to the leukaemia development. For example, a large number of cells (10^8) have been infused into those patients, a considerable percentage of which contained at least one copy of the MLV vector. Due to the SCID-X1 physiology, a high selection pressure is exerted on the modified cells that are parents of the functional B and T cells developing in the treated patients. There is also a possibility that the γc-chain gene transferred acts synergistically with the activated LMO2 gene to predispose the lymphocytic cells to neoplasia. Additional genetic alterations involved in the malignant cell transformation are supposed to play a role but have not yet been identified.

In conclusion, the leukaemia development following transfer of MLV vector-modified bone marrow stem cells to SCID-X1 patients shows that chromosomal integration of a single copy of an MLV vector expressing the γc- chain gene into blood progenitor cells may lead to insertional oncogenesis initiated by the insertional mutagenesis and activation of a proto-oncogene. The risk of chromosomal integration leading to insertional mutagenesis and activation of oncogenes must therefore be taken seriously [9]. The gene therapy example also shows that this risk, according to current knowledge, is not a general risk of all MLV vector treatment or all integrating vector usages although long-term follow-up will show the real frequency with which such events may be observed. The example given also shows that one activated oncogene may, under special circumstances, be sufficient for the development of some cancers.

Naked DNA Vaccination into Mouse Muscle Leads to Spontaneous Uptake of > 10^5 Copies into Cells in the Absence of Massive Chromosomal Integration

The 10 ng rule of the WHO says that vaccines manufactured in continuous cell lines (including tumorigenic cell lines) should contain not more than 10 ng residual cell DNA per vaccine dose. It is suggested, based on theoretical calculations, that keeping this limit will exclude any risk of oncogenicity attributable to residual cell DNA. This assumption is corroborated by clinical experience with licensed vaccines manufactured in tumorigenic cells where no tumours have been observed within a follow-up of more than 10 years or so. The theoretical risk assumption is mainly based either on insertional mutagenesis and activation of oncogenes or on direct oncogene transfer into human cells in vivo.

Preclinical toxicology tests performed to assess the risk of tumour development include inoculation of cellular DNA from the tumorigenic cell lines used for vaccine manufacture into mice. Often, the route of inoculation chosen for the preventive vaccines is also used for the oncogenicity test described. This seems appropriate according to current knowledge and has been accepted for licensing of some of the conventional vaccines on the market. However, such toxicology tests also suffer from a lack of appropriate positive controls included in the testing. It would therefore enhance the credibility of the toxicology testing performed if an animal test system including a positive control able to induce tumour development in vivo could be developed.

This issue has importance also because the nucleic acid and vectored vaccines currently tested in clinical trials are sometimes manufactured on tumorigenic cell lines such as HeLa cells or 293 cells. In some cases, as with adenoviral vectors, viral oncogenes like the adenovirus E1a gene are intentionally transfected into tumorigenic cell lines used for manufacture in order to complement the replication-defective adenoviral vector in these cells. Therefore, two classes of tumorigenic cells lines can be distinguished. One class includes continuously grown cell lines with a certain tumorigenic potential in mice for which the genetic principle which caused the cell transformation is unknown. The second class encompasses cell lines which contain and express known viral or activated cellular oncogenes. It is suggested to develop an appropriate animal model, in which the risk of tumour development in vivo following the inoculation of DNA form tumorigenic cell lines of both kinds, can be quantitatively assessed. It is not useful, at this point, to include such assays in batch release tests for vaccines manufactured in tumorigenic cell lines because they have not really been developed and tested for suitability. However, the suggested assay development would improve our knowledge of the real risk of tumour development due to residual cell DNA in vaccines and other medicinal products.

The qualification of tumorigenic cell lines as vaccine cell substrates may nowadays include inoculation of the cell DNA into two animal species such as, e.g., rats, hamsters or mice and an appropriate follow-up and histopathological examinations to exclude tumour formation. The relevant European document [10] suggests that a risk assessment be performed when tumorigenic cell lines are used as cell substrates. This could entail the experimental assessment of the oncogenicity of residual cell substrate DNA as just described.

Issues Related to the Preventive Use of Nucleic Acid Vaccines and Vectored Vaccines

Apart from the issue of oncogenicity of residual cell substrate DNA in preventive vaccines, additional issues related to the preventive use of nucleic acid vaccines and vectored vaccines include the risk due to chromosomal integration of the expression vector DNA itself and the risk theoretically attributable to other properties of tumorigenic cell lines such as, e.g., the presence of unknown viruses or microbes harboured by these cells. An issue not discussed here is the use of preventive vaccines in healthy volunteers intended to protect against an infectious agent that is not circulating in the geographical region in which such a clinical trial is planned. For all these issues, further discussion is needed to provide the necessary ethical and regulatory background for productive research and development of preventive vaccines based on gene transfer products. Many of the newer vaccines indeed belong to the group of gene therapy products. Some of these vaccines are desperately needed such as an HIV vaccine. The "CPMP Gene Therapy Working Party" at the European Medicines Agency EMEA may take some of these issues into account in order to support vaccine development.

References

1 Ledwith BJ, Manam S, Troilo PJ, Barnum AB, Pauley CJ, Griffiths TG 2nd, Harper LB, Schock HB, Zhang H, Faris JE, Way PA, Beare CM, Bagdon WJ, Nichols WW: Plasmid DNA vaccines: Assay for integration into host genomic DNA, in Brown F, Cichutek K, Robertson J (eds): Development and Clinical Progress of DNA Vaccines. Dev Biol. Basel, Karger, 2000, vol 104, pp 33-43.

2 Wang Z, Troilo PJ, Wang X, Griffiths TG, Pacchione SJ, Barnum AB, Harper LB, Pauley CJ, Niu Z, Denisova L, Follmer TT, Rizzuto G, Ciliberto G, Fattori E, Monica NL, Manam S, Ledwith BJ: Detection of integration of plasmid DNA into host genomic DNA following intramuscular injection and electroporation. Gene Ther 2004; 11 (8): 711-721.

3 Tenenbaum L, Lehtonen E, Monahan PE: Evaluation of risks related to the use of adeno-associated virus-based vectors. Curr Gene Ther 2003; 3 (6): 545-565.

4 Miller DG, Petek LM, Russell DW: Adeno-associated virus vectors integrate at chromosome breakage sites. Nat Genet 2004; 36 (7): 767-773.

5 Dreier G, Cichutek K: German Database of Gene Transfer Clinical Trials. Implementation of a central registry of gene transfer clinical trials in Germany. Bundesgesundheitsblatt Gesundheitsforschung Gesundheitsschutz 2004; 47 (6): 554-558.

6 Cavazzana-Calvo M, Hacein-Bey S, de Saint Basile G, Gross F, Yvon E, Nusbaum P, Selz F, Hue C, Certain S, Casanova JL, Bousso P, Deist FL, Fischer A: Gene therapy of human severe combined immunodeficiency (SCID)-X1 disease. Science 2000; 288 (5466): 669-672.

7 Fischer A, Abina SH, Thrasher A, von Kalle C, Cavazzana-Calvo M: LMO2 and gene therapy for severe combined immunodeficiency. N Engl J Med 2004; 350 (24): 2526-2527.

8 Bonini C, Grez M, Traversari C, Ciceri F, Marktel S, Ferrari G, Dinauer M, Sadat M, Aiuti A, Deola S, Radrizzani M, Hagenbeek A, Apperley J, Ebeling S, Martens A, Kolb HJ, Weber M, Lotti F, Grande A, Weissinger E, Bueren JA, Lamana M, Falkenburg JH, Heemskerk MH, Austin T, Kornblau S, Marini F, Benati C, Magnani Z, Cazzaniga S, Toma S, Gallo-Stampino C, Introna M, Slavin S, Greenberg PD, Bregni M, Mavilio F, Bordignon C: Safety of retroviral gene marking with a truncated NGF receptor. Nat Med. 2003; 9 (4): 367-369.

9 Gene Therapy Expert Group of the Committee for Proprietary Medical Products (CPMP): European Agency for the Evaluation of Medical Products. Insertional mutagenesis and oncogenesis: update from non-clinical and clinical studies. Gene Therapy Expert Group of the Committee for Proprietary Medical Products (CPMP), European Agency for the Evaluation of Medical Products — June 2003 meeting. J Gene Med 2004; 6 (1): 127-129.

10 CPMP Position Statement On the Use of Tumourigenic Cell Lines of Human Origin For the Production of Biological and Biotechnological Medicinal Products (CPMP/BWP/1143/00).

Dr. K. Cichutek, Paul-Ehrlich-Institut, Paul-Ehrlich-Strasse 51-59, DE-63225 Langen, Germany.
E-mail: cickl@pei.de

Petricciani J, Sheets R (eds): Vaccine Cell Substrates 2004.
Dev Biol (Basel). Basel, Karger, 2006, vol 123, pp 35-44.

Adaptation of the WHO Guideline for Residual DNA in Parenteral Vaccines Produced on Continuous Cell Lines to a Limit for Oral Vaccines

J.A. Lebron[1], P.J. Troilo[1], S. Pacchione[1], T.G. Griffiths[1], L.B. Harper[1], L.A. Mixson[3], B.E. Jackson[2], L. Michna[2], A.B. Barnum[1], L. Denisova[1], C.N. Johnson[1], K.L. Maurer[1], S. Morgan-Hoffman[1], Z. Niu[1], D.F. Roden[1], Z. Wang[1], J.J. Wolf[1], T.R. Hamilton[1], K.M. Laux[1], K.A. Soper[3], B.J. Ledwith[1]

[1] Biologics Safety Assessment, [2] Acute Toxicology, and [3]Biometrics Research, Merck Research Laboratories, Merck & Co., Inc., West Point, Pennsylvania, USA

Abstract: Although there is a WHO guidance for a limit on residual DNA for parenterally administered vaccines produced on continuous cell lines, there is no corresponding guidance for oral vaccines. To help determine an oral limit, we performed a study of Vero cell DNA uptake in rats, in which the relative uptake and persistence of Vero cell DNA administered orally was compared to its uptake when delivered intramuscularly (IM). The results of this study allowed the generation of an empirically derived IM versus oral factor (10^6) representing the relative inefficiency of DNA uptake by oral administration. This factor was then applied to the WHO recommended parenteral limit of 10 ng/dose to determine a corresponding upper limit on the level of residual Vero cell DNA for an oral vaccine of 10 mg. As a conservative approach, this empirically determined limit was reduced 100-fold to 100 µg. Thus, the results of this animal study, together with additional evidence in the literature, support a residual DNA safety limit of 100 µg per dose for an oral vaccine produced on a continuous cell line.

INTRODUCTION

An inherent consequence of vaccine manufacturing processes is the presence of residual cellular DNA from the host cells used for production. When the cells used are either primary cells or normal diploid cells, residual DNA is usually not an issue. However, when the cells are considered a continuous cell line, there is an inherent, but unlikely, risk that residual DNA from these cells might contain a cellular or viral oncogene, which in turn could be transmitted to the vaccine recipient [1]. Therefore, regulatory agencies and vaccine manufacturers closely monitor the amount of residual DNA in vaccines produced on continuous cell lines. For parenterally administered vaccines, the residual DNA limit was originally set at 100 pg/dose by a 1986 WHO Study Group [2]. Based on studies of polyomavirus DNA, this Study Group estimated that a 100 pg of cellular DNA would be equivalent to 0.5×10^{-10} of a tumour-producing dose, which for cellular DNA was calculated to be ~2 g of DNA. In 1998, the WHO Expert Committee concluded that levels of residual DNA up to 10 ng/dose are considered acceptable for parenterally administered vaccines produced on continuous cell lines [3]. This recent limit still provides a large safety margin (~2×10^8) when compared to the estimated cellular DNA tumour-producing dose. There is no guidance, however, on the limit of residual DNA in oral vaccines produced on a continuous cell line. In fact, the 1998 WHO Expert Committee concluded that the 10-ng limit for residual DNA does not apply to oral vaccines, and that acceptable limits should be set in consultation with national control authorities [3].

To establish a limit for residual DNA in an oral vaccine, we propose that the WHO's widely accepted limit for parenteral vaccines be adjusted by an appropriate factor that would reflect the relative inefficiency of DNA uptake by oral versus parenteral administration. Based on published studies of polyomavirus DNA infectivity [4], a conservative estimate for the reduced efficiency of oral versus parenteral administration would be 10^4. This factor was derived considering that supercoiled polyomavirus DNA was not infectious at the highest dose given orally (1 µg), while infectivity was observed at the lowest doses given parenterally (intra-peritoneal ID_{50} of 1.2×10^{-3} µg and subcutaneous ID_{50} of 1.3×10^{-4} µg). Thus, oral administration of polyomavirus DNA was at least (given negative results) 10,000-times less infectious than parenteral administration. When this 10^4 factor is applied to the WHO parenteral limit of 10 ng/dose, an oral limit of 100 µg/dose is estimated.

Although the experiments by Israel and his colleagues [4] shed light on what the approximate difference in DNA uptake efficiency might be for the oral and parenteral routes, these experiments used supercoiled polyomavirus DNA, not cellular DNA, and they used an immunological endpoint to quantify infectivity rather than directly quantifying the DNA. Thus, we felt that an additional comparative experiment would offer a more relevant oral versus parenteral factor.

Therefore, to generate this factor, we performed a Vero cell DNA uptake study in rats, in which the relative uptake of Vero cell DNA administered orally was compared to its uptake when delivered intramuscularly (IM). This study allowed the generation of an empirically derived IM versus oral factor (10^6) representing the relative inefficiency of DNA uptake by oral administration. When this factor is applied to the WHO recommended parenteral limit of 10 ng/dose, a new upper limit of 10 mg is generated for an oral vaccine. As a conservative approach, this empirically determined value was reduced 100-fold, resulting in a new residual DNA safety limit of 100 µg/dose of an oral vaccine produced on a continuous cell line.

Materials and Methods

Isolation of DNA from Vero cells

Vero cells used as the source of DNA were obtained at passage 135. DNA was extracted from these cells using a phenol : chloroform extraction procedure, which yielded a DNA preparation of excellent purity (A260/280 ~1.9) and high molecular weight (HMW, > 50 kb by agarose-gel electrophoresis).

Preparation of dosing solutions

Purified Vero cell DNA was prepared at 0.25 mg/mL in either an oral vaccine vehicle (containing sucrose, citrate, phosphate buffer, polysorbate-80, Ham's F12 and Medium 199) or in phosphate-buffered saline (PBS). Solutions were rocked overnight at room temperature to allow the viscous, HMW Vero cell DNA to form a homogeneous solution. DNA concentration was confirmed by spectrophotometry before dosing. The oral vaccine vehicle was supplied by Pharmaceutical Research & Development Department at Merck. PBS vehicle was obtained from Invitrogen Corp.

Administration of Vero cell DNA to rats

Sprague-Dawley rats (59 days-old at study start) were used in this study. On Study Day 1, two groups of 12 rats/sex received 125 µg of Vero cell DNA in the oral vaccine vehicle or in PBS by oral feeding or IM administration, respectively. During oral dosing, rats were carefully restrained and dosed very slowly, ensuring that the entire dose was administered into the oral cavity. Rats received 0.5 mL (total volume for oral feeding or 0.25 mL/quadriceps for IM injection) of the corresponding dosing solutions. Another group of four rats/sex remained undosed and was used as a control. All animal studies described here were approved by the Merck Research Laboratories Institutional Animal Care and Use Committee.

Necropsies and tissue collection

On Study Days 2, 3, 4, and 8, blood was collected from the orbital sinus of one untreated control rat/sex and three rats/sex each in the IM- and orally-dosed groups before necropsy. Blood samples were frozen in liquid nitrogen, and stored at approximately -70°C until processed for DNA isolation. Following blood collection and recovery from anaesthesia, rats were euthanized with CO_2. Before necropsy, each rat was treated with DNAZap™ (Ambion®) to remove any potential Vero cell DNA that may have contaminated the skin of the rats. As an attempt to increase the effectiveness of this treatment, the entire body of each of the rats was shaved the day before dosing and the day before the final necropsy (Study Day 7), to remove as much hair as possible. After DNAZap™ treatment, rats were exsanguinated and necropsied inside biosafety cabinets. Tissues were removed in the order listed in Table 1, individually rinsed in PBS, frozen in liquid nitrogen, and stored at approximately -70°C. Contamination control measures included the use of 10 % Clorox to decontaminate instruments after each tissue collection, trimming of tissues on clean weigh boats, frequent glove changes, and decontamination of the biosafety cabinets after each necropsy.

Isolation of DNA from tissues

Total genomic DNA was isolated from individual animal tissue samples. For all tissues, except blood, genomic DNA was prepared using the QIAGEN BioRobot EZ1 DNA Purification System (Qiagen). For blood samples, white cells were lysed, and the DNA was extracted with phenol : chloroform and precipitated with isopropanol. Each DNA preparation was analysed by spectrophotometry and agarose-gel electrophoresis to ensure its suitability for PCR. All DNA preparations were found to be of excellent purity (data not shown).

Quantitation of Vero cell DNA

TaqMan® PCR analysis targeting a 166-base pair (bp) long segment (Table 2) of the African Green Monkey (AGM) alpha satellite DNA (a 171 bp repetitive sequence found in AGM genomic DNA at a frequency of about five million copies/genome [5]) was used to quantitate Vero cell genomic DNA levels in all tissue DNA samples.

Before PCR, DNA samples were heat-denatured at ~95°C for ~15 minutes. Alpha-166A TaqMan® PCR reactions were performed in a final volume of 50 µL, containing TaqMan® Core Component Kit reagents (Applied Biosystems), 200 nM probe, and 180 nM of each primer. Amplification and real-time fluorescence detection were performed using the ABI Prism 7700 and 7900HT Sequence Detectors.

In each TaqMan® experiment, a standard curve consisting of 10-fold serial dilutions from 10^5 down to 0.1 fg of Vero cell DNA per µg of control rat DNA was assayed along with the test samples. The level of vector DNA in a sample was determined by interpolation of the sample's cycle threshold (Ct) value to the standard curve. This assay has a limit of detection of 0.001 fg (1×10^{-18} g) of Vero cell DNA per µg of rat DNA, and a limit of quantitation that is 10-fold higher (0.01 fg/µg). Spike experiments, in which known amounts of Vero cell DNA were added to sample DNA preparations before PCR, demonstrated the absence of PCR inhibitors in all the tissues assayed (data not shown).

Table 1 : Tissues collected during each necropsy.

Route of administration	
Intramuscular[a]	**Oral**[a]
Blood	Blood
Skin above quadriceps	Thymus
Quadriceps	Heart
Hamstrings	Oesophagus
	Lungs
	Liver
	Mesenteric lymph nodes
	Duodenum
	Jejunum
	Ileum
	Colon
	Caecum
	Stomach
	Tongue
	Mucosae[b]
	Inguinal lymph nodes
	Brain

a All tissues were collected on the control rats.
b Mucosae = Combined pharynx and oral mucosae.

Table 2 : TaqMan® PCR primers and probe sequences used for the quantitation of Vero cell DNA in rat tissues.

Segment	Oligonucleotide Primers/Probe	Name	Size
Alpha-166A	5'- GAA ACT GTT CTG TGT TCT GTT AAT TCA TC-3'	Alpha-MF5 (Forward Primer)	166 bp
	5'- 6FAM- ACA GTT ACA TCT TTC CC-MGBNFQ-3'	Alpha-MPr3 (Probe)	
	5'- AGC TTC TTT CCA GTT TTG AAG GG-3'	Alpha-MR5 (Reverse Primer)	
MGBNFQ = Minor groove binder, non-fluorescent probe from Applied Biosystems			

RESULTS

Following administration of a single dose on Study Day 1 of 125 μg of Vero cell DNA, given by either the oral or IM routes, rats were euthanized on Study Days 2, 3, 4, and 8 (1, 2, 3, and 7 days post-dosing, respectively), decontaminated with DNAZap™, and necropsied to collect the tissues listed in Table 1. Genomic DNA was then isolated from each of these tissues, and total uptake of Vero cell DNA was quantified using a highly sensitive TaqMan® quantitative PCR assay specifically developed to quantify AGM (Vero cell) DNA. TaqMan® PCR results for all the tissue DNA samples assayed in this study were then converted to the corresponding amount of Vero cell DNA per tissue. This conversion is needed to account for the difference in DNA extraction yields and tissue weights for each of the different samples collected, so that the average total amount of DNA per rat can be used when calculating the Vero cell DNA IM versus oral ratio. To obtain the amount of Vero cell DNA in each tissue, the TaqMan® PCR results were multiplied by the DNA extraction yield (expressed in μg of rat DNA per mg of tissue and assuming a 100 % recovery) for the corresponding tissue type, and by the average tissue weight (mg), as shown in Formula 1 below. The average DNA yield and the average tissue weights used for these calculations were generated during a previous study (data not shown).

Formula 1

$$\frac{\text{fg of Vero DNA}}{\mu\text{ of rat DNA}} \; \text{X} \; \frac{\mu\text{g of rat DNA}}{\text{mg of tissue}} \; \text{X} \; \text{Average tissue Wt (mg)} = \text{fg of Vero DNA in the tissue}$$

The DNA uptake level was markedly higher in the IM group than in the oral group at each of the four time points examined, and the levels decreased with time for both routes of administration (see Table 3 and Fig. 1). As shown in Table 3, the total average amount of Vero cell DNA detected in the tissues of orally-fed rats was only 1,169 fg on Study Day 2. Not surprisingly, ~95 % of this signal was detected in the oral cavity and the gastrointestinal (GI) tract tissues, as they represent the route of administration. The rest of the signal resulted from a single inguinal lymph node sample in one rat and a lung sample from another rat. When these values are averaged among the six rats in this time point, the resulting amount accounts for the other ~5 % of the total signal. It is possible that these two samples may have been positive due to a spurious contamination event since none of the blood samples from the orally-fed rats were positive at any time during the study, and since no other systemic organ was positive. However, even though this is the most likely explanation, as a conservative measure, these results have been included in the calculations. By Study Day 3, the average total signal detected was only 3.1 % of that detected on Study Day 2. This 33-fold reduction in the amount of DNA indicates that the DNA was being rapidly cleared. By Study Day 4, the DNA has been completely cleared, as no signal was detected either on Study Day 4 or on Study Day 8.

For the IM-dosed rats, the total average amount of Vero cell DNA detected in the tissues was 152,002,173 fg on Study Day 2, an amount that is over 130,000 times higher than that detected in the orally-fed rats at the same time point. By

Table 3: Summary of average Vero cell DNA levels in the various tissue groups for each time point and for each route of administration.

Route	Tissues	Average fg of Vero cell DNA in the tissues[d]			
		SD2	SD3	SD4	SD8
Oral	Oral cavity and GI tract[a]	1,107	36	0	0
	Systemic organs[b]	62	0	0	0
	Blood	0	0	0	0
	Total	1,169	36	0	0
IM	Injection site[c]	152,002,173	15,403,145	3,410,524	1,009,313
	Blood	0.25	0.45	0.02	0
	Total	152,002,173	15,403,145	3,410,524	1,009,313

a Oral cavity and GI tissues = Tongue, pharynx/oral mucosae, oesophagus, stomach, jejunum, duodenum, ileum, caecum, and colon.
b Systemic organs = Mesenteric lymph node, inguinal lymph node, brain, heart, liver, lungs, and thymus.
c Injection site = Skin over quadriceps, quadriceps, and hamstrings.
d Average value of N = 6 (3 rats/sex) for each time point and for each route of administration.
GI = Gastrointestinal.
IM = Intramuscular.
SD = Study Day.

Study Day 3, ~90 % of the DNA detected on the previous day had been cleared. Subsequently, this amount was further reduced by 78 % by Study Day 4, and then by an additional 70 % by Study Day 8.

The amount of Vero cell DNA across all tissues was then added to determine the average amount of Vero cell DNA per sex and the average of the combined sexes, per time point, for both routes of administration. Table 4 summarizes the results of these calculations.

These values were then used to calculate the safety limit for residual DNA from a continuous cell line in an oral vaccine. This safety limit was defined prospectively for this study as the product of the 10 ng WHO residual DNA limit for parenteral vaccines multiplied by the resulting ratio of the area-under-the-curve (AUC) for the last three time points (Study Days 3, 4, and 8) of the study for the IM route, divided by the AUC for the same time range for the oral route. Since IM and orally administered DNA will be either degraded, taken up systemically, or excreted, the last three necropsy time points were chosen to allow the DNA to be committed to one of these three paths, as it was expected that this might take more than 24 hours. The first time point was included in the study to evaluate the DNA clearance kinetics. However, for informational purposes, the AUC and the IM versus oral ratio were also calculated using all four time points, and were included in this report.

To calculate the AUC for either all four time points ($AUC_{[SD2-8]}$) or for only the last three time points ($AUC_{[SD3-8]}$) for both routes of administration, the values shown in Table 4 were used to generate area plots (Fig. 1) with a specialized mathematical computer software (Sigma Plot 8.0). This software was then used to determine the corresponding AUCs. The results of these calculations are summarized in Table 5.

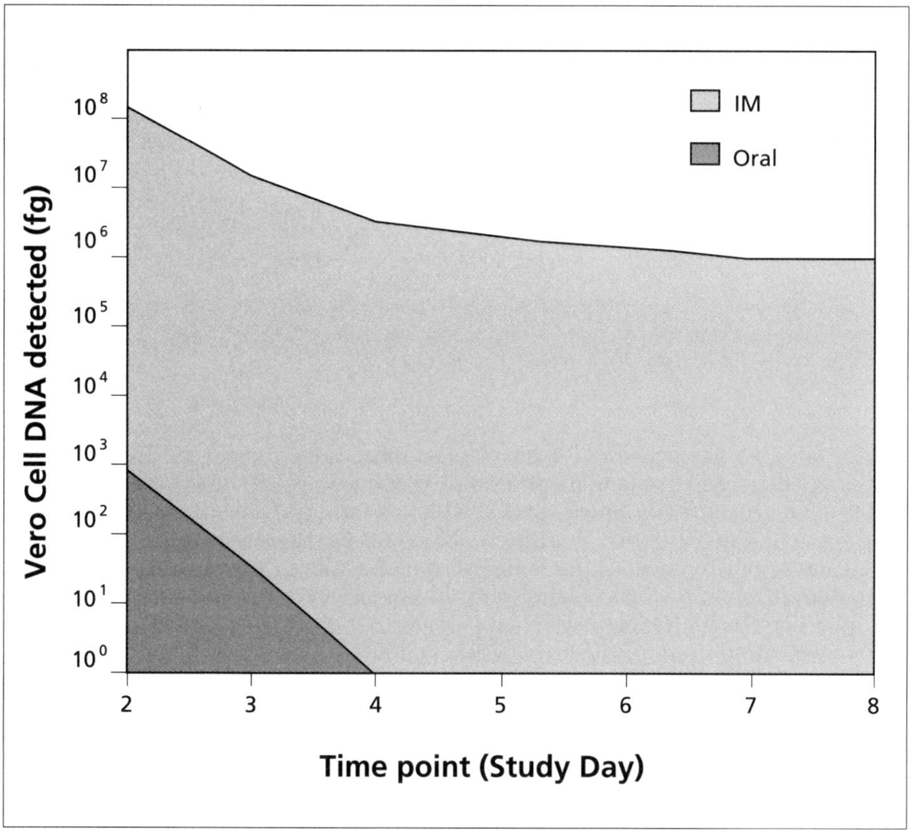

Fig. 1: Area plots generated for the IM and oral routes of administration during the eight-day comparative study. The area-under-the-curve (AUC) for the IM is 1 x 10^6- higher than the AUC for the oral route.

Table 4: Summary of the average total amount of Vero cell DNA detected in the rat tissues for each sex and for the combined sexes.

	Average total amount (fg) of Vero cell DNA					
	Intramuscular[a]				Oral[a]	
SD	**Female**	**Male**	**Combined**	**Female**	**Male**	**Combined**
2	238,616,074	65,388,271	152,002,173	896	1,441	1,169
3	15,461,715	15,344,574	15,403,145	28	44	36
4	3,510,459	3,310,589	3,410,524	0	0	0
8	1,786,204	232,421	1,009,313	0	0	0

SD = Study Day.
a Average of N = 3 for each sex, or N = 6 for the combined sexes.

Table 5: Summary of AUC values generated using either all four study time points or only the last three time points.

	Intramuscular		Oral	
Sex[a]	**$AUC_{[SD2-8]}$**	**$AUC_{[SD3-8]}$**	**$AUC_{[SD2-8]}$**	**$AUC_{[SD3-8]}$**
Female	147,118,306	20,079,412	730	14
Male	56,780,025	16,413,602	1,437	22
Combined	101,949,165	18,246,507	1,083	18

$AUC_{[SD2-8]}$ = Area-under-the-curve for all four time points in the study (Study Days 2, 3, 4, and 8).
$AUC_{[SD3-8]}$ = Area-under-the-curve for the last three time points in the study (Study Days 3, 4, and 8).
a Average of N = 3 for each sex, or N = 6 for the combined sexes.

The AUC values generated using the last three time points ($AUC_{[SD3-8]}$) of the study were then used to calculate the definitive IM versus oral ratio for each sex and for the average of the combined sexes, yielding a ratio of 1.4×10^6 for females, 7.5×10^5 for males, and 1.0×10^6 for the combined sexes. The results clearly show that there is no sex difference, as the female and male values differ by only 1.9-fold, a negligible difference when dealing with biological systems. When this factor is multiplied by the WHO recommended parenteral limit of 10 ng per dose, it results in an upper safety limit for the level of residual DNA from a continuous cell line in an oral vaccine of ~10,000 µg (10 mg) (Table 6).

Table 6: Summary of IM versus oral ratios determined using the data from either all four study time points (SD 2-8) or only the last three time points (SD 3-8).

	IM versus oral ratio[a]			Adjusted residual DNA
Time points	**Female**	**Male**	**Combined**	**safety limit (µg)**[b]
SD 2-8	2.0×10^5	4.0×10^4	9.4×10^4	940
SD 3-8	1.4×10^6	7.5×10^5	1.0×10^6	10,000

IM = Intramuscular.
SD 2-8 = Study Days 2 to 8.
SD 3-8 = Study Days 3 to 8.
a IM versus oral ratio for the corresponding AUC values generated using either N = 3 for each sex, or N = 6 for the combined sexes.
b Based on the combined sexes IM versus oral ratio, multiplied by the 10 ng WHO parenteral limit.

Discussion

Since the relative efficiency of DNA uptake by IM versus oral administration is central to the adaptation of the WHO limit, an animal study was conducted to determine this factor, specifically using Vero cell DNA as the test article and a highly sensitive and specific TaqMan® PCR assay for quantification of DNA uptake.

The results of this uptake study allowed the generation of an empirically derived IM versus oral factor (10^6) representing the relative inefficiency of DNA uptake by oral administration. These results are in agreement with published reports on DNA uptake efficiency developed using polyomavirus DNA, suggesting that oral DNA uptake is at least 10^4-fold less efficient than parenteral DNA uptake [4]. When this 10^6 factor is applied to the WHO recommended parenteral limit of 10 ng/dose, an oral limit of 10 mg is generated. As a conservative approach, this empirically determined limit was reduced 100-fold to 100 μg. Thus, the results of this animal study, together with additional evidence in the literature, support a residual DNA safety limit of 100 μg/dose for an oral vaccine produced on a continuous cell line.

While designing this IM versus oral study, we considered the possibility of administering the Vero cell DNA in the context of a vaccine, as this would mimic the conditions under which the residual DNA is administered to vaccinees. However, we decided to use purified DNA instead for various reasons. The WHO residual DNA limit for parenteral vaccines is based on data generated using purified polyomavirus DNA [2,3]. In fact, most, if not all, of the viral, cellular, and cloned oncogene DNA tumorigenicity studies considered by the WHO study group in drafting the residual DNA limit for parenteral vaccines also used purified DNA. Therefore, to draw a linear comparison between the data generated in this IM versus oral study and the WHO residual DNA limit for parenteral vaccines, which we adapted for oral vaccines based on the results presented here, we decided that it was more appropriate to use purified DNA. Additionally, by performing the IM versus oral Vero cell DNA uptake study using purified DNA, the data generated will be independent of any particular vaccine, making this a practical solution for developing generic guidances applicable to diverse types of oral vaccines. This approach is consistent with the one used by the WHO Expert Committee to set a limit for residual DNA in parenteral vaccines [2,3]. Just as this WHO generic limit has been widely applied without consideration to the vaccine component (virus/antigen), the animal data from this study would support an analogous limit that can be widely applied to oral vaccines. This is reasonable, given the large safety margins built into these limits, which would make any vaccine-specific effects on DNA uptake insignificant when compared to the overall risk.

In preparation for the definitive IM versus oral comparative study presented here, we performed five other studies of Vero cell DNA uptake in which the technical procedures were refined. The main technical difficulty we faced was the prevention of Vero cell DNA contamination of negative tissues. During the first Vero cell DNA uptake study (data not shown), we encountered contamination issues where several tissues (such as the skin, various muscles, and other internal organs) from orally dosed rats were found to be unexpectedly positive. We hypothesized that the reason these tissues were positive was because of an unforeseen contamination event of the skin resulting from the transfer of DNA by the normal grooming/licking behaviour of the rats following oral dosing or due to the difficulty of restraining rats during oral feeding, and that this external contamination may have cross-contaminated the underlying tissues during necropsy. Although this contamination was unforeseen, it is not necessarily surprising, given that rats are being orally dosed with micrograms of DNA, while the DNA is being quantified using an assay capable of detecting as little as 1 attogram (10^{-18} g) of Vero cell DNA. After conducting four follow-up exploratory studies (data not shown), we were able to identify improvements in the oral dosing techniques and to the contamination control measures. These measures allowed us to eliminate or greatly minimize contamination during and after oral dosing, and during necropsy. Of these techniques, the most stringent and effective is the decontamination of the skin of euthanized rats with DNAZap™ before necropsy. We also demonstrated that none of these contamination control measures interferes with the detection of "true positives."

Conclusions

To define the limit of residual DNA in oral vaccines produced on a continuous cell line, we performed a study of Vero cell DNA uptake in rats, in which the relative uptake of Vero cell DNA administered orally was compared to its uptake when delivered IM. As expected, the DNA uptake and persistence was markedly higher in the IM group than in the oral group, resulting in an IM versus oral ratio of 1.0×10^6. This ratio corresponds to an upper safety limit of about 10 mg/dose, after converting the WHO recommended limit for residual DNA from continuous cell lines in parenteral vaccines, to a corresponding limit for oral vaccines. As a conservative approach, this empirically determined value was reduced 100-fold to 100 µg. Thus, the results of this animal study, in combination with (i) published data on DNA uptake efficiency developed using polyomavirus DNA suggesting that oral DNA uptake is at least 10^4-fold less efficient than parenteral DNA uptake [4]; (ii) the lack of evidence that cellular DNA is tumorigenic when administered parenterally, even from highly tumorigenic cell lines [1]; and (iii) the WHO risk assessment that estimated that a theoretical cellular DNA tumour-producing dose with parenteral administration is ~2 g of DNA [2,3], supports the establishment of a new upper safety limit of 100 µg per dose for residual DNA from a continuous cell line in an oral vaccine.

Acknowledgements

We want to thank the Acute Toxicology group of Merck's Safety Assessment Department for their unconditional support.

References

1 Petricciani J, Loewer J: An overview of cell DNA issues, in Brown F, Lewis AM, Peden K, Krause P (eds): Evolving Scientific and Regulatory Perspectives on Cell Substrates for Vaccine Development. Dev Biol. Basel, Karger, 2001, vol 106, pp 275-282.

2 World Health Organization. Acceptability of cell substrates for production of Biologicals. Report of a WHO Study Group. Geneva, Switzerland, 1987 (World Health Organization Technical Report Series 747).

3 World Health Organization. WHO Expert Committee on Biological Standardization. Geneva, Switzerland, 1998 (World Health Organization Technical Report Series 878).

4 Israel MA, Chan HW, Hourihan SL, Rowe WP, Martin MA: Biological activity of polyoma viral DNA in mice and hamsters. J Virol 1979; 29 (3): 990-996.

5 Madhami HD, Leadon SA, Allen Smith C, Hanawalt PC: A DNA in African Green Monkey Cells is Organized into Extremely Long Tandem Arrays. J Biol Chem 1986; 261 (5): 2314-2318.

Dr. J.A. Lebron, Merck Research Laboratories, PO Box 4, WP45-325, West Point, PA 19486, USA.
E-mail: Jose_lebron@merck.com

Petricciani J, Sheets R (eds): Vaccine Cell Substrates 2004.
Dev Biol (Basel). Basel, Karger, 2006, vol 123, pp 45-53.

Biological Activity of Residual Cell-Substrate DNA

K. Peden, L. Sheng, A. Pal, A. Lewis

Division of Viral Products, OVRR, CBER, FDA, Bethesda, MD, USA

Abstract: Vaccines and other biological products manufactured in cells contain contaminating residual DNA derived from that production cell substrate, with the amount and form of this DNA depending mainly on the type of vaccine. The potential risk of this cellular DNA has been debated for over 40 years without resolution. Opinions on residual DNA have varied from it being considered an inert contaminant, and thus its presence should not be deemed to be a risk to vaccine recipients, to it being considered an important risk factor, particularly for vaccines manufactured in certain cell substrates, such as cells derived from tumours or cells that are tumorigenic. We are not of the opinion that DNA can be considered biologically inert, but whether or what risk residual cell-substrate DNA poses remains to be determined. In this paper, we discuss our approaches to address this issue and describe some preliminary work.

INTRODUCTION

The major risk for residual cell-substrate DNA has been considered to be due to the oncogenicity of the DNA, particularly if the cell substrate is tumorigenic or is derived from a human tumour. DNA oncogenicity has been considered to arise mainly through the introduction of a dominant oncogene, as oncogenesis arising through insertional mutagenesis has been considered to be an improbable event [1]. However, another potential safety concern is the risk that the genome of an infectious virus is present in this DNA, either as an integrated genome or as an extra-chromosomal element, and that, once inoculated, this genome could generate an infectious agent that could then establish a productive infection in the vaccine recipient. Such an infection could have pathogenic consequences. Thus, to assess potential risks associated with residual cell-substrate DNA, both biological activities (oncogenicity and infectivity) need to be considered and addressed.

Our approach to assessing risks has been presented previously as a Defined-Risks Approach [2] and involves, in part, developing quantitative assays for the assessment of each risk, as it is our view that an evaluation of risks needs to be based on quantitative experimental data. Furthermore, as long-term safety data from vaccinated people are usually unattainable, we feel that it is prudent to make risk assessments based on the most sensitive model system. To this end, we are developing sensitive quantitative assays to assess the oncogenicity of DNA in various animal model systems and the infectivity of DNA in cell-culture systems. Preliminary results from each of these assays will be presented, and extrapolations from these data will be made.

RESULTS

Development of in vivo assays to assess DNA oncogenicity

Currently, there are no validated animal models to assess DNA oncogenicity. In fact, few studies have attempted to measure DNA oncogenicity in vivo. Most of the published work has been with viral oncogenes or the DNA genomes of tumour viruses. The most complete studies have been done using polyoma virus DNA in mice and hamsters. Polyoma virus is a DNA virus of mice that both infects and induces tumours in mice and induces tumours in hamsters. The DNA genome of polyoma virus was also found to be oncogenic. For example, polyoma virus DNA induced tumours in newborn hamsters whether the viral DNA (0.5 μg) was injected via the intraperitoneal route [3], or whether the cloned polyoma virus genome (0.5 μg) was inoculated via the subcutaneous route [4].

The only published study using exclusively cellular oncogenes is the one done by Burns and colleagues in 1992 [5]. While this study was done predominantly to assess the risk associated with recombinant DNA for laboratory workers, it remains the only study where the oncogenicity of a single cellular oncogene has been assessed in vivo. The plasmid pHO6T1 (which contains the activated human T24-H-*ras* oncogene under the control of the Moloney murine leukaemia virus LTR and the SV40 enhancer) and the plasmid pT24C3 (which expresses the same T24-H-*ras* oncogene under its own promoter) were inoculated on to the skin of NIH Swiss mice by scarification. Vasoformative tumours histologically characterised as lymphangiosarcomas were induced by 12 weeks with both plasmids, demonstrating that 10 μg of an activated, dominant cellular oncogene, the human T24-H-*ras* oncogene, alone was sufficient to induce tumours. To our knowledge, experiments demonstrating that a single dominant cellular oncogene is capable of inducing tumours in normal mice have not been replicated.

The results from the studies reported to date indicate that DNA encoding dominant activated oncogenes, whether of viral or cellular origin, can be oncogenic in experimentally inoculated rodents, but that the efficiency of tumour induction is low. Amounts no lower than 0.5 μg of polyoma viral DNA were oncogenic in vivo [3,4]. Whether this is a consequence of the species or strain of the experimental animal, the route of inoculation, the particular oncogene, the promoter from which the oncogene was expressed, or the state of the DNA is not known.

Because DNA **can** be oncogenic, the issue, therefore, is whether cellular DNA containing activated oncogenes could also be oncogenic. To resolve this question, we reasoned that it would be necessary to establish a highly sensitive animal model to detect oncogenic activity. For the development of such an animal model system, appropriate positive-control reagents are required. These positive controls should be cellular oncogenes, because these oncogenes would be the main oncogenic concern with residual cell-substrate DNA from cell lines that are tumorigenic or are derived from human tumours that were not induced by viral oncogenes (e.g. E6 and E7 from human papilloma virus). These positive controls would be used to establish quantitative dose-response relationships for oncogenicity in different mouse models.

The overall aims of our DNA oncogenicity studies are: (i) to determine whether dominant cellular oncogenes, when expressed under a strong promoter, can be oncogenic in vivo; such expression plasmids would be our positive-control DNAs for subsequent studies; (ii) to evaluate several mouse strains for their susceptibility to tumour induction with our positive-control DNA; various mouse strains with different degrees of immunocompetence (e.g. newborn normal mice, athymic nude mice, and CD3 epsilon mice) and various transgenic and knock-out mice strains that have higher cancer incidence will be tested; (iii) to evaluate different ways to introduce DNA into animals (e.g. use of different transfection facilitators with needle and syringe, Gene Gun, Biojector); (iv) to determine which routes of inoculation are effective and to determine which is the most efficient; because vaccines can be inoculated via various routes, it is important to assess the ones that are used clinically (e.g. intramuscular, subcutaneous, intranasal, oral).

Development of positive-control reagents

In consultation with Don Blair, Stephen Hughes, and John Coffin at the Frederick Cancer Research Facility of the National Cancer Institute, we decided to use expression plasmids based on the Moloney sarcoma/leukaemia virus (MSV) long terminal repeat (LTR) and the cellular oncogenes c-*myc* and activated H-*ras*. Mouse c-*myc* and human T24-H-*ras* were the oncogenes chosen, because these two oncogenes complement one another in the transformation of primary cells in vitro [6]. Furthermore, these oncogenes have been associated with human cancers.

The expression plasmids for T24-H-*ras* and murine c-*myc* are based on pMSV/LS (Sheng and Peden, unpublished) and were generated in a series of steps from components isolated by PCR amplification from other plasmids. The structures of the plasmids are shown in Fig. 1 and were confirmed by DNA sequence analysis. Before initiating animal experiments, these plasmids were first shown to be biologically active in vitro using a focus-forming assay with NIH 3T3 cells and Rat-1 cells. For the assessment of in vivo oncogenicity, adult and newborn NIH Swiss mice and C57BL/6 mice were inoculated by the subcutaneous route with 10-fold serial dilutions of pMSV-T24-H-*ras* plus pMSV-c-*myc* or pMSV-T24-H-*ras* alone. The total amount of DNA inoculated per mouse was 25 µg in 50 µL; when necessary, the vector pMSV/LS was used to adjust the amount of DNA to 25 µg. The use of the vector DNA was the control for events caused by possible insertion of the LTR. As shown in Table 1, tumours were only induced when both the *ras* and the *myc* plasmids were injected and not with the *ras* plasmid alone and only at the highest dose (12.5 µg) of each plasmid. NIH Swiss mice were more susceptible than C57BL/6 mice, and newborn mice were more susceptible than adults. Tumours were only found at the site of injection. Histological examination of the tumours indicated that they were undifferentiated sarcomas. Cell lines were established from many of the tumours, and PCR analysis demonstrated that both the *ras* and *myc* genes were present in all lines. These results show that cellular oncogenes, when injected subcutaneously, can induce tumours in normal mice.

Development of in vitro assays to assess DNA infectivity

The DNA genomes of several types of viruses have been shown to be infectious when inoculated into susceptible cells in vitro. For example, the genomes of the polyomaviruses [7], adenoviruses [8], papillomaviruses [9], and herpesviruses [10]

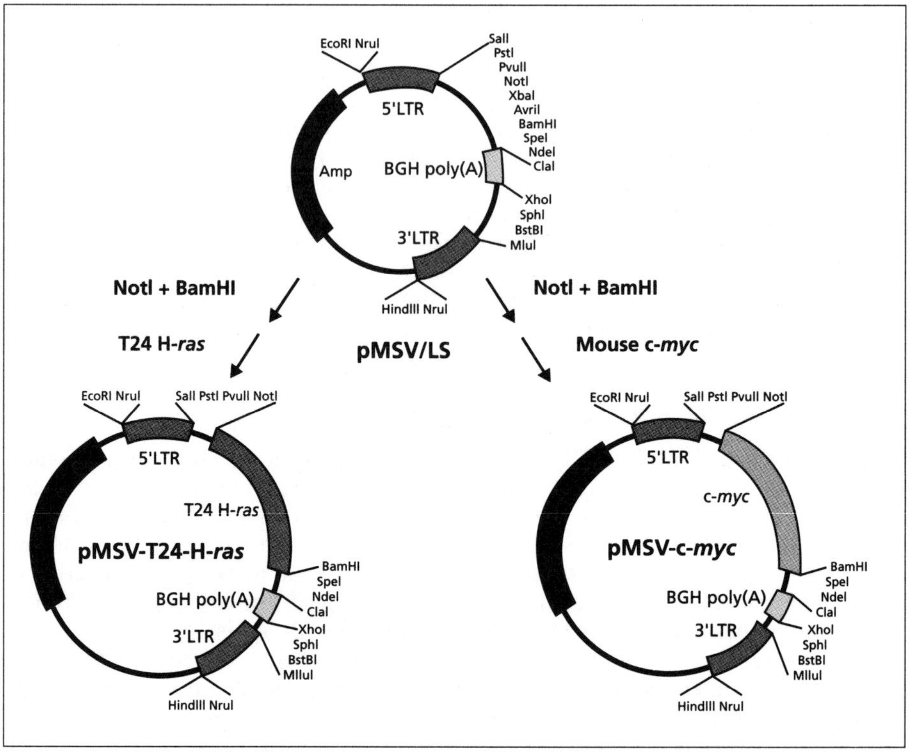

Fig. 1: Construction of pMSV-T24-H-*ras* and pMSV-c-*myc*.

Table 1: Tumor incidence in mice injected with *ras* and *myc* expression pasmids.

Cohort	*ras* plasmid (µg/mouse)	*myc* plasmid (µg/mouse)	NIH Swiss		C57BL/6	
			Adult	Newborn	Adult	Newborn
R6	12.5	0	0/10	0/10	0/10	0/8
R5	1.25	0	0/10	0/11	0/10	0/10
R4	0.125	0	0/10	0/9	0/10	0/10
R3	0.0125	0	0/10	0/8	0/10	0/10
RM6	12.5	12.5	**2/10**	**9/11**	**0/10**	**2/11**
RM5	1.25	1.25	0/10	0/10	0/10	0/10
RM4	0.125	0.125	0/10	0/10	0/10	0/10
RM3	0.0125	0.0125	0/10	0/10	0/10	0/10

Subcutaneous sarcomas developed at injection site at 4-7 weeks.

are infectious if transfected into the appropriate cells in vitro. The cloned proviral DNA of retroviral genomes can also be infectious [11].

DNA can also be infectious in vivo if inoculated into the appropriate animals. For example, polyoma virus DNA [12] is infectious in mice, and cloned retroviral DNA of simian immunodeficiency virus [13] or bovine leukaemia virus [14] can establish productive infections when inoculated into experimental animals.

That the DNA genomes of viruses can be infectious has consequences for the production of biologicals that are produced in mammalian cell substrates. [Although the RNA genomes of RNA viruses can also be infectious, we are not considering these a risk due to the lability of RNA as compared with DNA.] If that cell substrate contains the genome of a DNA virus, either integrated or extrachromosomal, or the proviral genome of a retrovirus, then the residual cell-substrate DNA in the vaccine will contain those genomes at a certain level and thus the vaccine recipient could receive infectious DNA. The viral genome could be that of a known virus or an unknown one. For the inoculation of DNA containing an infectious genome to be of consequence, the inoculated human cell would need to be permissive for the virus. The major concern would be if the virus once produced from the transfected cell could establish a productive infection in the human host. For example, one of the obvious ways infectious DNA could be important is in the case of an inactivated HIV vaccine made from wild-type HIV. While the virus would be inactivated, most likely through chemical means, and the extent of inactivation monitored, the inactivation of the proviral DNA, which could produce infectious HIV, may escape examination.

Because of the potential of amplification of the initial inoculum, the risk associated with an infectious DNA could be greater than that associated with an oncogenic event. We have approached the issue of infectious DNA in the following ways: (i) we have developed a transfection/co-cultivation assay to measure the infectivity of a proviral genome of a retrovirus; (ii) we have determined whether cellular DNA from infected cells is infectious, and (iii) we have explored ways by which the infectivity of DNA is reduced.

Establishing the transfection/co-culture system for retroviral DNA infectivity

The system we have established (Sheng and Peden, in preparation) involves the human immunodeficiency virus type 1 (HIV-1). Because HIV-1 is a cytopathic retrovirus, the sensitivity of detection should be high. Infectious molecular clones of HIV-1 exist, and stocks of HIV-1 are usually prepared by transfection of susceptible human cells such as 293 or 293T cells. We have used the LAI strain of HIV-1, as this virus is highly cytopathic [15]. Because 293 cells do not have CD4, the primary receptor for HIV-1, it is necessary to co-culture the transfected 293 cells with a permissive cell line to amplify the virus produced from the transfected cells. The cell line chosen was the Jurkat T-cell line, which efficiently replicates LAI. Co-culture is a highly efficient system to detect small quantities of virus.

After several variables were tested, the system selected was transfection of linear HIV-1 molecular clones into 293T cells using the Effectene transfection facilitator and co-culturing 24 h after transfection with T cells. Infection is monitored by measuring the reverse transcriptase activity in the medium; the higher the DNA dose, the sooner the appearance of virus. This assay is capable of detecting 1 pg of the cloned LAI DNA. In addition, we have shown that the HIV genome can be detected in 2.5 µg of cellular DNA from LAI-infected cells. These results show that the DNA genome of a retrovirus can be infectious down to 1 pg and cellular DNA, if containing an infectious retrovirus, can be infectious at the 2.5 µg level.

We have used this assay to explore ways to inactivate DNA infectivity. Methods used to eliminate DNA infectivity are directly pertinent to the elimination of DNA oncogenicity as well. Our initial experiments involved digestion of the DNA with a non-specific endonuclease and treating the DNA with β-propiolactone (BPL), reasoning that cleavage of the retroviral genome or chemical inactivation would result in the reduction and possibly elimination of infectivity.

For endonuclease digestion, we mixed pLAI, the infectious HIV-1 molecular clone, with an equal amount of Jurkat cellular DNA. This mixture was digested with the endonuclease for various times. The DNA was purified by phenol extraction and ethanol precipitation, and the size distribution of the DNA fragments was analysed by agarose gel electrophoresis. Each size class was evaluated for infectivity. Digestion to a mean size of 650 bp eliminated the infectivity of 0.15 µg of pLAI DNA, whereas larger DNA retained infectivity. For the BPL treatment, we incubated the pLAI/Jurkat DNA mixture with various concentrations of BPL at 4° C for 24, 48, or 72 h and assessed the infectivity of each sample. Treatment with 0.25 % BPL for 72 h at 4° C eliminated the infectivity of 25 ng of HIV DNA.

Discussion

In this paper, we describe our approaches to address whether residual cell-substrate DNA poses a risk to vaccine recipients. Because DNA can have an oncogenic activity and an infectivity activity, we are developing assays to measure both of these biological activities.

With respect to DNA oncogenicity, we have generated expression plasmids for the human T24-H-*ras* and the mouse c-*myc* genes and have inoculated these plasmids via the subcutaneous route into newborn and adult NIH Swiss and C57BL/6 mice. Tumours were induced only when mice were inoculated with both the T24-H-*ras* and the c-*myc* plasmids and only at the highest dose of each plasmid (12.5 µg). Newborn NIH Swiss mice were the most sensitive system tested so far. These results demonstrate that plasmids expressing cellular oncogenes can be oncogenic in vivo. These plasmids will serve as positive control reagents for future studies.

The in vitro transfection/co-culture system we have established can detect 1 pg of the cloned proviral DNA of HIV-1 and 2.5 µg of DNA derived from HIV-infected cells. With this assay, we have explored methods of reducing the infectivity of DNA. Our results to date indicate that digestion of the DNA with a non-specific endonuclease or reacting the DNA with the chemical β-propiolactone (BPL) can eliminate the infectivity of HIV DNA. Digestion to a mean size about 650 bp eliminates the infectivity of 150 ng of HIV-1 DNA, and incubation with 0.25 % BPL for 72 h at 4°C eliminated the infectivity of 25 ng of HIV-1 DNA.

Estimates of safety factors based on these results

We define the "safety factor" as the multiplicative factor by which the biological activity is reduced. Therefore, the safety factor is analogous to "clearance" of adventitious agents, which includes both removal and inactivation.

To make estimates for safety factors concerning DNA, a number of assumptions are necessary.

(i) For any given DNA, the level of the response of a cell to that DNA is proportional to the amount of that DNA. This assumption can be documented in many in vitro systems, and we assume that this would also be true in vivo.

(ii) The activity of a gene/viral genome integrated in chromosomal DNA or as part of plasmid or phage vector is equivalent or lower. Few experiments have been done to test this assumption, although our ability to detect the infectivity of integrated HIV DNA suggests that it is true.

(iii) The amount of uptake of a given gene/viral genome by a cell and the expression of this gene/virus in the cell is related to the concentration of the gene/viral genome in the DNA.

(iv) The activity of a given gene/viral genome inoculated as chromatin is the same or lower than when the same gene/viral genome is inoculated as protein-free DNA. This assumption has not yet been tested.

Several other factors need to be considered to make estimations of risk arising from DNA. For example, if we assume that a single-copy gene is between 3,000 and 30,000 base pairs (bp), then this gene represents between 10^{-6} and 10^{-5} of the haploid mammalian genome (3×10^{9} bp); this means that the amount of DNA corresponding to a single-copy gene is 10^{6}-fold to 10^{5}-fold less abundant for equivalent amounts of cellular DNA compared with a plasmid DNA expressing the same gene. Therefore, the amount of mammalian genomic DNA equivalent to 1 µg of plasmid DNA is between 1×10^{6} µg, or 1 g, and 0.1 g. Conversely, if a single oncogene is present in 1 µg of mammalian DNA, the equivalent amount of the same oncogene, if cloned in a plasmid vector, is between 1×10^{-6} and 1×10^{-5} µg, or 1 pg, and 10 pg.

With these assumptions, and considering the respective sizes of the cellular genome versus the size of a cellular gene and viral genome, we can estimate safety factors.

Estimate from HIV-infected cellular DNA

When the DNA from HIV-infected cells was tested for infectivity, 2.5 µg of this DNA was infectious. Therefore, in the **absence** of any inactivation (chemical, enzymatic, radiation), residual cell-substrate DNA at 10 ng would have a safety factor of 250-fold.

Estimate from cloned HIV DNA

Cloned HIV DNA was infectious at 1 pg. Because the size of the HIV proviral genome represents 1.67×10^{-6} of the diploid cell genome, the amount of cellular DNA that would correspond to an infectious HIV dose is: 1 pg $\div$ (1.67×10^{-6}), or 0.6 µg (600 ng).

Thus, in the absence of any inactivation (chemical, enzymatic, radiation), 10 ng of cellular DNA provides a 60-fold safety factor. [We use the diploid size for a viral genome, as there is usually a single copy of the provirus per cell.]

The difference between the estimates made from the results of the infected cellular DNA and the cloned HIV DNA is likely due to the fact that not all cells were infected in the HIV-infected culture, although it could be due to lower infectivity of viral DNA as part of the cellular DNA.

Estimate from endonuclease digestion of cloned HIV DNA

Digestion of DNA to a mean size of 650 bp or smaller resulted in the loss of biological activity of 0.15 μg of pure viral DNA. Based on the proportion of a retroviral genome in the cell, 150 ng of viral DNA corresponds to: $150 \div (1.67 \times 10^{-6})$ of cellular DNA, or 90×10^6 ng, or 90 mg. Therefore, for 10 ng of cellular DNA with a single provirus, the safety factor is 9×10^6.

Estimate from β-propiolactone treatment of cloned HIV DNA

Treatment of cloned HIV DNA with 0.25 % β-propiolactone (BPL) for 72 h at 4° C eliminated the activity of 25 ng of the HIV DNA. Again, based on the proportion of a retroviral genome in the cell, 25 ng of viral DNA corresponds to: $25 \div (1.67 \times 10^{-6})$ ng, or 15×10^6 ng, or 15 mg. Therefore, for 10 ng of cellular DNA with a single provirus, the safety factor is 1.5×10^6.

Estimate from in vivo oncogenicity studies

The amount of plasmid DNA that induced tumours in mice was 12.5 μg of each of *ras* and *myc* plasmids. Assuming that the average size of an oncogene is 10^4 bp, and the size of haploid genome is 3×10^9 bp, then an oncogene is present in the genome at: $10^4 \div (3 \times 10^9) = 3.33 \times 10^{-6}$. Extrapolating to two oncogenes, since both *myc* and *ras* were required to induce a tumour, the amount of cellular DNA that could induce a tumour is: 25 μg $\div (3.33 \times 10^{-6}) = 7.5 \times 10^6$ μg. If the amount of cellular DNA were reduced to 10 ng, then this represents a safety factor of 7.5×10^8. If there are 200 oncogenes in the mammalian genome, as has been estimated, then the safety factor is reduced to 3.8×10^6. Similarly, if the oncogene has been amplified and is present in multiple copies, this would also reduce the safety factor. Again, these estimations do not take into account any DNA inactivation steps.

The calculations of safety factors are based on the data that we have available now and will probably change as more data accrue and as more sensitive models are developed. Nevertheless, we think that the exercise is worthwhile in that the methods used to reduce DNA activity have demonstrated that the safety factor can be increased significantly over just lowering the amount of residual cell-substrate DNA. Therefore, using a combination of reducing the amount of DNA and chemical inactivation and/or size reduction, safety factors of at least 10^7 should be readily attainable. The methods chosen will depend on the type of vaccine.

We are not aware of a consensus on what an acceptable safety factor should be for a given risk, although risks of 1 in 10^6 or greater have been considered unacceptable [2]. Therefore, attaining safety factors of more than 10^7 for the elimination of the biological activities of DNA would seem to be prudent, even recognising that the risks associated with DNA are at this stage unknown.

Acknowledgements

This work was supported in part by grants from the FDA Office of Science HC and the National Vaccine Program Office. L.S. and A.P. were initially supported by a fellowship funded by the National Vaccine Program Office. We thank Hana Golding, Phoebe Mounts, Bill Egan, Jerry Weir, and Phil Krause for comments on the manuscript.

REFERENCES

1 Kürth R: Risk potential of the chromosomal insertion of foreign DNA. Ann N Y Acad Sci 1995; 772: 140-151.

2 Lewis AM, Jr., Krause P, Peden K: A defined-risks approach to the regulatory assessment of the use of neoplastic cells as substrates for viral vaccine manufacture, in Brown F, Lewis AM, Peden K, Krause P (eds): Evolving Scientific and Regulatory Perspectives on Cell Substrates for Vaccine Development. Dev Biol. Basel, Karger, 2001, vol 106, pp 513-535.

3 Israel MA, Chan HW, Hourihan SL, Rowe WP, Martin MA: Biological activity of polyoma viral DNA in mice and hamsters. J Virol 1979; 29: 990-996.

4 Israel MA, Chan HW, Martin MA, Rowe WP: Molecular cloning of polyoma virus DNA in Escherichia coli: oncogenicity testing in hamsters. Science 1979; 205: 1140-1142.

5 Burns PA, Jack A, Neilson F, Haddow S, Balmain A: Transformation of mouse skin endothelial cells in vivo by direct application of plasmid DNA encoding the human T24 H-ras oncogene. Oncogene 1991; 6: 1973-1978.

6 Land H, Parada LF, Weinberg RA: Tumorigenic conversion of primary embryo fibroblasts requires at least two cooperating oncogenes. Nature 1983; 304: 596-602.

7 DiMayorca GA, Eddy BE, Stewart SE, Hunter WS, Friend C, Bendich A: Isolation of infectious deoxyribonucleic acid from SE polyoma-infected tissue cultures. Proc Natl Acad Sci U S A 1959; 45: 1805-1808.

8 Burnett JP, Harrington JA: Infectivity associated with Simian adenovirus type SA7 DNA. Nature 1968; 220: 1245.

9 Watts SL, Ostrow RS, Phelps WC, Prince JT, Faras AJ: Free cottontail rabbit papillomavirus DNA persists in warts and carcinomas of infected rabbits and in cells in culture transformed with virus or viral DNA. Virology 1983; 125: 127-138.

10 Fleckenstein B, Bornkamm GW, Ludwig H: Repetitive sequences in complete and defective genomes of Herpesvirus saimiri. J Virol 1975; 15: 398-406.

11 Fisher AG, Collalti E, Ratner L, Gallo RC, Wong-Staal F: A molecular clone of HTLV-III with biological activity. Nature 1985; 316: 262-265.

12 Atanasiu P, Orth G, Rebiere JP, Boiron M, Paoletti C: Production of tumors in the hamster by inoculation of desoxyribonucleic acid extracted from tissue cultures infected with polyoma virus. C R Hebd Seances Acad Sci 1962; 254: 4228-4230.

13 Letvin NL, Lord CI, King NW, Wyand MS, Myrick KV, Haseltine WA: Risks of handling HIV. Nature 1991; 349: 573.

14 Willems L, Portetelle D, Kerkhofs P, et al: In vivo transfection of bovine leukemia provirus into sheep. Virology 1992; 189: 775-777.

15 Peden K, Emerman M, Montagnier L: Changes in growth properties on passage in tissue culture of viruses derived from infectious molecular clones of HIV-1_{LAI}, HIV-1_{MAL}, and HIV-1_{ELI}. Virology 1991; 185: 661-672.

Dr. K. Peden, Building 29A, Room 3D08, CBER, FDA, 29 Lincoln Drive, Bethesda, MD 20892, USA.
E-mail: peden@cber.fda.gov

Petricciani J, Sheets R (eds): Vaccine Cell Substrates 2004.
Dev Biol (Basel). Basel, Karger, 2006, vol 123, pp 55-73.

PANEL I DISCUSSION

Potential Oncogenicity of Cellular DNA

Panel Members: *John Petricciani (Panel Chair), Klaus Cichutek, Johannes Doehmer, Phil Minor, Keith Peden, Alan Shaw, Alastair Strain, Giuseppe Vicari, David Wood*

Panel Questions

Question 1:
Based on currently available data, what range of cellular DNA derived from cells in which there is no evidence of viral elements should be considered essentially risk-free (i.e., less than 1 in 10^6) for an oncogenic event?

Question 2:
Should cellular DNA derived from cells in which there is no evidence of viral elements be considered an impurity rather than a risk factor when it is reduced to acceptable levels?

Question 3:
Is DNA at any level from any cell type of any concern when an inactivating agent such as BPL is used in the manufacturing process and data are available to demonstrate the inactivation of DNA as well as the vaccine virus (e.g. rabies)?

Question 4:
Should distinctions be made between the amount of cellular DNA that should be allowed as an impurity in parenteral versus oral products?

Question 5:
Are there different levels of risk associated with DNA from autologous, homologous, or heterologous sources?

Question 6:
Are positive control(s) appropriate or feasible in any future risk assessment studies on cellular DNA in view of the absence of data to demonstrate that cellular DNA is oncogenic?

Question 7:
What sensitive and validated methods can be applied in cell substrate characterization studies to evaluate the potential of cellular DNA to be oncogenic?

Question 8:
Can an alternative approach to laboratory testing, such as risk assessment, be used to address the risk of cellular DNA in biological products?

Dr. Petricciani: Taking question one, based on the available data, is there a range of cellular DNA from cells in which there is no evidence of viral elements that can be considered essentially risk free?

Dr. Cichutek: The 10 ng rule established by WHO is currently the basis for evaluation of the quality of medicinal products in the competence of the Paul-Ehrlich-Institut. 10 ng cell substrate DNA per dose of a vaccine are therefore acceptable. The European "points to consider" document which explains the harmonized view of European regulators calls for a more intensive evaluation of risk without explicitly naming a specific amount of foreign DNA that is acceptable. When viral elements are present in foreign DNA of a product administered to humans, the risk may be considered to be increased versus the risk associated with a DNA missing viral LTR elements.

I would like to re-examine what the experimental data underlying the establishment of the 10 ng rule were to understand what it really means. What I heard from the presentations so far is that the calculations to estimate the theoretical risk vary by a considerable factor. Some of the underlying assumptions may have to be discussed again in view of new data. For example, we know now that muscle takes up DNA spontaneously and expresses it. Some of the original theoretical calculations assumed an uptake frequency of 10^{-6} which is probably much higher. We need to rediscuss the original basis leading to the establishment of the 10 ng rule.

Second, the 10 ng rule was based on a theoretical risk assessment. As I showed you, for humans the risk of insertional oncogenesis is real. Two SCID-X1 patients treated with retrovirally modified autologous bone marrow cells suffered from leukaemia caused by insertional mutagenesis and activation of the LMO2 oncogene. It must be borne in mind that this was due to special circumstances such as treatment at a very young age, the underlying disease SCID-X1 giving a selection advantage to corrected bone marrow cells after treatment, and others. This is described in a January 2003 report of the EMEA/CPMP Gene Therapy Expert Group. Also, we have today the opportunity to analyse the real risk of insertional oncogenesis in animal models. For example, Baum and others described an animal model where leukaemias in mice can be readily induced following retroviral modification of bone marrow cells and in vivo passage. This gives us the opportunity to analyse the genes involved, the susceptibility of certain cell types, the regulatory elements involved, and other factors. Other models using various routes of administration could potentially be established in a similar way. I think what Keith Peden told us about such estimations is a very good beginning, and also the Vero cell data are very good for looking at the risk that could be there. I think there can be models that can be built up. I am challenging the data discussed for the sake of evaluation. When you say that for 20 years we have had no progress in estimating the real risk associated with DNA administration to humans and now the time has come to progress, I would like to hear about the experiments that have been done during this time to be able to assess the basis for the risk assumptions discussed. Obviously, based on the human and animal data that I presented, the design of new experiments addressing the risks should be discussed. I would like to see more than sporadic experiments that show that the basis for the 10 ng rule still has a solid basis, and I can see that some of these experiments and experiences we learned from clinical gene transfer and pre-clinical research could lead to new experiments looking at the risk of insertional oncogenesis in animals. We have now a chance to introduce positive controls into these experiments.

Dr. Strain: I would like to make a couple of comments. The first is that we ought to be considering the various studies, of course. In particular the study that Dr. Cichutek described about the IM administration is clearly important but it may be just one example, and it is important to consider more studies than just that one.

The second point that has not really been discussed is the recent data just presented. The detection and presence of DNA is one thing. What about the fragment size of that DNA? I do not know if the PCR studies can address that topic, but clearly the size of the residual DNA is one of the issues that needs to be discussed.

Dr. Vicari: I agree with this comment because if the size of DNA is below the dimension of a functioning gene, its presence should be less relevant. In addition, when we established limits for DNA in the past, we were considering recombinant proteins which were highly purified, and we could fix limits which were quite low. Now we are talking about vaccines which cannot be purified to the same degree. Therefore, we should pay more attention to some elements of the nature of this DNA, such as whether it has been submitted to treatment such as beta-propiolactone. Also, one should consider whether the DNA is in the presence or the absence of vectors or facilitators as well as its size. All these elements should be considered in relation to the risk to patients.

Dr. Doehmer: We have seen from the last speaker that the uptake of DNA by intramuscular injection is far better than other routes by which some of the DNA is applied. As a toxicologist, I feel familiar with risk assessment, and one of the major issues in risk assessment is exposure. Intramuscular injection is a very specific exposure which should not be overstated and should be put into perspective in relation to other applications.

The 100 pg of residual DNA that we concluded can be tolerated in a dose of product in our WHO Study Group in 1986 was based on the experimental conditions we were familiar with at that time to get DNA into a cell and integrated functionally into genomic DNA. IM injection is a very special situation because it is cloned homogeneous DNA, and the recipient cell is very specific. It is a big cell with approximately 10 nuclei, in the same environment.

Again, I think we should not overstate those results. What is equally important is the situation where injection is given IV or where the product is taken up orally. So, I think the results seen with IM injection on DNA uptake do not change the situation. Rather, it is just highlighting a very special situation and should not be taken as a generalization for any risk assessment.

Dr. Wood: To provide some further information on where the 10 ng level came from in the WHO specifications, as Johannes has just said, this derives from the additional work of the 1986 study group and the results of that study group were published in full in the WHO technical report series number 747. That sets out the background information that was available at that time, the basis on which the initial 100 pg per dose was assessed. This was subsequently revised about 10 years later when the WHO looked at this again. On the basis of the accumulation of information in the meantime, that is where the decision was made to revise the limit to 10 ng per dose and is published in the technical report series, number 878.

Dr. Petricciani: I think it is clear from the discussion that there are three important variables and it is difficult to come to a generalized statement: (i) quantity; (ii) size, and (iii) the route of administration. It is clearly different if dosing is subcutaneous versus intramuscular, so it is more complicated than just a simple, straightforward assessment.

Dr. Minor: I have a great deal of difficulty in this area but I wonder if there may be other problems as well in terms of the specifics. For example, the assumption from this slide is that the actual nature of the DNA does not matter, but it seems to me that if you use the gene therapy vector that Klaus described without the virus, i.e. the DNA, and put it into SCID people, it would have a different risk assessment than if you took a different whole cell DNA and put that into healthy babies.

The issue about what counts as risk-free is something that needs to be considered as well. If it is 1 in 10^6, I am not sure that would be acceptable for a childhood vaccine, for example; the recipient may actually affect risk assessment.

I am still not clear whether the nature and quality and the actual specificities of the DNA influence matters, because it is not clear to me where the burden of proof lies. Is it necessary for people who are uneasy about this to prove that it is dangerous, or is it necessary for people who are quite happy about it to prove that it is safe?

Dr. Cichutek: First of all, I really think it is time to move that subject forward. The difficulties that arise are currently that many vectors are moving into the field of preventive vaccination. Clinical trials have been approved, including trials in healthy volunteers, using "new" vectors. Some of these vectors may have a certain known level of chromosomal integration.

The DNA issue could definitely be moved forward. I am trying, for example, in the EMEA/CPMP Gene Therapy Expert Group, to produce new considerations concerning the risk of insertional oncogenesis in the light of new experiences in the field of gene therapy. I fully agree with most of what has been said. Now is the time to revisit some of these vectors. One important statement that has to be made is that nobody is doubting that the 10 ng rule by itself is acceptable. Nobody has a scientific basis for lowering the amount of cell substrate DNA acceptable per dose.

If there are non-diploid cell lines used for vaccine production that are known to contain specific oncogenes we, as regulators, would always like to see data to show that the specific oncogenes can no longer be detected in the amount of cell substrate DNA to be administered in a vaccine dose. So, at this point the interpretation of the guidelines is that there should not be an activated oncogene in the DNA. During the discussion of the quality of gene transfer medicinal products, the experts from the Paul-Ehrlich-Institut may, for example, ask sponsors to look by PCR for the presence or absence of such known and defined oncogenes, and I should like to discuss whether that is a good scientific question or not. The fragment size of the cell substrate DNA is also taken into account because only a full-length oncogene could possibly be biologically active.

To move forward, we should consider new experiments. We started well, but to describe the risk of tumorigenicity properly, we need positive controls that show tumour formation in animal experiments. It is relevant to look at the route of administration for the perceived vaccine use. Then you have to look at the specific DNA from the relevant cell substrate. All these need to be taken into account. I think we should move forward by looking at what kind of experiments could be done; what kind of data showing absence of a risk of insertional oncogenesis have been collected during the past few years. Then we may proceed with developing a summary of such data and a collection of proposals about what needs to be done experimentally in the future.

Dr. Peden: I just have a couple of points on what Klaus just said. In my experience it is not clear that people do not want to lower the 10 ng. I think for certain cell substrates it is viable. We may want to lower it, say for example HeLa cells.

The other thing is that it is all very well to look at oncogenes present in your product, but as we found out from your presentation on gene therapy, we did not know that the common gamma chain was an oncogene in itself. There are always new oncogenes that we do not know about. Cellular oncogenes are rather more of a problem. I think you are right that if you want to do oncogenicity studies you need positive controls, which is why we are starting to develop those. It is all very well to tell sponsors that they should look at 100 mcg of DNA in a mouse, but unless you know what the positive control is for that I think it is not so useful to do those experiments without positive controls.

Finally, there may be a problem with mouse studies. They may not of course reflect the human case. Injecting into a mouse may not be as sensitive as in humans, or it may be more sensitive. We do not know what the relative sensitivities of the animal strains are. Our aim is to try to develop the most sensitive system using animal models. To that end, we are starting to look at different genetically modified strains such as those with which David Onions is familiar, that have genetic changes in them, thus making them more sensitive.

That is not just a way of getting to the most sensitive system. In fact, humans do have mutations. There are humans who carry BRCA1 mutations that may make them more sensitive to receiving certain oncogenes in vaccines. So, it not just theoretical; it may in fact have human practical relevance.

Dr. Shaw: To add a little more historical context to some of the naked DNA arguments, back in the early 1990s, when we were looking at naked DNA as a vaccine delivery system, we did some extensive studies, and Brian Ledwith is probably the best qualified to talk about these, but the general outcome was that super-coiled plasmid DNA is much more efficiently taken up by muscle cells than is either nicked, relaxed circles or linearized DNA.

Since what we are talking about here, other than DNA vectors, is sheared DNA of some sort which would be in linear form, the uptake of that in muscle cells would probably be far less effective.

Dr. Peden: There are data from polyoma that say that linear, in fact, is sometimes more efficient. Again it depends on the route and the virus. I think you are right in terms of IM injection with plasmid DNA but there are experiments reported in the literature that say linear polyomavirus DNA is more infectious subcutaneously than circular. Again, that is a replicating system but it is one of the matters we are discussing.

Dr. Robertson: One thing I have always wondered about is the nature of the DNA. Although people were talking about it this morning, only one person mentioned chromatin.

The situation, as far as I see it, is going to be a vaccine produced in a mammalian cell line where the contaminating DNA is not a plasmid DNA, so those studies are good, but of debatable relevance. It is not necessarily gene therapy we are looking at now, it is vaccines. The nature of the DNA that is going to be contaminating is presumably going to be bound up with protein.

Are there any data on exactly what happens to chromatin DNA if you inoculate that rather than looking at purified DNA, be it cellular or plasmid?

Dr. Ledwith: We have not done any studies of chromatin versus purified DNA but the study that was mentioned in Dr. Petricciani's presentation showed that about a milligram of T-24 DNA in Rhesus monkeys over eight years was with chromatin. At least there is some negative tumorigenicity in those data.

Dr. Petricciani: That is the only chromatin study that I know of.

Dr. Peden: In the infectivity model, as I mentioned, we want to address that. It is not clear whether chromatin would be more or less infectious, less or more active. One feels that it is going to be less biologically active but we really do not know the answer to that.

It could be argued that it could be taken up faster because it has proteins attached to it and the cell would take it up and protect it, but I don't think the answer is known. That is why we are planning to do those experiments.

Dr. Cichutek: Infectivity experiments are highly relevant. I think they should be expanded to do some in vivo work because, for example, there have been data — maybe in 1988 or 1989 – where in vivo inoculation into monkeys of plasmid DNA specifying a full-length viral genome of SIV produced a lentiviral infection. That is a relevant model showing that biologically active DNA can produce an effect after in vivo inoculation. The spontaneous uptake in vitro is one thing and then overlaying the transfected cells also with some infectable and permissive cells is another. It adds to the level of complexity. The real-life situation could be mimicked by some in relevant animal models.

Dr. Peden: There are data on that; Phil Krause and Andrew Lewis referred to it in their paper on SIV in macaques.

In the 1999 cell substrate meeting, Damian Purcell talked about their model using a gene gun. They can get an infection in a monkey with as little as 10 ng of DNA. In the previous studies it was in the hundreds of micrograms to establish an infection in a monkey. Again, it depends on the route of inoculation. It is relevant because vaccines go in with different facilitators and different routes.

We need to know what those numbers are. But the problem, of course, with SIV is that you need macaques, which should not be treated for doing dose responses. It is better with mouse models that have been used for retroviral DNA.

Dr. Egan: This is obviously a highly difficult and complicated question. Perhaps it is why we are not hearing very definitive numbers in response to it. There are many variables, including the quantity of DNA, its size, the route of administration, and also a number of variables relating to the quality of the DNA, chromatin DNA, naked DNA, circular DNA, linear DNA, and so on. Perhaps we could have some discussion about some of the things Keith proposed, trying to look for the most sensitive system, making estimates of that, and then trying to estimate roughly the factors that present margins of safety. For example, if you had some estimate of the oncogenicity/tumorigenicity of some DNA, how is that affected by whether it is given intramuscularly or orally? One could then back off the estimate. And, we just heard some lovely data that the difference is about 10^3 or 10^4. Then one could look at size. Maybe larger sizes are very tumorigenic, and as it is reduced, it becomes less tumorigenic or oncogenic by two, three or four orders of magnitude.

I think there is something to be said for starting with the most sensitive system that one can devise and then trying to evolve some idea about what the safety margins are for the various factors.

Dr. Cichutek: I completely agree with the comment. It is a complicated subject; however, I think you highlighted exactly the point that experimental data are really needed, and what we learned from the gene transfer field is that we now have, for example, a leukemia model in mice usable for a platform study where you could look at the transfer of specific genes and the possibly resulting leukaemia development. That, of course, is not yet relevant for the IM vaccine inoculation situation so I think everything we said is summed up by: "Let us do the relevant experiments and look at all these variables in an animal model."

Dr. Alain: I have two questions. Since 2001, FDA has recommended that manufacturers reduce the size of DNA and evaluate this size for vaccine development produced on Vero cells. EMEA recommends a similar approach for tumorigenic cell lines, and in 2001, with an additional recommendation regarding the nature of viral sequence.

The first question I have is what is the consensus from the scientific community regarding the DNA size that we need to reach to reduce oncogenic risk?

The second question is more related to a basic testing problem. Until recently the enzymatic method known as the threshold method was considered as a reference method for residual DNA quantification. Following the FDA and EMEA recommendation, quantitative PCR methods have been implemented for the evaluation of DNA content and DNA size. These new methods are particularly useful since differences are observed with the threshold method for DNA fragments below 80 base pairs.

So, the second question is: what are the recommendations of the scientific community and maybe from regulatory agencies regarding testing methods for the evaluation of DNA content and size?

It would be fruitful if this meeting were to make some progress regarding harmonization of recommendations and also of testing.

Dr. Peden: As to the size, I addressed that. We are trying to determine at what size you lose biological activity and at what level.

I think those data will be available within the next few years. I mean, logically it makes sense that you reduce it to the smallest size possible.

At 650 base pairs and above, we found activity. We need to know the size and also the distribution. We do not know to what level it should be reduced, but it is obvious that it should be reduced to less than 650 base pairs, and our general feeling is that it should be around 100. But that is not a hard and fast number – it seems likely to be in that range. As for the methods, I do not know yet.

Dr. Alain: Regarding the methods, I think that new methods are really relevant, particularly Q-PCR to evaluate the DNA content and also to evaluate the size of DNA, and what will be your recommendation for this?

Dr. Peden: Speaking for the FDA, we do not have recommendations. So, you come to us with the method and we evaluate it. People have used quantitative PCR to look for size.

Dr. Cichutek: I could say something very similar for Europe. I do not think there is a clearly established harmonized view in that respect. Generally, PCR methods are welcome. Southern blots are still also used to look at the DNA. There is no clear size limit that we can give at the moment.

Dr. Doehmer: The concern is residual cellular DNA and its oncogenic potential. To make judgments on that, we should rely on the experience that we have with experimental conditions of gene transfer. If you want to know something about the size of fragments that are potent enough to be functionally integrated into genomic DNA, then we have to go back to the experience that we have with gene transfer, where we actually constructed plasmid DNA to have it functionally integrated.

In the beginning of the 1980s I started doing transfer experiments, expressing the growth hormone gene, and the experience at that time was that linear DNA is more efficient than circular DNA. It was stated that nucleases in the cells are crucial in the potency of the DNA to become integrated. My experience is that you have random linearization of plasmid DNA when you offer plasmid DNA in a circular form. On the other hand, you have chewing on the ends if you offer linear DNA. My experience was to make a stretch of DNA functionally integrated with a selective marker, if the gene of interest and the promoter in the range of 2KB, should be protected at least by 1,500 base pairs on either end. This represents quite a large piece of DNA being needed to have a good chance of becoming functionally integrated, i.e. a cloned DNA.

We should be aware of the experience that we have with gene transfer experiments but, at the same time, we should focus on cellular DNA. The experience that we have in different fields of gene transfer can have an influence, but our focus should be on the infectivity of the cellular DNA in a preparation and the chance that it has oncogenic potential. Anything else that is being discussed here surrounding this question can be used as a source of experience. We should come back not to cloned DNA or any plasmid DNA or gene therapy or anything like that. The issue is residual cellular DNA and its oncogenic potential.

Dr. Robertson: We have tremendous experience in the use of diploid cell-derived vaccines. Presumably these types of vaccines will have the same level and type of DNA contamination as we are concerned about with continuous cell lines. Diploid cells contain a great number of proto-oncogenes, much the same as the continuous cell will. With the continuous cell perhaps you are able to identify the presence of active oncogenes. But does the presence of proto-oncogenes in diploid cell DNA and thus in contaminating DNA vaccines, not say that that type of DNA is not an issue, is not a problem? What is our concern with the DNA coming from continuous cell lines? Because proto-oncogenes were to integrate, could they not be switched on with the same probability as an active oncogene entering and integrating into the recipient cell also being expressed?

Dr. Petricciani: I think that is precisely the point that Howard Temin was making when he commented about the inherent exposure that we all have to certain levels of proto-oncogene from basic cell turnover.

Dr. Cichutek: I fully agree with the comment. If you meant these activated oncogenes would, of course, impose a different risk compared with proto-oncogenes, theoretically all kinds of regulatory elements in DNA could pose a risk, but that is a theoretical discussion.

I also agree with the comment of Dr. Doehmer. I think it is important to return to assessing the risk by doing the experiments. To talk put residual DNA into animals which does not result in tumours does not give us the real scientific answer needed to address a possible risk of insertional oncogenesis quantitatively.

We need positive controls to assess numbers correctly. It does not help to inject residual DNA intramuscularly in some kind of model without knowing whether this is a predisposition model and a relevant positive control. In many cases in the gene transfer and also in the vaccine field, predisposition models are used and they help to assess risks in a quantitative way.

Dr. Robertson: But have the experiments not been performed by the use of millions, if not billions, of doses of diploid cell DNA? That DNA is being inoculated all the time with no apparent adverse events.

Dr. Strain: How much do we actually know of the nature of residual contaminant DNA which is present in these vaccine products? The fundamental question is that both the nature and size of the DNA is crucial. If meaningful experiments are to be done, they should be designed on the basis of the possible nature of the DNA, that being the residual contaminant DNA in question, rather than the experimental, convenient models that are being used or being discussed and described.

Dr. Wood: I want to endorse Klaus' comments that we should base our decisions on experimentation. In this regard, I think the experiments that Keith has initiated are very important because they are the first new data in this field. We can use some of the transgenic model systems we have available which we can divide into five or six major complementation groups so that we can know – (clearly, MYC and RAS complement each other in different groups) and use these genetically primed mice in all the issues we have been talking about: for size, the involvement of chromatin, and the delivery route; these are the models we should be evaluating.

Dr. Coffin: Two points on two issues. As Keith pointed out in his talk and I think also at the 1986 meeting, I find it disappointing that only FDA took the initiative, with extremely limited resources, to do something about defining some of these qualitative issues and work towards developing quantitative assays. I challenge all of you to build on this, before we return in 50 years and find this is still the only experiment we have. Funding within the government for these kinds of studies can be extremely hard to come by.

Secondly, the diploid cell experience regarding proto-oncogene uptake is not very relevant because oncogenes require some activating event. While there is a very small probability that an introduced DNA fragment will be put next to a strong promoter in some sort of integration event, that is a rare event on top of a rare event and does not really tell you what would happen should the contaminating DNA contain an already activated oncogene. Experiments such as this are the only way to resolve it. I request everyone interested to do some experiments on the subject.

Dr. Petricciani: This goes back to the issue mentioned earlier. The example was rabies vaccine in which there is an inactivating agent that not only inactivates the virus but is also a well-described inactivating agent for DNA. This is an important question, probably not so much for new vaccines that are under development as it is for the improvement of old vaccines. The issue is: under such conditions do we really need to worry about DNA at all?

Dr. Minor: If you consider how you would apply this question to a virus validation study, the question would not be that you treated it with BPL; therefore, there is no problem. It is a question of how effective the BPL is in inactivating the DNA. If the assumption is that you give it BPL and then all the DNA is killed, clearly it is a question to which you know the answer. To what extent does BPL, under the conditions used, actually inactivate the DNA?

Dr. Petricciani: That is absolutely correct. This assumes a validation package to support the manufacturing process.

Dr. Wood: This is, in fact, what WHO recommendations state, that they allow for substantially more levels of DNA per dose provided you have this validation of the specific manufacturing conditions. So, at least as far as WHO is concerned, we already are there in terms of this question.

Dr. Minor: But has anybody actually done the validation studies and what form do they take? I mean, I think there is a great culture out there about virus inactivation validation type studies which are quite hard. I am not sure how you do it for DNA and how confident you would be with the answers.

Participant: There have been a number of studies done, certainly for amines. What you can do is a series of nested PCRs looking at the frequency of inactivation per 100 base pairs. Those data have been used to justify the use of amines.

Dr. Peden: But it may be good to use a biological assay too that is more sensitive, and compare them.

Participant: Well, I am not sure about that, Keith. At least with amines, you have an inactivation progression of the reaction so you start to lose longer-term amplifications in your reaction. It is quite precise. That can then be applied to a specific target. It could be demonstrated that you have inactivated an insert.

Dr. Peden: It would have to be validated for your particular inactivation of your vaccine internally.

Participant: The same principle could be applied to beta-propiolactone.

Dr. Krause: I am very impressed with Keith's experiments where he also looked at BPL in a biological system but, of course, he was inactivating full-length HIV DNA and it may be that the kinds of genes that are of interest to be able to prove you have inactivated are smaller than that. So, I think that one really has to look, as David is saying, at this from a number of different perspectives and also at precisely what it is you are concerned with inactivating. Keith was also able to show 6 logs of inactivation of the HIV DNA using fairly stringent BPL inactivation conditions. I assumed that more inactivation with BPL would occur but, again, I think it gives a way of starting to look at this because what this field needs are data. I really applaud Keith's efforts.

Dr. Minor: I am not sure what the readout was in the HIV study. Was it infectivity? It is the readout that would worry me with this. It is reasonably clear to me that if you treat with BPL you are going to be better off than if you do not, in terms of any safety risk which exists. There is no safety risk which makes it extremely difficult. It may make it better but I am not sure how this could be quantified. I know you put a decent figure on that to say this really has succeeded, whatever the risk.

Dr. Cichutek: I agree but first there is a risk if tumorigenic cell lines are used in manufacturing. You could have some real oncogenes in it that could be transferred. Second, they can be used as a surrogate. If you look at the defragmentation of those specific oncogenes, at least there is a surrogate for what happens to other potential oncogenes in a cell.

Dr. Goldman: some of this question is addressed in your answer. The point of how to demonstrate the inactivation of the DNA by BPL or other chemical means, if you do have a designer cell where you have something that you know is causing the transformation of the cell line you can use PCR and specific methods such as that, but what if you do not? What does the panel think of general DNA activity type of assays, such as polymerase activity, hybridization and even the threshold system, which requires specific hybridization? If the DNA cannot be hybridized, it is not functional and cannot integrate. So, what does the panel think about that either in a Southern blot, threshold-type assay or just a general polymerization assay where the DNA cannot act as a substrate?

Dr. Peden: This is an interesting question. I like the idea of what David said about PCR of a certain size because it is hard to believe that if you cannot amplify any cellular gene after BPL treatment, there will be some gene that is hidden from BPL treatment. That has some merit. There is also the biological assay, whatever it is, which does not have to be HIV infection the way we do it. It can be a plasmid DNA because high levels of transformation are possible. In answer to Phil Minor's point, this effect can be quantified. If it is mixed with your vaccine, a plasmid DNA, you treat with BPL under the inactivation conditions you use for your vaccine, isolate the plasmid DNA and transform, and you have a control. You have plus or minus inactivation so you know what the level is. That gives you some safety factor.

Hybridization I would worry about, more so than PCR, because it just seems that is a less sensitive assay but, again, I have not seen the data for that. If you convinced me you can see the same sort of sensitivity of clearing 10^9 or something using hybridization that would be very interesting too. Does that help?

Participant: One thing I did not hear which is important to keep in mind in assays in the removal of sequences by cleaving DNA or by BPL, or by some other treatment, is the relative target size. The HIV assay is nice but you are inactivating a very large DNA sequence (~10 KB), whereas a small oncogene such as RAS is 10-fold smaller. If you cut the DNA to 600 base pair lengths there will be a great deal of RAS left even if HIV is down 9 logs. It is important to keep in mind the target size of the sequences that you might be interested in.

Dr. Petricciani: The next question reverts to one of the issues discussed earlier, the route of administration and in particular the oral route versus other routes. Any comments from the panel on this?

Dr. Cichutek: Definitely I see room for the distinction of a possible risk of insertional oncogenesis according to the specific route of administration. As we heard from the data presented today, there can be a large quantitative difference. For example, at some point plasmid DNA has also been used orally to vaccinate animals against SIV. When the data were examined closely, success could be achieved in most cases only if additional transfection reagents were mixed. Therefore, non-viral vectors, but not naked DNA, could be efficiently taken up by the oral route. A distinction should really be made in the risk assessment.

Dr. Wood: The current WHO recommendations do make that distinction. They specifically say that the upper limit of 10 ng does not apply to a product given orally. But then there is the usual sort of "let-out" clause saying that the acceptable limits should be set in consultation with the national control authority. In other words no guidance of a particular limit was given in that document. I think that was because there were no data available at that time, so it is very interesting to see the data being presented this morning. As far as I am aware, it is the first attempt to generate some data to support a limit for orally delivered products. In principle there is a difference, and it is good to see that data are being generated.

I would also add that there is much interest in other routes of administration. The intranasal route of administration of vaccines is something which is very much under development at the moment, and we should be thinking of other routes as well, not just the oral versus parenteral.

Dr. Peden: It is extended but the Malcolm Martin, Mark Israel, Wally Rhodes experiments of feeding polyoma DNA also came to the same conclusion, which is probably a more sensitive system than using PCR, but it is very nice to see that those experiments agree.

Dr. Petricciani: to the next question, number five: is there a reason to believe that there are different levels of risk depending on whether DNA is coming from autologous, homologous or heterologous sources? For example, is DNA from a murine cell such as BHK21, more or less of an issue than DNA from HeLa cells?

Dr. Cichutek: If heterologous sources mean xeno, then I would think this issue of transferring another xeno organism into humans could be an issue. Maybe there is an additional level of risk in that respect. In terms of autologous versus homologous sources, I cannot see the point to differ there. However, I would like to see what the scenario is.

Dr. Petricciani: Well, let me just repeat the example I gave. Is vaccine produced in BHK21 where there is residual cellular DNA different in terms of the perceived risk than the same vaccine produced in HeLa cells, which is human, whereas the other is murine?

Dr. Cichutek: My answer would be yes with respect to infectious agents that could be transferred with the DNA. Xenogeneic cell transfer could potentially pose a specific risk of introducing a foreign pathogen, e.g., a murine retrovirus, but, otherwise, in terms of the oncogenicity due to chromosomal insertion, I don't see a clear difference.

Dr. Petricciani: Now we are reverting to the issue of positive controls. Positive controls have been mentioned a number of times, and I should like the panel to discuss these in assessing cellular DNA as an element that may lead to an oncogenic event. What is the appropriate positive control? I am not aware of any data, certainly in animal systems, where cellular DNA has actually produced a tumour. What are you going to use for the positive control?

Dr. Minor: I think the original WHO discussions were based on polyoma. It is a highly potent oncogenic DNA, and the figure that looked most promising by whatever route was 100 pg per dose.

Clearly that is an overestimate of the amount that would be tolerated; far more could be tolerated because polyoma DNA is not being injected. If the answer is that DNA levels can be reached which give you a high level of confidence, then you are probably happier that way. But if you take a polyoma DNA which results in the conclusion that all the DNA has to be cleaned out, then I think you may not be happy with it, in which case you would want to consider a more appropriate positive control perhaps than a polyoma. I think it depends not only on the answer you receive but also on the answer you want. This is one of the matters in this area that concerns me.

Dr. Sheets: It should be borne in mind that many of these tests for vaccine cell substrate characterization are done at testing vendors, who are quite reluctant to take polyoma DNA into their animal facility and contaminate it. I think that infectious DNAs are going to be almost an absolute "cannot be done" for the routine kind of characterization of a cell substrate. I think we need to be thinking in terms of positive controls that will not contaminate an animal facility.

Dr. Minor: I was not necessarily specifying – polyoma was mentioned because that was the discussion at WHO. But it seems to me that you could have a highly oncogenic DNA that you would use as your positive control and polyoma was one that was used initially. Maybe there are others.

Participant: I endorse Phil's comments in terms of whether you want a tough oncogene – actually, we have had one demonstrated today, if you use an RAS-MYC combination that is obviously highly potent. One of the things about polyomavirus is that you have considerable transformation but for the base, and certainly the first 200 amino acids at the amino terminus of the T-antigen, that is transforming although not as efficiently as the whole T-antigen. So, that almost confirms Keith's kind of target for around 650 base pairs. It is a very sensible target because we know that that is near the minimum size of the transforming element. It seems a good model to continue to use.

Dr. Peden: Well, 650 was not a target but is what we had at that point. As a general rule, I think we need positive controls, otherwise you do not know whether the experiments worked. But finding those positive controls is incredibly difficult. We also did what Steve used and John Coffin and Don Blair said, that they would recommend using the RAS and the MYC, and they turned out to be a positive control. As you saw from the experiments, you needed both so it either requires both oncogenes to form a tumour in a primary cell, which agrees with the in vitro data, but you may also notice that if you go down 10-fold in efficiency you lose the activity. So, what we have done now is put those two genes on the same plasmid to see whether we increase the sensitivity of the assay.

We are not going to stop at the RAS and the MYC. The idea is to develop a series of positive controls to find out which oncogenes score positive, both in the normal mouse and in the transgenic mouse, because that is the aim of it. Then we should be able to have some idea of what positive controls are available and what they mean.

Dr. Cichutek: I am very happy about the question because, to turn it around, it tells us that we all agree that an experiment without using a positive control does not tell us anything at the moment. It only leads to a certain level of confidence. Seeing that DNA can be introduced is one thing, a much better thing is to use a positive control DNA resulting in tumour induction in the experimental setting.

To return to the gene transfer experience, there is the leukaemia model where you saturate bone marrow stem cells with about five copies of a retroviral vector containing LTRs. These vectors integrate but they do not transfer a specific tumour-promoting gene or an oncogene. It is only the LTR-containing vectors that integrate. Then, due to one, two or three activating events in specific proto-oncogenes, 100 percent of the animals develop leukaemia. That is an excellent model. From there you can start and then go back. I am not saying that we have to use exactly this model to assess the risk of insertional oncogenesis following intramuscular inoculation of DNA; I am just saying that such platform models would be very valuable and maybe your model can be developed into something similar. One may also think of using a transgenic model where you have a certain probability of cancer development and you may be able to increase this by introducing DNA. Those kind of experiments could be thought of.

Dr. Doehmer: In toxicology, the assessment starts with identifying the hazard. The hazard with residual cellular DNA is oncogenic potential. The question for the positive control, therefore, is to come as close as possible to the genetic configuration that causes oncogenicity in residual cellular DNA. We have experience with that. We know what sort of genes and promoters are necessary and what sort of genetic configuration is necessary to induce oncogenicity.

I can think of plasmid constructs that are positive in NIH-3T3 cells or 293 cells. I think this is a suitable positive control for checking the oncogenic potential of residual cellular DNA.

Dr. Shaw: I have a more practical problem with this approach. The systems that have been described today about positive controls using MYC and RAS and SV40 or polyoma are excellent systems for demonstrating that the method that you are using for putting nucleic acids into your test system are capable of producing a biological effect. But if we then turn it around to the most likely scenario, where we are looking at something like a Vero cell where we do not really know why it grows, but it grows reasonably well, it is not necessarily tumorigenic under the conditions in which we use it, but it potentially might have something in there that could become tumorigenic at some point. How do we bridge from the positive controls that we can construct using known materials either to confirming or refuting the notion that the DNA that comes as part of a heterogeneous mixture in sizes and compositions from a broken-up cell can be introduced into an animal system to produce a tumour? Because it is difficult to make a link between a heterogeneous mixture of things that we do not know much about and well-defined positive controls.

Dr. Wood: I think the answer to that, Alan, is really a subsidiary point, which is moving on to the transgenic animals. If you take the transgenic line and prime with MYC, and you know that those animals die within 90 days, usually a 100 percent death rate over 90 days, you add a second genetic insult to those. They can die in 60 days and a third will die in 20 days. So, you have fairly tightly based statistical arguments for looking at secondary and tertiary oncogenic hits. These are being used as an internal positive control and should then be challenged with different kinds of DNA. If you challenge them with an active oncogene they die very quickly and I suspect if you challenge them with Vero cell DNA it does nothing at all. But so far that experiment has not been done.

Dr. Shaw: And here we are five years later. I take your point very seriously; the knockout models that have been proposed will address complementation situations for oncogenes that we know about under specific circumstances. But your sort of regulatory sceptical agency is likely to reply that you have defined that particular lack of risk but what about the others? That is where I think we run into a problem.

Dr. Cichutek: We are not at the point where the 10 ng rule is challenged for cell lines that have been used in vaccine production because there is plenty of experience of human use not pointing to any tumour risk. There are new developments such as AAV vectors, adenoviral vectors, and others, that are produced on tumorigenic cell lines containing no viral oncogenes and this is an additional new challenge here.

So, in that respect there may be models where you could increase or decrease the probability of tumour development, e.g. if you use a predisposition transgenic model, and this is what we have to take into consideration. For example, think of all the HIV vaccines that are currently being developed. There is no longer a conventional one. The conventional ones have unfortunately failed. So, all the others may be associated with a certain level of chromosomal integration of the vector and, in addition, some of them will be produced in tumorigenic cell lines because otherwise you do not achieve sufficient vector titre. That is the important future issue. We need to discuss what would be an acceptable level of integration of vectored vaccines and how to decrease the amount of potential tumorigenic DNA and oncogene in the vaccine preparation.

Dr. Shaw: I think we have two different issues here. One is relatively simple, and that is where you have a cell that has been transformed with a specific set of genes so you think you know why it is transformed. If you have a series of genes for adeno-E1A and E1B for example, that can be tested for, you can in a relatively straightforward manner exclude that material from your product by testing, purification or other method.

The more difficult problem is the thing that you do not know about in other kinds of cell lines that will probably be used for something or other. I was using Vero as an example although it is well accepted now, but the issue is other cell types, including potentially human cell lines that do not have a frank oncogenic basis that is well described. I think that presents a bigger challenge.

Dr. Cichutek: Yes. I would like to mention two things. First of all, I had one slide showing a proposal where applicants have come forward to agencies and propose how to qualify such cell lines by looking in nude mice or other animals at the risk of tumour induction of cell lysates, the cellular DNA etc. of such cell lines. They could also propose methods to decrease DNA fragment size or show fragmentation of the specific oncogenic DNA under discussion. That is a suggestion on which I would like to receive some feedback from other regulatory agencies. The other thing that we have not even tackled is that some of these new vaccines actually integrate knowingly. There are animal data showing that some new vaccines under experimental research in animals produce a certain level of integration and that should be discussed as well. Is that acceptable at all? Can we define an acceptable level of chromosomal integration? How do we go about assessing such a level?

Dr. Petricciani: Because time is short, I am sorry that we cannot start that discussion.

Dr. Lewis: I want to endorse what Alan said about the practical experience and what the positive control may or may not tell us. I do not think it is ever going to be possible to develop positive controls at the level of actually defining or eliminating the possibility of risk. It is practical. The examplc I would use is that there is a situation in animals called injection site sarcoma in cats. This is a situation in which felines were given repetitive vaccines on a yearly basis and after a period of time approximately 1/5,000 cats will develop sarcoma at the inoculation site. This is not an uncommon event but it occurs at a frequency that would be almost impossible to measure.

I think any kind of testing on those vaccines and products would probably not pick up the possibility that this could occur. But we are always limited by the state-of-the-art so I think the best we can do at this point until we understand these phenomena better is to do what we can, develop positive controls that work and then work our way down the scale as time progresses and improve on what we have.

Dr. Petricciani: The final question, as opposed to laboratory testing, is it possible to approach risk assessment as has been done in the past, based on certain assumptions? As Klaus has pointed out, some of the original assumptions may or may not be valid and there are some additional data. So, any quick comments on this one?

Dr. Cichutek: Just a general one that we have been presented with theoretical calculations. They are important and give you some level of reassurance. Experiments are always better for assessing a risk. So, theory plus experiments are better than theory.

Consensus Statement

The range of cell substrates that could be used to produce biological products, including vaccines, is broad, and includes the following categories:

1. Primary cells.
   Examples: monkey kidney, chick embryo fibroblasts.
2. Diploid cell lines (human and non-human primate).
   Examples: WI-38, MRC-5, FRhL-2.
3. Continuous cell lines.
   a) Tumorigenic
      i) In vitro "transformation" during subculture (animal)
         Examples: BSC-1, LLC-MK2.
      ii) Transformed in vitro by whole virus or viral element(s) (animal and human).
         Example: 293, PerC.6.
      iii) Derived from tumour tissue (human and animal)
         Examples: Namalwa, HeLa, T-24.
   b) Non-tumorigenic (animal)
      Example: VERO at passages < 200.

However, the range of cell substrates that are widely accepted for vaccine production is limited to categories 1, 2, and 3b in part due to concerns about residual cellular DNA. DNA from cells in category 1 and 2 is not considered to pose a risk and there is no restriction on the level of residual cellular DNA in vaccines derived from them.

Based on theoretical considerations, residual cellular DNA from any cell substrate might give rise to: (i) an infectious event or (ii) an oncogenic event. An oncogenic event could arise through insertional mutagenesis, resulting in activation, inactivation, up-regulation, or down-regulation of critical genes, or inactivation of tumour suppressor genes. An oncogenic event could also arise via the introduction of a dominant, activated oncogene. Because insertional mutagenesis has been considered by several study groups and advisory committees to be a very low probability event, because insertional mutagenesis would be expected to occur with any cellular DNA and is not a unique feature of continuous cell lines, and because DNA vaccines, which contain a highly efficient promoter, have been permitted at levels up to 4 mg per dose for intramuscular injection, oncogenesis via insertional mutagenesis has not been considered a significant risk to vaccine recipients.

The one theoretical risk that may be specific to the DNA of continuous cells lines (Category 3) is the potential for a dominant, activated cellular oncogene to be transmitted in a biological product and then to cause a tumour in recipients of that product. Therefore, the considerations that follow apply only to category 3 cells.

The potential for risk from an infectious agent is significantly reduced, if not eliminated, in those cases where the cell substrate has been well-characterized to demonstrate the absence of viruses and viral sequences that are of concern because of their potential to cause disease in humans. In these cases, the theoretical risk associated with the DNA is therefore limited to an oncogenic event.

Many previous conferences have considered these issues, and various positions have been taken on the acceptability of cell substrate DNA at levels ranging from 10 pg to 10 ng. However, a consideration of the entire body of information available has not been undertaken recently, including human experience that relates to the issue of DNA risk. No information has been published on the use of new sensitive models to assess the oncogenic potential of residual cellular DNA.

Based on the presentations at this conference and the Panel discussion, we have reached a consensus on the following points:

1. No published study to date has demonstrated that tumour cell DNA can cause tumours in animal models or in humans. Quantitative risk estimates by five groups, based on various assumptions, suggest that if there is a risk of oncogenesis from cell substrate DNA in biological products, it is extremely low [Ref. 1-5].

2. Rigorous cell substrate and product characterization studies should be undertaken to identify risk factors associated with cellular DNA, and the manufacturing process should address them by achieving a level of cellular DNA well below a theoretical point at which DNA might be expected to be oncogenic. The acceptable level of residual cellular DNA per human dose has evolved as the assessment of risks has evolved from 10 pg (FDA 1984), to 100 pg (WHO 1986), to the current value of 10 ng (WHO 1994).

3. The risk of residual cell substrate DNA can be reduced to negligible levels when a DNA-inactivating method or a nucleic acid fragmentation method is used in the manufacturing process of vaccines, and data are available to validate these methods, and to demonstrate the consistency of the manufacturing process.

4. WHO should establish a working group to recommend studies designed to answer specific questions relating to theoretical risks associated with residual cellular DNA. Such studies should be supported by appropriate government agencies. Among the studies that might be recommended are:
   a) Platform studies to address the risk of oncogenesis by residual cellular DNA in appropriate models, including the use of relevant positive controls.
   b) Dose-response studies to determine the relationship of DNA dose to biological activity.
   c) Studies to determine whether there is less risk from DNA derived from non-tumorigenic continuous cell lines than from those that are tumorigenic. The 1986 WHO Study Group recommendation of 100 pg of DNA was meant to apply to both tumorigenic and non-tumorigenic continuous cell lines. The subsequent 1994 WHO recommendation by the Expert Committee on Biological Standardization (ECBS) of 10 ng of DNA also was meant to apply to all continuous cell lines.
   d) Risk reduction strategy studies to determine the impact of reducing the size of DNA fragments to various lengths, and to assess the effect, if any, of the configuration of the DNA (naked DNA, chromatin DNA, etc.) on risk.
5. Agreement among major regulatory agencies and WHO should be reached on levels of cellular DNA that can be considered risk-free. WHO could facilitate reaching a consensus on this point and should be encouraged to do so by undertaking a review of the DNA issue, as was done by the 1986 Study Group and the 1994 ECBS, when sufficient new data become available to warrant such a meeting.

References

1 Report on the meeting of the study group on biologicals. WHO Technical Report Series, 1987, No. 747.

2 Petricciani JP, Regan PJ: Risk of neoplastic transformation from cellular DNA: Calculations using the oncogene model, in Petricciani JC, Hennessen W (eds): Cells -Products – Safety. Dev Biol Stand. Basel, Karger, 1987, vol, 68, pp 43-49.

3 Temin H: Overview of biological effects of addition of DNA molecules to cells. J Med Virol 1990; 31: 13-17.

4 Kurth R: Risk potential of the chromosomal insertion of foreign DNA. Ann NY Acad Sci 1995; 772: 140-151.

5 Krause PR, Lewis AM Jr: Safety of viral DNA in biological products. Biologicals 1998; 26: 317-320.

Petricciani J, Sheets R (eds): Vaccine Cell Substrates 2004.
Dev Biol (Basel). Basel, Karger, 2006, vol 123, p 75.

SESSION II

Potential Oncogenicity of Viruses Latent in Cells

Petricciani J, Sheets R (eds): Vaccine Cell Substrates 2004.
Dev Biol (Basel). Basel, Karger, 2006, vol 123, pp 77-88.

Molecular Methods for Virus Discovery

R.F. Jarrett, D. Johnson, K.S. Wilson, A. Gallagher

LRF Virus Centre, Institute of Comparative Medicine, University of Glasgow, Glasgow, U.K.

Abstract: The potential use of novel cell substrates from diverse animal species raises concerns about the transmission of hitherto unknown viral agents. Viruses that do not cause a cytopathic effect in cell culture may escape detection by conventional methods and molecular methods may therefore prove useful for screening for hitherto unknown viruses. This review describes currently used molecular methods for virus discovery, including degenerate PCR assays, representational difference analysis and rolling circle amplification, and summarises the advantages and disadvantages of each technique.

INTRODUCTION

Characterizing potential virus contaminants in vaccines is a complex and constantly evolving problem. Contaminants have been associated with the use of primary cell substrates and of non-irradiated materials such as trypsin and serum. The move towards using clonally derived cell lines, irradiated serum and recombinant trypsin have been major steps in improving the safety of the new generation of vaccines. At the same time there has been renewed interest in cell substrates, some of them tumorigenic, from species such as dogs, African green monkeys and Syrian hamsters. Of particular concern is that these cells might harbour a virus which, while non-pathogenic in the species of origin, might be tumorigenic in a human host. Virus infections in these cell substrates may not always be revealed by traditional assays that depend on a cytopathic effect, and this has driven the need to implement new technologies for virus detection.

Molecular techniques for virus discovery have been developed over the last ten to fifteen years and now have a proven track record. Early attempts to identify novel viruses by molecular means lacked sensitivity and specificity but modern amplification techniques, such as the polymerase chain reaction (PCR), have revolutionised the field of virus discovery. Currently used methods include degenerate PCR assays, subtractive hybridisation including representational difference analysis (RDA) [1], and the more recently described rolling circle amplification (RCA) technique for detection of closed circular DNA genomes [2]. This review will summarise these techniques and their advantages and disadvantages. Additional methodologies, such as screening cDNA expression libraries with immune serum, have been successfully used in the past [3] but are less relevant to the analysis of vaccine cell substrates and will not be considered further.

Degenerate PCR Assays

Degenerate or redundant PCR assays use pools of primers that include most or all of the possible nucleotide sequences encoding a conserved amino acid motif. They have been used extensively to identify genes encoding proteins related to known proteins, for example new members of protein families or functional homologues in different species. In the context of virus discovery, these assays are used to detect novel viral genomes related to known genomes; they are therefore useful for the identification of new members of virus families or subfamilies.

The first step in the design of degenerate assays is to identify proteins that are well conserved among different members of a virus family and to perform an alignment of all the available amino acid sequences. Conserved blocks of sequence are then identified; in our experience conserved pentapeptides are ideal for primer design. A degenerate nucleotide sequence, containing all the possible sequences encoding the pentapeptide, is then derived from the amino acid sequence. The level of degeneracy (i.e. the number of different primers in the primer pool) will depend on the number of codons for each of the amino acids. Since sensitivity is inversely proportional to degeneracy, it is preferable to select conserved peptides that contain amino acids such as methionine, with only one codon, rather than amino acids such as serine, with six; unfortunately, this is not always possible. The degenerate nucleotide sequence corresponding to a pentapeptide is 14 bases long, since the terminal degenerate nucleotide is omitted from the primer. To achieve a reasonable level of sensitivity a primer with a degeneracy of 128-fold or less is desirable.

To increase PCR efficiency, a non-degenerate clamp sequence is usually added to the 5' end of the primer; this facilitates annealing of the primer to PCR products generated during the reaction. In our laboratory, the clamp sequence is derived from the consensus **nucleotide** sequence at the relevant position of the viral gene, with some modifications to ensure a reasonably high primer Tm. Since this clamp sequence has homology to the viral target template, it will stabilise primer binding during the early cycles of the PCR. We also usually add the sequence GTTT to the 5' ends of primers to ensure optimal cloning efficiency using TA cloning methods [4]. The first rounds of thermal cycling are performed using a low annealing temperature (e.g. five cycles with annealing at 44 °C) to allow the degenerate sequence to anneal to the template and then the annealing temperature is raised to facilitate specific amplification of PCR products.

A similar strategy for degenerate PCR design has been independently developed in Tim Rose's laboratory and designated CODEHOP (COnsensus-DEgenerate Hybrid Oligonucleotide Primer) [5]; a software programme is available for primer design at http://blocks.fhcrc.org/codehop.html. The main difference between this method of primer design and that described above is that the clamp sequence is derived from amino acid sequence, using codon preferences based on species.

Degenerate PCR assays are inherently non-specific and lack sensitivity compared to conventional PCR assays. Probes are generally not available and therefore positive results cannot be confirmed by hybridisation; identification of novel viral sequences is therefore dependent on the identification of DNA fragments in the anticipated size range and subsequent sequence analysis. For these reasons, we prefer to analyse PCR products using capillary electrophoresis and GeneScan™ analysis (Applied Biosystems, Warrington, U.K.). This gives improved sensitivity (up to 100-fold) compared to conventional agarose or polyacrylamide gel electrophoresis, and the superior fragment size resolution helps distinguish non-specific amplification products, present in control samples, from fragments of interest [6].

The major focus of research in our laboratory is the search for a novel viral agent in Hodgkin lymphoma [7]. This malignancy presents many challenges as the malignant cells are fragile, difficult to purify and scarce, usually representing less than 1 % of the total cells in the tumour. To detect viral genomes present in the malignant cells, our degenerate PCR assays have to be able to detect a single copy genome present in 1 % of the sample. Since 1 μg of tumour DNA is generally used as template in our assays, we aim to detect 1.5×10^3 viral genomes within a background of high molecular weight DNA. The epidemiology of Hodgkin lymphoma suggests that the disease may result from delayed infection with a common infectious agent. We have therefore focused our efforts on the search for new members of virus families that are widespread in the human population and infect most individuals at a young age. For these reasons, we have designed degenerate PCR assays for the detection of herpesviruses and polyomaviruses, as described below.

PCR assays for detection of herpesvirus genomes

Degenerate PCR assays have been used successfully to identify many new members of the herpesvirus family [8,9]. Assays are generally based on the polymerase and glycoprotein B proteins, as these are well conserved across family members. Alignment of the available polymerase sequences reveals several blocks of amino acid sequence that are well conserved in all herpesviruses, with the exception of the fish viruses (Fig. 1). We have designed an assay based on the YPSII and IYGDT amino acid motifs (Fig. 1); although YPSII is not absolutely conserved, the substitution of the final isoleucine (I) with methionine (M) does not affect the primer design. Since 144 primers are required to cover all possible nucleotide sequences encoding the pentapeptide YPSII, we have split the synthesis of this primer into two, to maintain sensitivity. An internal primer, based on the YGF/VTG motif, allows the assay to be run in a semi-nested format (Fig. 1); again, the synthesis of this primer has been split into two, with one primer covering the codon usage of phenylalanine (F) and the other valine (V). Clamp sequences have been added to the 5' ends of the primers, essentially as described above (Table 1).

This assay is extremely sensitive and robust [6,10]. Using the outer primer sets we can detect 10^3 copies of a herpesvirus genome present in a background of 1 μg DNA and this sensitivity increases to $< 10^2$ copies using the semi-nested primers. We have validated the assay using a wide range of templates (Fig. 1), and have amplified *pol* gene sequences from porcine cytomegalovirus, a known virus for which no sequence was available at the time of analysis. The assay is relatively simple to perform and can therefore be used for screening purposes. Similar assays have been used by other laboratories to identify novel herpesvirus sequences in a

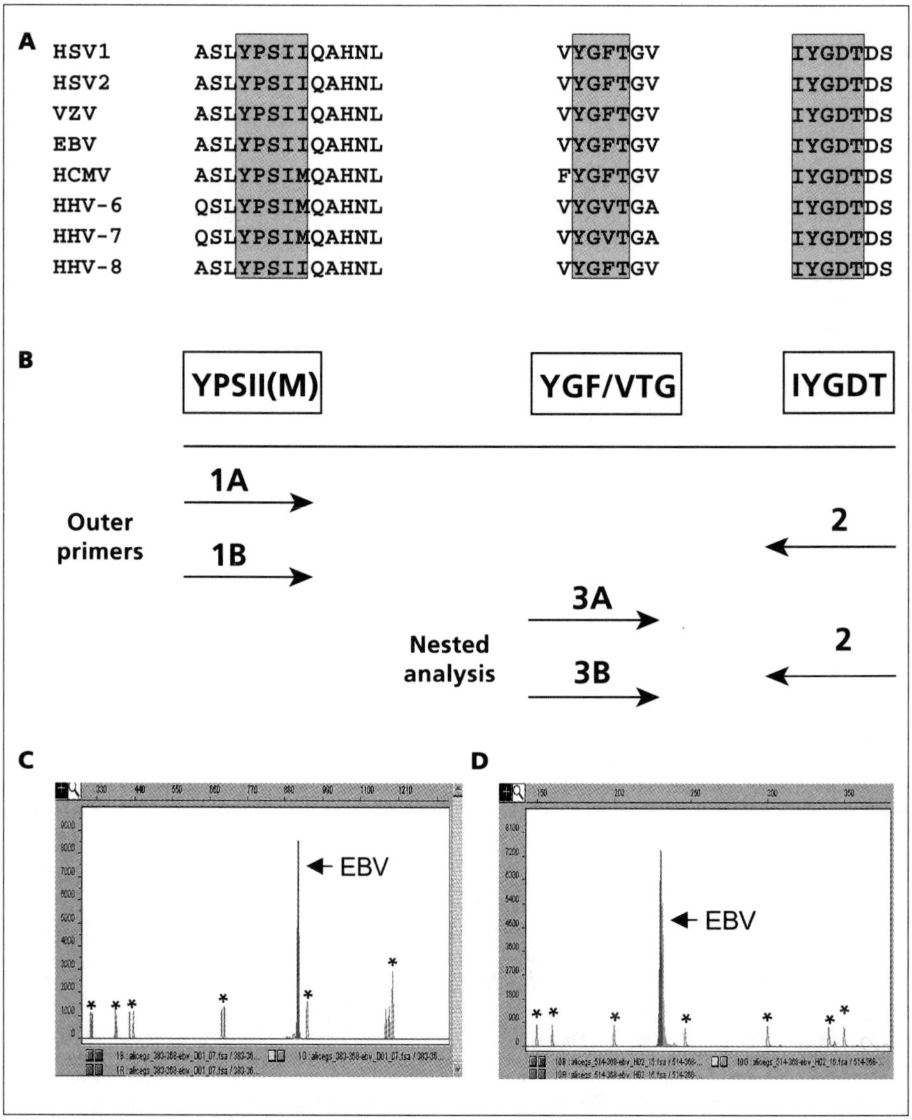

Fig. 1: **Degenerate PCR assay for detection of herpesviruses.** Panel a: Amino acid sequences of the herpesvirus polymerase proteins were aligned and blocks of conserved sequence identified; the alignment of the human herpesviruses is shown but these sequences were well conserved across herpesviruses of all species except fish. Pentapeptides were then selected (boxed sequences) on the basis of codon usage, to minimise degeneracy of the corresponding primer. Panel b: Three outer primers and two inner primers were synthesised; primer 1 was split into two syntheses, which differed in their codon usage for serine, and primer 3 was split in two syntheses with primer 3A including all nucleotide sequences encoding phenylalanine (F) and 3B including those encoding valine (V). The assay is run in a semi-nested format using the indicated primers. Panel c: GeneScan analysis showing amplification of Epstein-Barr virus (EBV) specific *pol* gene sequences using outer primers 1A and 2. * indicates size markers. Panel d: GeneScan analysis of products of semi-nested assay showing detection of EBV *pol* sequences using primers 3A and 2. * indicates size markers.

Table 1 : Primers used in degenerate PCR assays.

Number	Sequence
Herpesvirus	
1A	*GACTTTCCAAGTTTC*TAYCCNAGYATHAT
1B	*GACTTTCCAAGTTTC*TAYCCNTCNATHAT
2	*ACAAACATACAGTCC*GTRTCNCCRTADAT
3A	*GTTTGATGCCGACCT*TAYGGNTTYACNGG
3B	*GTTTGATGCCGACCT*TAYGGNGTNACNGG
Polyomavirus	
4	*GTTTATGGTTGTCTTT*GANGAYGTNAARGG
5A	*GTTTCGGTGTTTCTTT*TCNARRTTNACYTT
5B	*GTTTCGGTGTTTCTTT*TCNARRTTNACNGG
6	*GTTTGAGGCACAGAA*TAYTCRTTCATNGT

Primer numbers refer to those used in Figures 1 and 2. Clamp sequences are shown in italics.
N = A + C + G + T ; Y = A + T ; R = A + G ; D = G + A + T ; H = A + T + C.

wide range of animal species including non-human primates, pigs, and turtles [8,9,11]. Novel herpesvirus sequences were not detected in DNA samples from Hodgkin lymphoma biopsies using this approach [6].

PCR assays for detection of polyomavirus genomes

Polyomaviruses have well conserved amino acid sequences, particularly in the gene encoding the virus capsid protein VP1 [12]. However, tumours associated with polyomaviruses frequently contain integrated viral genomes which have lost these sequences, although the gene encoding the T antigen is retained. We therefore elected to design an assay based on the large T antigen, which is less well conserved than the structural proteins. Sequences from SV40, particularly those at the 5' end of the large T gene, are frequently included in plasmid vectors and contamination of DNA samples with these vectors can lead to false positive results in PCR assays for SV40 and related viruses [13]. It is therefore prudent to avoid the 5' part of the T antigen gene when designing primers.

Alignment of the available large T antigen sequences did not reveal any conserved pentapeptides although there were several tetrapeptides that were conserved across all viruses, with the exception of the Kilham strain of mouse polyomavirus (Fig. 2). We selected the E/DDVKG motif for design of the 5' primer, since codons for E (glutamatic acid) and D (aspartic acid) only differ at the third position, and the VNLE motif for design of the 3' primer. Synthesis of the latter primer was split into two, with separate primers covering KVNLE and PVNLE (Fig. 2). An additional 3' primer derived from the amino acid sequence TMNEY was also synthesised; although this sequence is not conserved in all polyomaviruses, the respective primer has only 16-fold degeneracy and was predicted to give greater sensitivity than the VNLE primers. Clamp sequences were added to all primers, as described above.

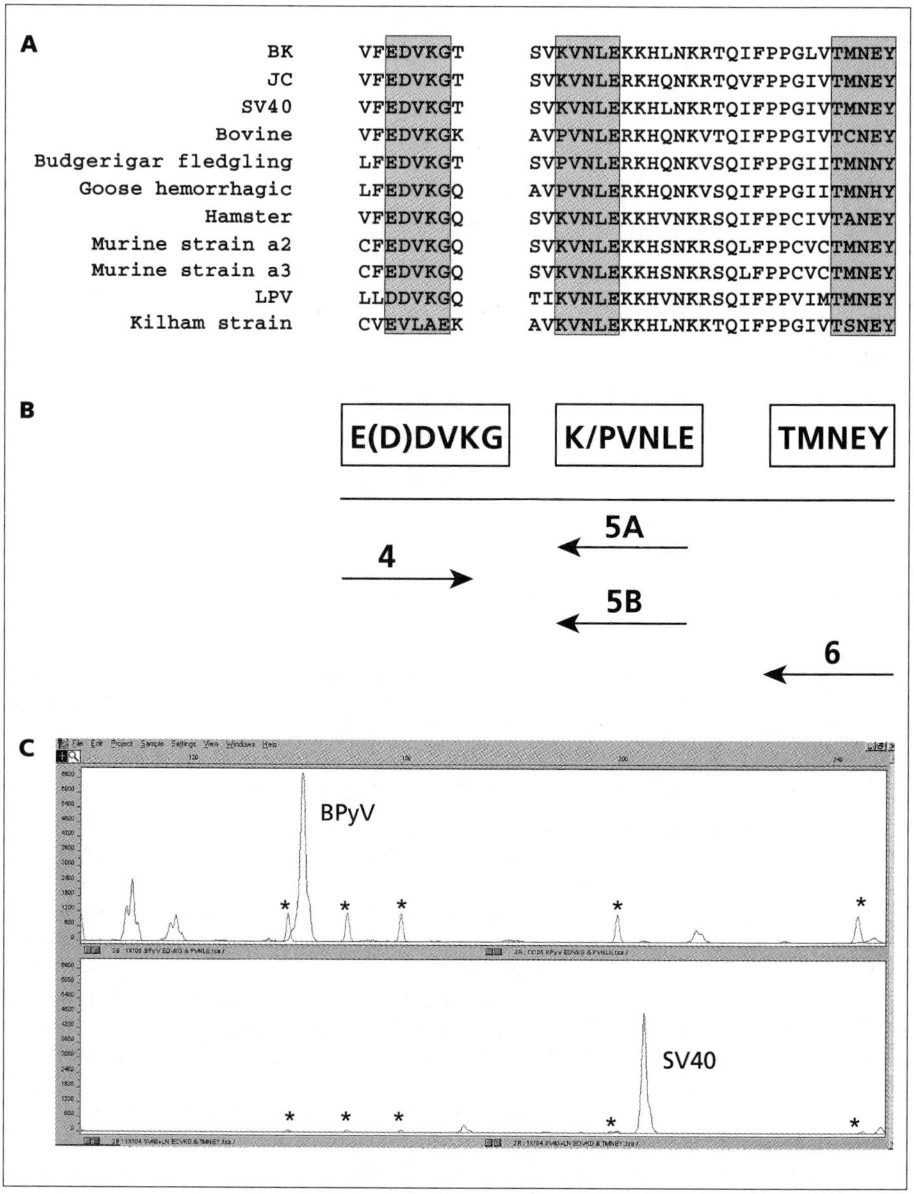

Fig. 2: **Degenerate PCR assay for detection of polyomaviruses.** Panel a: Amino acid sequences of the polyomavirus large T antigen were aligned and blocks of conserved sequence identified (boxed areas). Panel b: One 5' primer and three 3' primers were synthesised; primer 5A covers all possible codons for KVNLE and primer 5B covers all possible codon usage for PVNLE. Primer 6 is derived from the sequence TMNEY; this sequence is not conserved across all polyomaviruses but the corresponding primer has a degeneracy of only 16-fold, resulting in a sensitive PCR reaction. Panel c (upper electropherogram): GeneScan analysis showing detection of bovine polyomavirus (BPyV) using primers 4 and 5B. Panel c (lower electropherogram): GeneScan analysis showing detection of SV40 sequences using primers 4 and 6. * indicates size markers.

The assay was validated using a range of templates including the human viruses JC and BK, the non-human primate virus SV40, hamster polyomavirus and bovine polyomavirus (Wilson et al, submitted). Primer sets had the expected specificities. In addition, sequences from a rabbit polyomavirus, a virus for which no sequence information was available, were amplified (Wilson et al, submitted). Using the KVNLE and PVNLE primers we were able to detect $10^3 – 5 \times 10^3$ viral genomes in a background of 1 µg DNA; as predicted, the assay using the TMNEY primer was more sensitive, detecting $< 10^3$ viral genomes in a 1 µg sample.

Detection of polyomavirus sequences in human lymphoma

The detection of SV40 in samples of non-Hodgkin lymphoma has become a controversial issue with some groups detecting viral genomes in > 40 % of cases [14], while others have failed to detect evidence of the virus [15]. We screened DNA samples from 152 cases of non-Hodgkin lymphoma and 35 cases of Hodgkin lymphoma for the presence of SV40, BK virus and JC virus genomes using specific TaqMan™ assays with negative results [16]. A previously described consensus PCR assay, based on nucleotide sequences conserved between JC, BK and SV40 genomes, was also used to screen 128 of the non-Hodgkin's lymphoma samples and 18 of the Hodgkin lymphoma samples; again only negative results were obtained. The degenerate assays described above were used to investigate 35 cases of Hodgkin lymphoma but we found no evidence of polyomavirus genomes (Wilson et al, submitted). These results suggest that it is unlikely that SV40 is involved in the pathogenesis of commonly occurring non-Hodgkin lymphomas, and polyomaviruses are unlikely to play a direct role in Hodgkin lymphoma.

The above description provides examples of degenerate PCR assays designed to search for novel DNA tumour viruses. However, these assays can be used to detect a wide range of viruses including RNA and DNA viruses, viruses causing lytic and latent infections, and viruses with circular and linear genomes. For example, degenerate PCR was used successfully to identify a new hantavirus associated with an acute respiratory distress syndrome [17]. These assays are sensitive, versatile, robust and relatively easy to perform and can therefore be used for screening purposes. The main disadvantage is that they require knowledge of the viral agent that is likely to be involved and can only be used where sufficient sequence information, from related viruses, is available. Assays will be most reliable when primer design is based on the alignment of a large number of viral genomes, as is the case with the herpesvirus assays.

Subtractive Hybridisation Techniques

Subtractive hybridisation is used to isolate differences between two nucleic acid species. Since viral infection constitutes the addition of exogenous nucleic acid sequences to a cell, this methodology can be used to detect RNA or DNA viruses. These assays are more difficult and cumbersome to perform than degenerate PCR assays but have the advantage that no a priori knowledge of the potential viral agent is required.

Subtractive hybridisation involves two samples of nucleic acid – the **tester** sample, which includes the **target** sequence of interest, and the **driver** sample. The tester and driver samples are hybridised, allowing complementary strands of nucleic acid to anneal, and then various strategies are used to rescue either single-stranded tester molecules or tester: tester hybrids. The driver and tester samples must be as closely matched as possible, to minimise differences in nucleic acid composition. For subtractive hybridisations using DNA samples, the driver sample should be derived from a non-affected sample from the same individual, whereas in cDNA subtractions the driver should be from the same cell type (at the same stage of differentiation) as the tester sample. Driver samples are always used in excess in the hybridisation reaction.

Subtractive hybridisation works best when nucleic acid complexity is low and the hybridisation reaction approaches completion. Body fluids, such as serum, and cell culture supernatants, probably represent the least complex samples for virus hunting. RNA is considerably less complex than genomic DNA; RNA/cDNA subtractions are therefore simpler to perform than DNA subtractions and a wider choice of methodologies is available. Strategies for subtractive hybridisation using genomic DNA samples generally lack sensitivity because the complexity of the DNA prevents the hybridisation from reaching completion. The representational difference analysis (RDA) protocol circumvents this problem by using only a representation, or defined part, of the human genome in each hybridisation reaction [1]. Representations are defined by digestion with a restriction enzyme and subsequent size selection. This results in a reduction of complexity of the DNA sample, which varies depending on the enzyme used.

Following the generation of tester and driver representations, adaptor sequences are ligated to the 5' ends of the tester sample. The tester and driver samples are then hybridised and a 'fill-in reaction' is performed. This results in the addition of nucleotides, complementary to the adaptor molecule, to the 3' ends of all molecules that have hybridised to tester molecules. As a result, tester molecules that hybridised to other tester molecules (tester: tester hybrids) will have adaptor sequences, or their complements, on each end and will therefore be amplified exponentially using the adaptor as primer. Molecules that formed driver: driver hybrids will not have adaptor sequences on either end and will not be amplifiable using the adaptor primer, whereas molecules that were part of tester: driver hybrids will have an adaptor sequence at only one end and will be amplified linearly. Similarly, single-stranded tester molecules will be amplified linearly. Each hybridisation and 'fill-in' is therefore followed by a PCR reaction, using the adaptor as primer, which results in the selective enrichment of double-stranded tester molecules. Following amplification, the adaptor sequence is changed and further rounds of hybridisation and amplification are performed. The first round of RDA is largely subtractive but, as a result of the second order kinetics of reassociation, subsequent rounds have an increased kinetic component.

The chance of detecting a viral genome using RDA increases with genome size since there is a greater chance of fragments from larger genomes being present in the representation. We have estimated that there is a 99 % probability of detecting the presence of a herpesvirus genome in a single representation, but two independent representations are required to have a 95 % probability of detecting an adenovirus genome and three to four are required for the detection of a retrovirus genome.

The RDA protocol requires meticulous bench work but, in our experience, the technique can reliably detect viral genomes present at single copy level. By monitoring the procedure using quantitative PCR, we have shown that fragments of the EBV genome can be amplified > 1 million-fold using three rounds of RDA. In our hands the technique does not work reliably when the target fragment is present at lower levels and when the starting DNA samples are not of good quality.

The main advantage of RDA is that no a priori knowledge of the infectious agent is required. It is also versatile in that it can be used to detect linear and circular genomes and modified protocols can be used for the detection of RNA viruses and viruses present in body fluids. The need for matched samples, multiple representations and experienced laboratory workers are the main disadvantages. The technique is not ideal for screening samples and a high index of suspicion is required before embarking on RDA. Nevertheless, RDA has a good track record for virus discovery including the identification of human herpesvirus 8 [18], the elusive agent in Kaposi's sarcoma, and the circovirus TTV [19].

Rolling Circle Amplification

Rolling circle amplification (RCA) is a technique used to exponentially amplify single- or double-stranded circular DNA templates. It is widely used as a laboratory method for amplifying small circular molecules, such as plasmids, and has only recently been described as a method for virus discovery [2]. RCA uses bacteriophage ϕ29 DNA polymerase, a high-fidelity enzyme with a strong strand-displacing capability, high processivity and proofreading activity. The processivity, coupled with the strand-displacing capability, enables the generation of tandem-repeat concatemers of circular DNA molecules. Elongation of displaced single-stranded products results in exponential amplification of the template DNA. Random hexamers are used to prime the reaction, eliminating the requirement for custom primers. RCA can therefore be used to amplify complete, circular DNA genomes without any knowledge of viral genomic sequence.

Rector et al (2004) first reported the use of the RCA technique in virus discovery. Using a commercially available kit (TempliPhi 100, Amersham Biosciences, Amersham, U.K.) they were able to selectively amplify complete genomes from human papillomavirus 16 DNA. Assay sensitivity was improved by increasing the concentration of nucleotides in the reaction. Optimised conditions were then used to isolate and sequence a novel variant of bovine papillomavirus type 1 and a novel papillomavirus from a Florida manatee [2,20]. We have used this protocol to amplify complete papillomavirus genomes successfully from equine sarcoid tumours (Fig. 3) and complete polyomavirus genomes, including the genome of the rabbit polyomavirus mentioned above. The technique is relatively simple to perform and is suitable for screening purposes. The major disadvantages of RCA are: (i) it can only be used to selectively amplify circular genomes and (ii) assay sensitivity is decreased when genomic DNA samples are used as a template, since the background DNA is also amplified, albeit with lesser efficiency, during the multiple displacement amplification reaction.

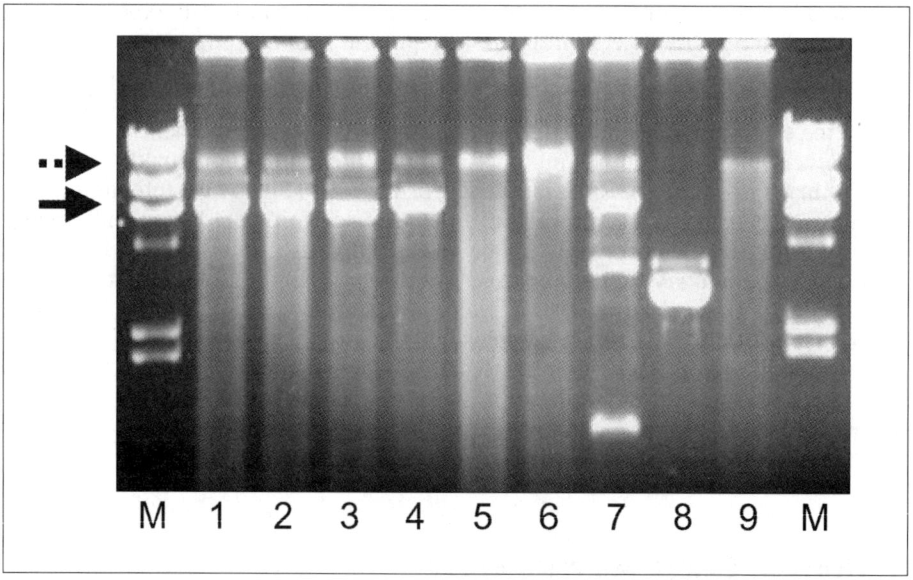

Fig. 3: **Analysis of equine sarcoid tumour samples using rolling circle amplification.** DNA samples were subjected to rolling circle amplification followed by restriction enzyme digestion with *Hin*dIII and agarose gel electrophoresis. Solid arrow indicates the band representing the complete papillomavirus genome; dashed arrow indicates non-specific products. Lane M, *Hin*dIII-digested lambda phage DNA size marker; lanes 1-5, samples from equine sarcoid tumour samples – full length papillomavirus genomes are detected in four of the samples; lane 6, equine DNA; lane 7, cloned genome of bovine papillomavirus type 1 in a DNA background; lane 8, plasmid control; lane 9, water.

Conclusions

The last fifteen years have witnessed the dawn of the molecular era of virus discovery. Several complementary methods for the identification of novel viruses have been developed, each with its own pros and cons. Degenerate PCR has the advantage of sensitivity and simplicity but can only be used to detect related members of known virus families. Subtractive hybridisation techniques, such as RDA, are technically more complex and time-consuming and are less sensitive but do not require any knowledge of the potential infectious agent. The recently described RCA technique can only be used to amplify circular DNA genomes but, like RDA, does not depend on any a priori knowledge of the sequence of the potential virus. Both degenerate PCR assays and RDA have a proven track record and RCA holds promise for the future.

The choice of method for use in virus hunting will depend on: (i) the evidence suggesting involvement of a particular virus family or type; (ii) the index of suspicion; (iii) the reagents or samples that are available and (iv) the predicted level of infection and therefore the required assay sensitivity. Of the assays described above, degenerate PCR assays are the most straightforward to perform and have the most applicability to screening vaccine cell substrates. These assays may prove useful in the investigation of novel cell substrates, such as clonal cell lines that are potentially tumorigenic.

Acknowledgements

We would like to thank Freda Alexander for performing the power calculations used to determine the required number of RDA representations for virus discovery and David Onions for helpful discussion. Work in our laboratory is supported by a Leukaemia Research Fund specialist programme grant.

References

1 Lisitsyn N, Lisitsyn N, Wigler M: Cloning the differences between two complex genomes. Science 1993; 259: 946-951.

2 Rector A, Tachezy R, Van Ranst M: A sequence-independent strategy for detection and cloning of circular DNA virus genomes by using multiply primed rolling-circle amplification. J Virol 2004; 78: 4993-4998.

3 Choo QL, Kuo G, Weiner AJ, Overby LR, Bradley DW, Houghton M: Isolation of a cDNA clone derived from a blood-borne non-A, non-B viral hepatitis genome. Science 1989; 244: 359-362.

4 Brownstein MJ, Carpten JD, Smith JR: Modulation of non-templated nucleotide addition by Taq DNA polymerase: primer modifications that facilitate genotyping. Biotechniques 1996; 20: 1004-1010.

5 Rose TM, Henikoff JG, Henikoff S: CODEHOP (COnsensus-DEgenerate Hybrid Oligonucleotide Primer) PCR primer design. Nucleic Acids Res 2003; 31: 3763-3766.

6 Gallagher A, Perry J, Shield L, Freeland J, MacKenzie J, Jarrett RF: Viruses and Hodgkin disease: no evidence of novel herpesviruses in non-EBV-associated lesions. Int J Cancer 2002; 101: 259-264.

7 Jarrett RF: Viruses and Hodgkin's lymphoma. Ann Oncol 2002; 13 (S1): 23-29.

8 Ehlers B, Ulrich S, Goltz M: Detection of two novel porcine herpesviruses with high similarity to gammaherpesviruses. J Gen Virol 1999; 80: 971-978.

9 Rivadeneira ED, Ferrari MG, Jarrett RF, Armstrong AA, Markham P, Birkebak T, et al: A novel Epstein-Barr virus-like virus, HV (MNE), in a Macaca nemestrina with mycosis fungoides. Blood 1999; 94: 2090-2101.

10 MacKenzie J, Gallagher A, Clayton RA, Perry J, Eden OB, Ford AM, et al: Screening for herpesvirus genomes in common acute lymphoblastic leukemia. Leukemia 2001; 15: 415-421.

11 Quackenbush SL, Work TM, Balazs GH, Casey RN, Rovnak J, Chaves A, et al: Three closely related herpesviruses are associated with fibropapillomatosis in marine turtles. Virology 1998; 246: 392-399.

12 Volter C, zur HH, Alber D, de Villiers EM: A broad spectrum PCR method for the detection of polyomaviruses and avoidance of contamination by cloning vectors, in Brown F, Lewis AM (eds): Simian Virus 40 (SV40): A Possible Human Polyomavirus. Dev Biol Stand. Basel, Karger, 1998, vol 94, pp 137-142.

13 Lopez-Rios F, Illei PB, Rusch V, Ladanyi M: Evidence against a role for SV40 infection in human mesotheliomas and high risk of false-positive PCR results owing to presence of SV40 sequences in common laboratory plasmids. Lancet 2004; 364: 1157-1166.

14 Vilchez RA, Madden CR, Kozinetz CA, Halvorson SJ, White ZS, Jorgensen JL, et al: Association between simian virus 40 and non-Hodgkin lymphoma. Lancet 2002; 359: 817-823.

15 Brousset P, de A, V, Gascoyne RD: Immunohistochemical investigation of SV40 large T antigen in Hodgkin and non-Hodgkin's lymphoma. Int J Cancer 2004; 112: 533-535.

16 MacKenzie J, Wilson KS, Perry J, Gallagher A, Jarrett RF: Association between simian virus 40 DNA and lymphoma in the United kingdom. J Natl Cancer Inst 2003; 95: 1001-1003.

17 Nichol ST, Spiropoulou CF, Morzunov S, Rollin PE, Ksiazek TG, Feldmann H, et al: Genetic identification of a hantavirus associated with an outbreak of acute respiratory illness. Science 1993; 262: 914-917.

18 Chang Y, Cesarman E, Pessin MS, Lee F, Culpepper J, Knowles DM, et al. Identification of herpesvirus-like DNA sequences in AIDS-associated Kaposi's sarcoma. Science 1994; 266: 1865-1869.

19 Nishizawa T, Okamoto H, Konishi K, Yoshizawa H, Miyakawa Y, Mayumi M. A novel DNA virus (TTV) associated with elevated transaminase levels in posttransfusion hepatitis of unknown etiology. Biochem Biophys Res Commun 1997; 241: 92-97.

20 Rector A, Bossart GD, Ghim SJ, Sundberg JP, Jenson AB, Van Ranst M. Characterization of a novel close-to-root papillomavirus from a Florida manatee by using multiply primed rolling-circle amplification: Trichechus manatus latirostris papillomavirus type 1. J Virol 2004; 78: 12698-12702.

Dr. R.F. Jarrett, LRF Virus Centre, Institute of Comparative Medicine, University of Glasgow, Bearsden Road, Glasgow G61 1QH, U.K.
E-mail: r.f.jarrett@vet.gla.ac.uk

Petricciani J, Sheets R (eds): Vaccine Cell Substrates 2004.
Dev Biol (Basel). Basel, Karger, 2006, vol 123, p 89.

Significance of Retrovirus Contaminants of Cell Substrates

J.M. Coffin

Department of Molecular Biology and Microbiology, Tufts University, Boston, MA, USA

Abstract: Retroviruses are perhaps the most difficult of potential contaminants of bioproducts, to deal with including vaccines. They are widespread in nature and in the laboratory, and many are the cause of serious disease in humans and experimental animals. Clearly, the presence of any amount of any infectious retrovirus in cell substrates used for the production of vaccines or other products cannot be tolerated. We now have very sensitive non-specific (PERT) and specific PCR assays capable of detecting a few virions in cell culture supernatants. The problem is that such assays do not readily distinguish infectious virions from non-infectious virus-like particles (VLPs).
Retrovirus contamination of cell cultures can arise from a number of different sources, including infection by an exogenous virus of the host from which the cell cultures were obtained; expression of germline inherited (endogenous) proviruses; contamination subsequent to isolation (for example, human tumours passed through nude mice often become infected with endogenous murine leukaemia virus [MLV]); and replication-competent recombinants between endogenous and exogenous viruses or deliberately introduced vector constructs. All mammalian and avian cells contain large numbers of endogenous proviruses; some have the potential to produce infectious virus; many more are capable of producing harmless VLPs. Distinguishing one from the other can present significant problems, and there are no good general rules.

Dr. J. Coffin, Tufts University, 136 Harrison Avenue, Boston, MA 02111, USA.
E-mail: john.coffin@tufts.edu

Petricciani J, Sheets R (eds): Vaccine Cell Substrates 2004.
Dev Biol (Basel). Basel, Karger, 2006, vol 123, pp 91-97.

Practical Assay Issues with the PERT/PBRT Assay: A Highly Sensitive Reverse Transcriptase Assay

A. Chang, S. Dusing

Biologics Testing, BioReliance, Invitrogen Bioservices, Rockville, MD, USA

Key words: Biosafety testing, reverse transcriptase, PCR

Abstract: Product safety testing for retroviruses can be achieved by a panel of screening assays, including electron microscopy, viral gene specific PCRs, virus propagation, and detection of reverse transciptase activity. The application of PCR-based reverse transcriptase assays (PERT) that are approximately a million-fold more sensitive than conventional nucleotide incorporation assays in the testing of biologicals is described. Use of PERT assays can be applied to three areas: (i) screening for adventitious retrovirus contamination; (ii) detecting and quantifying endogenous viral particle load and (iii) monitoring levels of infectious retrovirus generation in cell lines that contain endogenous retroviruses.

INTRODUCTION

Retroviruses comprise a large number of viruses known to infect many species, including humans, and have been associated with neoplasia, immunodeficiencies, leukaemias, and lymphomas [1]. A feature of the retrovirus life cycle is the integration of the newly reverse-transcribed viral DNA into the host cell genomic DNA, forming the provirus. The presence of the provirus can activate or inactivate cellular genes, possibly with genetic consequences. In addition, retroviruses have high mutation and recombination rates that can potentially alter the host range and virulence. Safety testing of cell substrates used to produce biologicals for viruses is an important and regulated issue. The most frequent viral contaminant found in these cell substrates are retroviruses [2,3].

Results and Discussions

PERT and detection of adventitious retrovirus contamination

Currently, the biotechnology industry uses four assays for the detection of retroviruses from cell substrates and biological products: (i) transmission electron microscopy; (ii) infectivity and amplification of virus on sensitive cells; (iii) viral gene specific PCR assays and (iv) detection of reverse transcriptase (RT) activity. The detection of RT represents an important part of this panel of assays as it has a broad spectrum and can detect all retroviruses as by definition all retroviruses contain reverse transcriptase within the virion.

Numerous existing RT assays measure the incorporation of radiolabelled deoxyribonucleoside triphosphates (e.g. ^{3}H-TTP) in DNA using RNA as the template [4]. These assays have been used for measuring RT activity in many laboratories. The incorporated radiolabel is measured by a scintillation counter or by autoradiography. These assays are often limited in their ability to detect low levels of RT activity.

Several polymerase chain reaction (PCR) based RT detection assays (PERT: **P**roduct **E**nhanced **R**everse **T**ranscriptase assay, PBRT: **PCR**-**B**ased **R**everse **T**ranscriptase assay) that are 100,000-fold more sensitive than the conventional ^{3}H-TTP incorporation-based assays have been described [5-7]. These assays are similar fundamentally in that they detect RT activity by amplifying the newly synthesized cDNA from an RNA template via PCR (Fig. 1a)

The PERT assay has been examined for sensitivity by diluting purified RT from various sources (HIV, MuLV, AMV). All RT's were normalized to the RT unit defined by manufacturers as "the amount of enzyme required to incorporate 1nmol of deoxynucleotide into acid-insoluble material in 10 minutes at 37°C using poly (rA)-oligo (dT)$_{50}$ as template-primer". All the RT tested had similar limits of detection of 10^{-9} units. A typical sensitivity experiment using M-MLV RT is shown in Figure 1b.

As with all assays that detect RT, it has been shown that the PERT assay is capable of detecting RT activity from different retrovirus subfamilies (oncovirus, lentivirus, and spumavirus), such as foamyvirus, human T-cell leukaemia virus type I, squirrel monkey retrovirus, equine infectious anaemia virus, avian myeloblastosis virus, human immunodeficiency virus, Gibbon ape leukaemia virus and potential unknown or variant viruses. Because of the high sensitivity and broad spectrum the PERT assay represent an important assay in the panel of assays used to demonstrated product safety. On December 18th 1998, CBER posted a "*letter to viral vaccine IND sponsors – Use of PCR-based Reverse Transcriptase Assay*" requesting the use of these improved reverse transcriptase assays as part of the battery of assays to demonstrate the absence of adventitious retroviruses in viral vaccines.

PERT and RVLP enumeration

Recently, advances in quantitative PCR have introduced a powerful and sensitive method that can determine not only the presence of the PCR target, but determine how much is present in the initial template concentration [8]. This technology measures PCR product accumulation during the log phase of the reaction and is referred to by several names, (Real-time, TaqMan™, 5'nuclease assay) and the reaction includes a fluorogenic-labelled sequence specific oligonucleotide in the

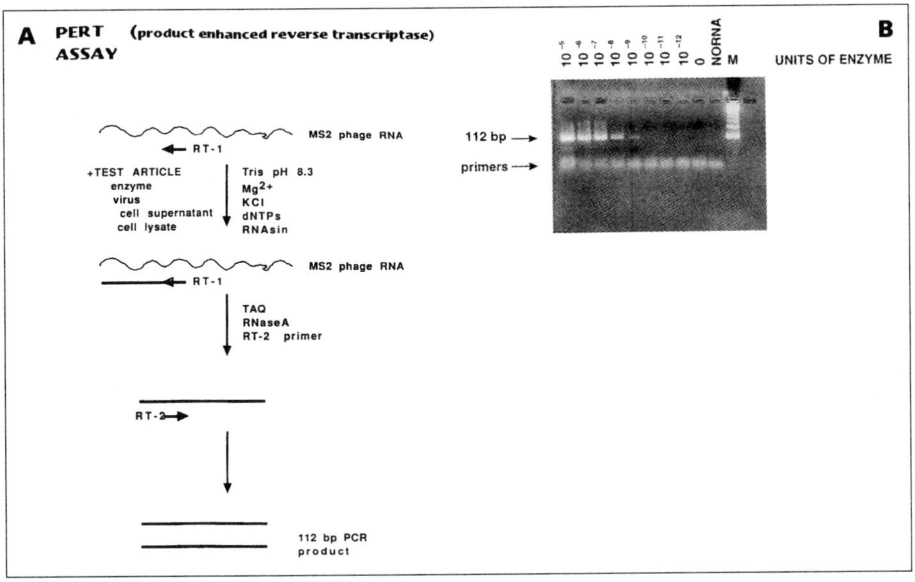

Fig. 1: **A** Schematic diagram of PERT assay. The thin wavy line represents the RNA template annealed with the oligonucleotide primer. The test article is incubated in the presence of RT reaction components. **B** Titration of purified RT (M-MLV). The numbers above each lane refer to the total units of RT tested. The samples were electrophoresed on a high resolution agarose gel. The 112-bp PCR product amplified from the reversed transcribed cDNA is indicated by the upper arrow. NORNA: no RNA negative control M: molecular weight marker.

PCR reaction. The amount of fluorescence generated can be directly correlated to the threshold cycle (Ct) at which the signal is first noted. The second generation PERT assays were adapted to this quantitative format. By including standards, either as units of RT activity, viral particles, or infectivity titres, it is now possible to quantify the amount of RT in a sample relative to the included standards.

As non-infectious entities possess RT (e.g. defective retrovirus, endogenous retroviruses, ancient retrovirus, retrovirus-like particles, retrotransposons), it is expected that these will produce a positive result in the PERT assay. Cell lines commonly used to produce biological products contain non-infectious retroviral-like elements (e.g. CHO, Sf9, avian origin cells) and will test positive in the PERT assay. Of particular interest is the endogenous Type C retroviral-like particles (RVLP) observed in Chinese Hamster Ovary (CHO) cells. Regulatory agencies worldwide require sponsors to quantify these Type C RVLPs and calculate the potential retroviral load of the bulk harvest before the purification process and thus determine the clearance capacity required to guarantee that the final product is free from potential retroviral contamination. The existing electron microscopy (EM) method, the current industry standard, is relatively insensitive and variable, and thus not an ideal tool for accurate quantification. Experiments comparing PERT and a CHO-specific gene PCR with the EM method have been performed and the data show that the molecular assays demonstrate improved accuracy, sensitivity, and quantification capability than EM (Fig. 2) [9].

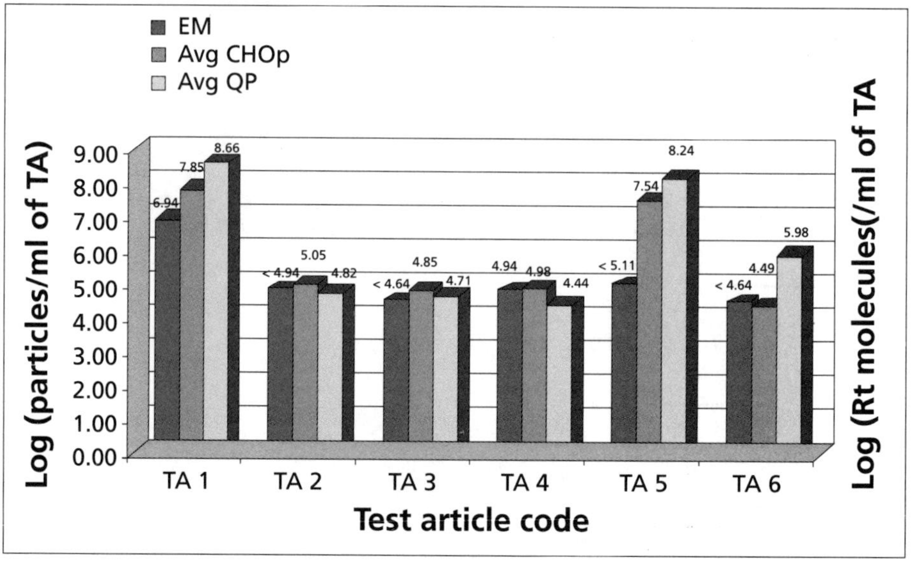

Fig. 2: **Comparison of CHO particle enumeration assays:** Six CHO Cell lines were assayed for the presence of CHO Type C RVLP by EM by direct particle count (■), Q-PCR to CHO RVLP RNA, by copies of RNA (■), and Quantitative PERT by total molecules of RT (□). All values have been converted to total particles (log) per ml of sample (test article: TA). A conversion factor of 2 RNA molecules and 30 RT molecules was used to equate one particle for the molecular assay. Values that below the limit of detection are reported as a less than value.

PERT and infectious retrovirus assays

There are only two retroviral subfamilies that are known to be pathogenic for humans [10,11]. These are the leukaemia viruses, HTLV-I and HTLV-II and the immunodeficiency viruses, HIV-1 and HIV-2. However, it is clear that there are many retroviruses that possess a wide host range and may pose a threat to the human population. The viruses may be either primate or non-primate in origin such as the murine amphotropic and xenotropic retroviruses [1]. More recently, documented cases of the porcine endogenous retrovirus (PERV) capable of infecting human cells in culture (293 human cells) have raised concerns in the xenotransplantation community [12,13]. As the presence of PERV in the 293 cells is non-cytopathic, the infection was monitored by the PERT assay. The assay not only detected the presence of the PERV, but demonstrated virus propagation through the passages via the increase in RT molecules over passages (Fig. 3).

The ability of PERT to monitor non-cytopathic retrovirus infection can be expanded to assay for the presence of an infectious retrovirus in cells that have a background RT level due to the presence of endogenous non-infectious retrovirus. Figure 4 demonstrates how PERT is used to differentiate the endogenous RT signal from that of an infectious retrovirus. Turkey embryo fibroblasts (TEF) were infected with avian leukosis type J (ALV-J) retrovirus and maintained in culture for 28 days with seven subcultures. The level of retrovirus present in the culture medium harvested from these cells at subculture was determined by PERT (dashed line) and by an ELISA for viral p27 protein (solid line). The uninfected TEF cells

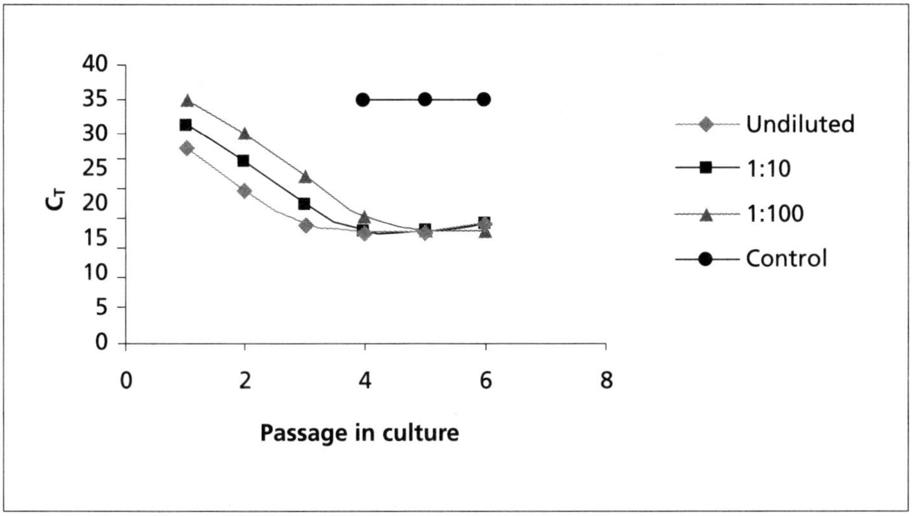

Fig. 3: **PERT analysis of PERV-infected 293 cells.** 293 cells were inoculated with undiluted (◆) and 10^{-1} (■) and 10^{-2} (▲) dilutions of PERV and maintained in extended culture. Infection with and replication of PERV was monitored by PERT with results reported as C_t (threshold cycle) values. Values below C_t 35 are considered to be positive. The level of RT detectable by the PERT assay increased with the time in culture in a dose-dependent manner.

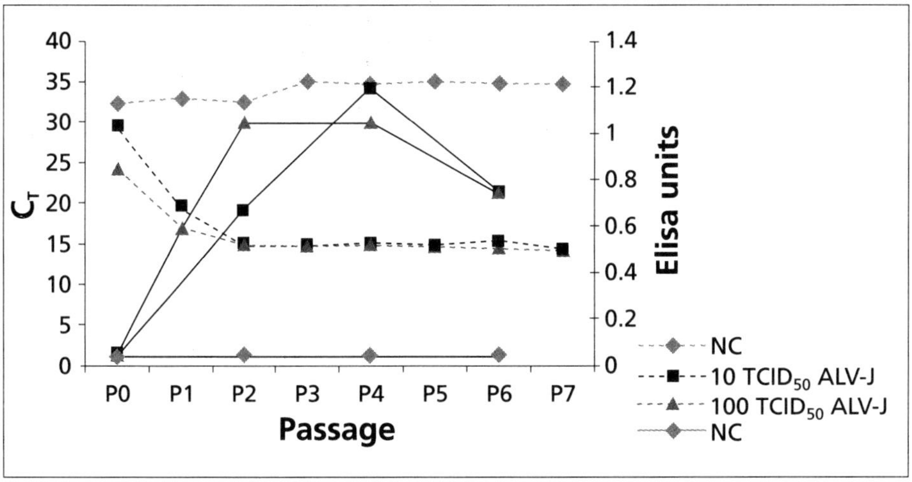

Fig. 4: **PERT analysis of ALV-J infected TEF cells.** TEF cells were inoculated with 10 (■) or 100 (▲) $TCID_{50}$ of ALV-J and the course of the infection was monitored by PERT assay (dashed line) or ALV p27 ELISA (solid line). Mock-inoculated TEF cells (◆) were a control for the assay. The level of RT activity in the culture supernatant as determined by PERT assay is reported as the C_t (threshold cycle) value depicted on the left vertical axis. Values below the background RT level detected in the negative control culture are considered to be positive for RT. The ELISA units of ALV p27 protein are depicted on the right axis. Note the increase of RT and ALV p27 antigen signal early in the assay demonstrating virus amplification and the continued presence of the virus throughout the remainder of the cultivation.

(diamonds) expressed a background level of RT activity, as established by C_t values ranging between 32 and 35, and this signal remained constant over the seven passages. In contrast, the TEF cells inoculated with 10 or 100 $TCID_{50}$ of ALV-J produced RT at a level that increased logarithmically through the second passage and subsequently remained at a high level, indicating the presence of an active retroviral infection in the TEF cells. A corresponding increase in the amount of ALV p27 verified that the increased RT activity detected by the PERT assay was, indeed, attributable to the ALV-J infection. These data clearly demonstrate the presence of the infectious ALV-J retrovirus in a cell line that contains endogenous retrovirus.

SUMMARY

The successful analysis of the retroviral status in biological samples depends on the interpretation of data generated from several assays (EM, virus propagation, PCR, and RT). The PERT assay uses the extreme sensitivity of PCR to detect retroviruses by the presence of their enzymatic RT activity. As noted, the assay possesses a broad spectrum and can detect all retroviruses that harbour reverse transcriptase activity. The quantitative version can also be used to enumerate endogenous retroviruses by calculating the number of RT molecules present. Historically, EM is used to quantifying retroviral particles, and although EM is an excellent assay for determining cellular morphology or cellular localization, it suffers technically as a quantitative assay. In addition to the rapid turn-around time, the molecular-based PERT assay offers better reproducibility and accuracy than EM. Finally, the PERT assay can be a useful endpoint in retroviral cell culture propagation assays, especially when the presence of the retrovirus does not produce a cytopathic effect. By calculating the amount of RT presence over passages, one can monitor the presence of a productive infection, even in the presence of a background RT level. With the ease of performing the molecular analysis, all three applications of the PERT assay are useful in establishing product safety with regard to the presence of retroviruses.

REFERENCES

1 Coffin JM: Retroviridae and their replication, in Fields BN, Knipe DH (eds): Fields Virology. New York, Raven Press, 1990, pp 1437-1500.

2 Moore WA: Experience in cell line testing, in Brown F, Esber EE, Williams M (eds): Continuous Cell Lines – An International Workshop on Current Issues. Dev Biol Stand. Basel, Karger, 1992, vol 76, pp 51-56.

3 Losikoff AM, Poiley JA, Raineri R, Nelson E, Hillesund T: Industrial experience with the detection of retrovirus, in Brown F, Esber EE, Williams M (eds): Continuous Cell Lines – An International Workshop on Current Issues. Dev Biol Stand. Basel, Karger, 1992, vol 76, pp 187-200.

4 Roy-Burman P, Dougherty M, Pal BK, Charman HP, Klement V, Gardner MB: Assay for type C virus in mouse sera based on particulate reverse transcriptase activity. J Virol 1976; 19: 1107-1110.

5 Silver J, Maudru J, Fujita K, Repaske R: A RT-PCR assay for the enzyme activity of reverse transcriptase capable of detecting single virions. Nucleic Acids Res 1993; 21: 3593-3594.

6 Pyra H, Boni J, Schupbach J: Ultrasensitive retroviral detection by a reverse transcriptase assay based on product enhancement. Proc Natl Acad Sci U.S.A. 1994; 91: 1544-1548.

7 Heneine W, Yamamoto S, Switzer WM, Spira TJ, Folks TM: Detection of RT by a highly sensitive assay in sera from persons infected with HIV-1. J Infect Dis 1995 151: 1210-1216.

8 Heid CA, Stevens J, Livak KJ, et al: Real Time Quantitative PCR. Genome Methods 1996; 6: 986-994.

9 Brorson K, Xu Y, Swann PG, Hamilton E, Mustafa M, de Wit C, Norlin LA, Stein KE: Evaluation of a quantitative product-enhanced reverse transcriptase assay to monitor retrovirus mAb cell-culture. Biologicals 2002; 30: 15-26.

10 Cann AJ, Chen ISY: Human T-cell leukemia virus types I and II, in Fields BN, Knipe DH (eds): Fields Virology. Raven Press, New York. 1990, pp 1501-1528.

11 Hirsch MS, Curran J: Human Immunodeficiency Viruses, in Fields BN, Knipe DH (eds): Fields Virology. Raven Press, New York, 1990, pp 1545-1570.

12 Patience C, Takeuchi Y, Weiss RA: Infection of human cells by an endogenous retrovirus of pigs. Nature Medicine 1997; 3: 282-286.

13 Wilson CA, Wong S, Muller J, Davidson CE, Rose TM, Burd P: Type C Retrovirus Released from Porcine Primary Peripheral Blood Mononuclear Cells Infects Human Cells. Journal of Virology 1998:72: 3082-3087.

Dr. A. Chang, Biologics Testing, BioReliance, 14920 Broschart Road, Rockville, MD 20850, USA.
E-mail: achang@bioreliance.com

Petricciani J, Sheets R (eds): Vaccine Cell Substrates 2004.
Dev Biol (Basel). Basel, Karger, 2006, vol 123, pp 99-107.

Antibodies to SV40 in Sera from Humans and Macaques

P. Minor, P.A. Pipkin

National Institute for Biological Standards and Control, Potters Bar, U.K.

Background

Poliovaccines manufactured in the 1950s were inadvertently contaminated with the simian polyoma virus SV40, which is a common silent infection of the Rhesus macaque monkeys which supplied the kidneys used to generate the cell cultures used in production. The poliovaccines used were the formalin-killed poliovirus preparations developed by Salk, but SV40 is less sensitive to formalin than poliovirus and it is estimated that 10 to 30 million people were exposed to live SV40 as a result [1]. While other vaccines such as an experimental vaccine against respiratory syncytial virus, and a vaccine used in the military against adenovirus were also contaminated with SV40, the major iatrogenic exposure is from poliovaccines. The virus does not cause a visible cytopathic effect in Rhesus macaque cell cultures, and was not reported until 1960. By 1962 it is likely that most poliovaccines in use were free of the virus as requirements were introduced to restrict production to animals free of antibodies to SV40, and cultures producing a cytopathic effect after infection were used to screen for it. Subsequent studies using PCR to investigate old retained stocks [2,3] supported the view that poliovaccines used in the U.S.A. and the U.K. were free of SV40 from at least the 1970s. SV40 is known to cause tumours in experimental animals, mainly hamsters [4], and epidemiological studies were undertaken at the time which generally failed to reveal any excess risk of cancer in vaccine recipients [5]. However, the studies tended to be short term. In one a risk to the child was indicated when pregnant women were immunised [5]. In the 1990s, however, SV40 sequences were reported in a number of human tumours, including mesothelioma, osteosarcoma, ependymomas and choroid plexus tumours [6-10]. More recently such sequences have been reported in non-Hodgkin's lymphomas [11,12]. In some cases virus has been isolated [9] suggesting that the patients are indeed infected with SV40. The extent and origin of such infection, if genuine, are matters of great interest. In 1972 Shah reviewed studies on the incidence of antibodies to SV40 in a number of categories of human subjects [13]. Titres were determined by neutralisation assay and were low, no greater than one in 128 and typically one in four to one in eight in dilution end point titre assays. Nearly 20 % of the group consisting of children who had received inactivated polio vaccines possessed antibodies. Furthermore 27 % of the group of monkey handlers from Northern India who were exposed regularly to infected animals and 8.7 % of other residents of Northern India were also seropositive. Other groups, including cancer patients from Northern India or the U.S.A., children born after vaccines were cleared of SV40 and individuals bled in 1951 or 1954, before poliovaccines were introduced, all had positive rates of 2 to 6 %. This suggested that exposure to polio vaccine can induce antibodies to SV40, but that there was also a low level of positivity of roughly the same titre in unexposed populations.

Over recent years we have undertaken an examination of human sera from various sources to establish the current prevalence of antibodies to SV40 and attempt to determine the origin of the positives found. The situation is complicated by the occurrence of human polyoma viruses including JC and BK which cross-react with SV40 to some extent. Thus the positive sera found could be the result of cross-reaction rather than infection with SV40. This paper summarises the findings and also describes studies of a monkey breeding colony where SV40 infections are common which was studied to examine natural transmission of the virus in a species known to be susceptible to infection.

Seroprevalence in Populations with Different Poliovaccination Histories

Panels of sera were available from a number of sources [14], including the United Kingdom, where previous studies had shown that all batches of stored vaccine found were free of SV40 sequences from 1970 onwards. Panels were also available from measles vaccination campaigns in Africa, where the individuals had probably received vaccine of a similar origin to that used in the U.K., and from a poliomyelitis outbreak in Morocco, where the individuals were presumably not vaccinated at all, and from Poland, where the vaccine used was manufactured in the former USSR, and is believed to have been strictly SV40 free rather later than in the other countries. Three panels of sera from the United Kingdom collected in 1985, between 1989 and 1994, and in 1995 respectively, were assayed by a rapid screening neutralisation method, in which a one to eight dilution of sera was incubated with virus and cells over 10 days and examined for cytopathic effect. Positive samples where the cell sheet was not affected were then titrated in full. The incidence of positives was about 4.5 %, as shown in Table 1. Moreover the titres were low, most sera being positive at a dilution of no higher than one in eight, and none higher than one in 64. This is consistent with the data reported by Shah for populations not knowingly exposed to SV40 [13]. Fewer sera were available from Africa, but the overall positivity and titres were very similar to those found in the United Kingdom at about 3.3 % and typically one in eight respectively, as shown in Table 1. The one exception was the collection of sera from Morocco collected in response to an outbreak of poliomyelitis, where all 29 sera were positive, although with a similar distribution of titres to the positives in other panels. This high rate of positivity remains

Table 1: Serum antibodies to SV40 in sera from the U.K., Africa and Poland.

Source	**Number tested**	**Number with titre**				**Percent positive**
		≥ 8	≥ 16	≥ 32	≥ 64	
UK						
NBTS Liverpool (1985)	800	34	12	5	1	4.5
NBTS Edgware (1989-1994)	1000	47	2	0	0	4.7
PHLS	254	12	7	4	3	4.7
Zaire	184	6				3.2
Sierra Leone	62	2				3.2
Morocco	419	14	2			3.3
(Morocco)	29	29	22	13	1	100
Poland						
1997 Measles survey – healthy people	264	1				0.4
a) 1992-1995 Polio survey – healthy people	274	13	7	1	1	4.7
b) 1996 Influenza survey – healthy people	146	7	2	1	1 (1/128)	4.8
c) 1997 Asbestos area – healthy people	201	3	2	1	1 (1/256)	1.5
d) 1992-1997 Neuro-infection – herpes infection	38	2				5.3

Antibody titres were determined by end-point dilution in a neutralisation assay.

unexplained but cannot be due to the use of polio vaccine in view of the origin of the sera. Finally a collection of sera from Poland had an overall rate of positives of 2.8 %. The findings were biased by the low positivity in one collection of sera taken to estimate the level of antibodies to measles. As the assays were performed at the same time and repeated the difference is real, but unexplained. It was concluded that while there were differences between the different panels examined they could not be explained by exposure to polio vaccines from different origins. All collections were of generally low titre and the typical rate of positives was very similar to that reported by Shah at 3 to 5 %.

Origin of Antibodies Reacting with SV40: Age-Dependent Seroconversion

It was known that the titres of antibodies in sera from old world primates assayed in the same neutralising assay were high in contrast to those seen in the human sera tested, and that positivity rates were also high in at least one breeding colony tested [14-16]. There has been and continues to be speculation about the origin of the antibodies to SV40 found in human sera and many have concluded that they are induced against a human polyoma virus. Brown et al (1975) examined a number of sera from isolated communities and concluded that the antibodies reported were induced by BK virus. Studies were performed on the age dependent positivity in a collection of 2435 sera from the United Kingdom in which the neutralising antibodies to SV40 were measured as above, and antibodies to BK and JC by a haemagglutinin inhibition assay. All assays were therefore based on the measurement of biological activity rather than ELISA reaction. The results are shown in Figure 1. The seropositive rate to BK rose rapidly to a maximum of about 90 % by age 5-9, and then declined slowly to 68 % at 60-69. The kinetics of seroconversion to JC were very different, rising gradually but uniformly over the entire age range studied. Finally, antibodies to SV40 were found in 3-5 % of sera over the entire age range and did not change at all. One possible explanation for the presence of antibodies to SV40 in human sera is that the virus was introduced into the population in the 1950s from polio vaccines. Some of the tumours in which virus sequences have been found have been in children and it is possible that parents pass the virus on to their offspring. If this were the case it might be expected that the cohort receiving the vaccine and their children would have been of higher seropositivity than the cohorts in between. In fact there is no age-dependent variation in positivity, suggesting that this straightforward scenario is incorrect. The titres of the antibodies to SV40 were low, as in the previous study, while the titres to BK and JC were higher, with a median of 160 and 80 respectively. While there was no obvious correlation between the presence of antibodies to the three viruses in any one serum, only one of the SV40 positive sera was negative, and in two others the titres were low to both BK and JC in the presence of antibodies to SV40. The origin of the SV40 reactive antibodies remains unclear from this study. It is possible that they are cross-reacting antibodies, probably induced by BK. On the other hand the absence of an age-dependent change in seropositive rates for SV40 in contrast to BK suggests that this may not be the case. Renal transplant patients frequently reactivate

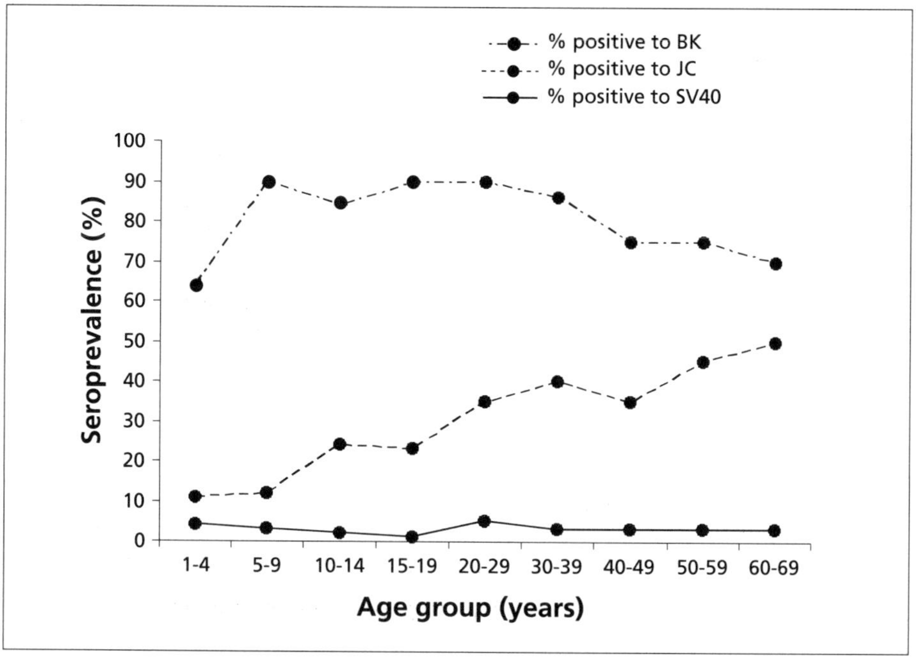

Fig. 1: Age-dependent seropositivity to polyoma viruses in U.K.-sourced sera. Antibodies to BK and JC were determined by H1 assay, and to SV40 by neutralisation assay.

JC and BK as a result of the immunosuppressive therapy they undergo. The level of immunosuppression is insufficient to prevent a humoral antibody response occurring, and in two such patients the levels of antibodies to SV40, BK and JC were followed, as shown in Figure 2 [14]. The titres to SV40 did not follow those to either of the other viruses; in one the titre to JC rose slightly after that to SV40 while the BK titre stayed constant, and in the other the titre to BK rose several weeks after the titre to SV40 had risen, the titre to JC staying constant. The significance of these findings also remains unclear as they are consistent with either cross-reactive antibodies of differing properties in different patients, or with the existence of a third virus which could be SV40 or another cross-reacting infectious agent. A possible way to approach the problem is to use cross-adsorbed sera or competition assays. If all SV40 reactive antibodies can be removed by adsorption with either BK or JC it suggests that they are not induced by SV40 itself. Assays for antibodies to SV40, BK and JC were developed using an ELISA format with a recombinant expressed virus-like particle (VLP) as the antigen for each of the three viruses [17]. The assay for SV40 antibodies was validated with monkey sera whose neutralising antibody titre to SV40 was known, and it was shown that the ELISA and neutralising antibody titres correlated for sera with high and very low titres. Some of the sera with intermediate titres by neutralisation assay were negative by ELISA, and some sera had low titres in ELISA for BK and JC antibodies. The different assay formats are therefore measuring different populations of antibodies. However, all sera that were positive for SV40 in the ELISA were also positive for JC and BK. Moreover, in a

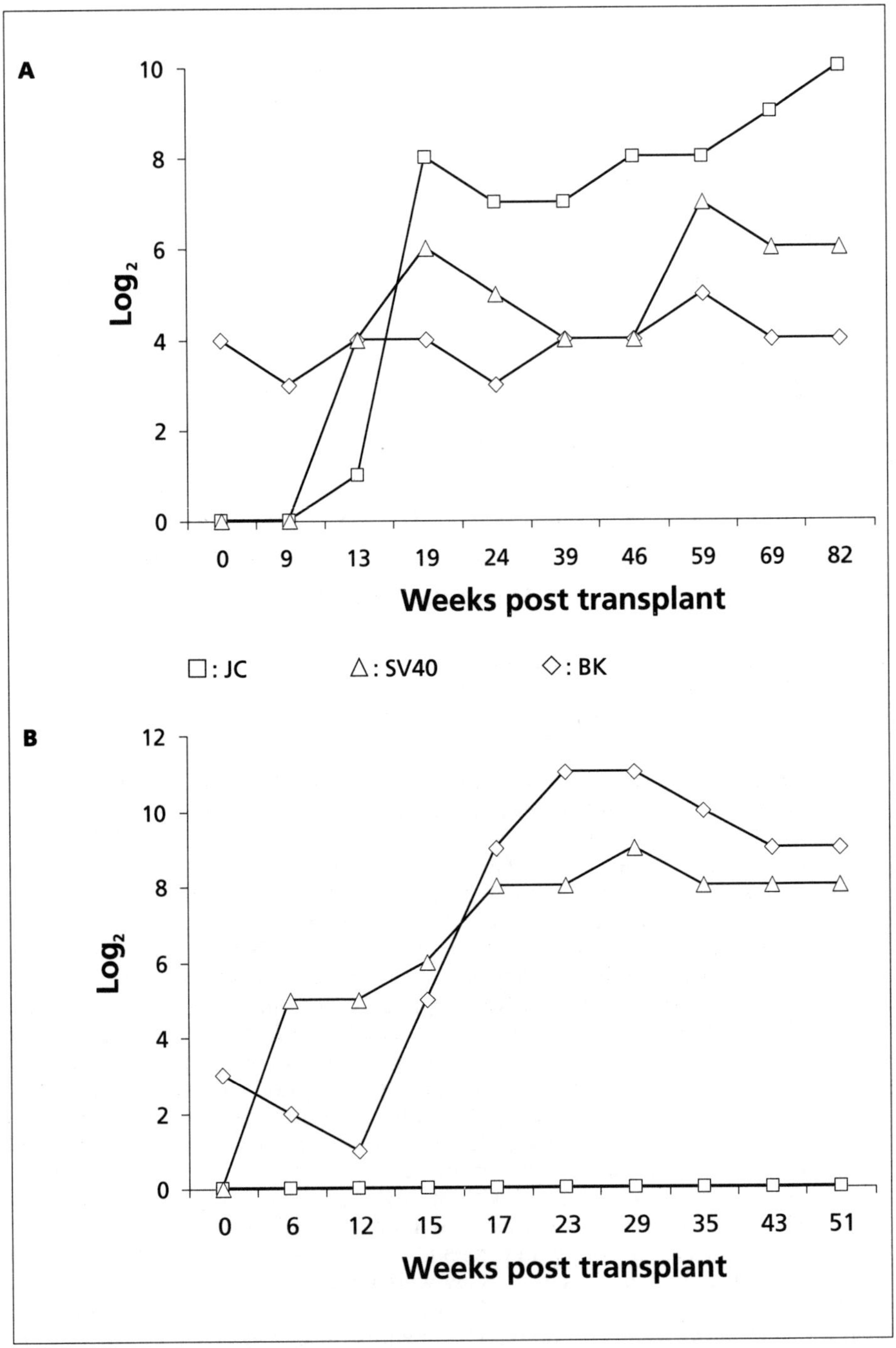

Fig. 2a, b: Antibody titres to JC, BK and SV40 in two renal transplant recipients.

competition format, all reactivity to SV40 could be blocked by the addition of either JC or BK VLPs. For none of them could the reaction with BK or JC be completely blocked by the presence of SV40. This indicates strongly that the ELISA reactive antibodies at least are cross- reactive, and not induced by SV40.

Natural Infection and Transmission of SV40 in a Breeding Colony of Macaques

The existence of human infections with SV40 remains controversial. The prevalence is low based on serum surveys and the titres are lower than might be expected of a natural infection. On the other hand there is credible evidence of the presence of viral sequences in certain human tumours. It is possible that examination of monkeys where transmission is clear might help in understanding the situation in humans. In particular it has been suggested that transmission from parent to offspring might occur in humans, thus explaining the occurrence of SV40 in certain childhood tumours. The existence of such transmissions in monkeys where infection is not in question was therefore of interest [15]. Blood samples have been taken as part of the routine health screening of a colony of cynomolgous macaques, and were examined for antibodies to SV40. The animals are held in breeding groups of about twenty in each, and the babies are removed to a separate gang on weaning, typically at older than six months, and sometimes at over one year. Blood samples were taken from the time of weaning, and from the mothers. Despite the high titres of SV40 antibodies in the mothers, none of the infants showed any serological signs of infection at the time of ganging, as shown in Table 2. The number of animals seroconverting varied depending on the cohort examined but in all cases examined, the mean time to seroconversion, if it happened, was about 35 days. Antibody titres in positive animals were high, typically greater than 4000. It is possible that the infants are infected by the mothers (who were with one exception highly seropositive) but that the infection is suppressed by antibodies in the milk so that the baby only seroconverts post-weaning. This is hard to reconcile with the differences between cohorts. In some cases the baby is rejected by the mother and is then hand-reared. In no case examined was such a baby found to be positive, suggesting that if the babies were infected by the mother, it was not at or before birth. The differences between the cohorts are not yet explained, but it seems likely that they arise from differences in husbandry. One possibility is that the social structure of the gang of weaners may be significant. In one of the earlier cohorts, the animals were all of a similar age and the oldest was 10 months. In a later gang where seroconversion was less complete, some of the animals were very much older than the rest. It is at least conceivable that the transmissions occur during the early stages of the formation of the groups, when social hierarchies are being formed, and that the presence of older animals means that the hierarchies are formed more quickly and with less fighting. This hypothesis is untested. The studies of transmission in the monkey colony suggest strongly that natural transmission is not from mother to child, and that infection occurs horizontally after weaning. By analogy, if transmission occurs in human populations at all, it seems unlikely that the main route is from mother to offspring.

Table 2a: Seroconversion to SV40 in a colony of cynomolgous macaque monkeys.

Animal	Age at weaning (monthly)	Colony	8 days	43 days	88 days	Final days	Titre of mother 43 days after infant ganging
350BCE	8	C25	< 8	> 4096	8192	2048	4096
398BBD	9	C25	< 8	> 4096	8192	4096	> 4096
039C1	10	C32	< 8	4096	4096	4096	> 4096
577CH	9	C33	8192	Died			> 4096
085EE	10	C37	< 8	32	2048	nk	256
651FC	6	C42	< 8	< 8	2048	4096	> 4096
845H	7	C46	< 8	> 4096	Died		4096
868F	7	C46	< 8	< 8	2048	nk	8
902F	8	C48	< 8	512	8192	nk	> 4096
930F	7	C48	< 8	64	1024	4096	512
109HH	7	G16	< 8	128	2048	1024	> 4096
141HA	9	G16	< 8	< 8	8192	2048	> 4096

Table 2b.

Animal	Age at weaning (monthly)	Colony	27 days	35 days	90 days	256 days	Titre of mother 27 days after infant ganging
812DA	9	C31	< 8	< 8	< 8	< 8	8192
602BA	12	C32	8192	8192	4096	8192	8192
816BB	8	C32	< 8	< 8	< 8	< 8	2048
237DL	9	C42	< 8	< 8	< 8	< 8	8192
548BJ	18	C42	128	512	128	512	4096
558GC	8	C51	< 8	< 8	< 8	< 8	4096
817CD	11	C51	< 8	< 8	4096	8192	1624
523FF	11	C54	< 8	< 8	< 8	< 8	> 8192
028HC	8	C55	< 8	< 8	< 8	nk	1024
176AFD	8	C55	64	< 8	< 8	< 8	4096
357CCE	8	C55	< 8	8192	8192	8192	> 8192
417HC	8	C55	< 8	Died	–	–	8192
894BC	6	C55	< 8	< 8	< 8	< 8	8192
030KE	11	C60	2048	8192	4096	> 8192	8192
040LF	10	C63	< 8	< 8	< 8	< 8	> 8192
631BDE	9	C63	< 8	< 8	< 8	< 8	2048

Conclusion

Antibodies to the simian polyoma virus SV40 can be found at comparable titres and rates in sera from around the world. They may be cross-reactive antibodies induced by the human polyoma viruses JC and BK, although this has not been proved. It seems unlikely that vertical transmission occurs from mother to baby in macaque monkeys, where infection occurs, or in humans where it is controversial.

References

1 Shah K, Nathanson N: Human exposure to SV40: review and comment. Am J Epidemiol 1976; 103: 1-12.

2 Sangar D, Pipkin PA, Wood DJ, Minor PD: Examination of poliovirus vaccine preparations for SV40 sequences. Biologicals 1999; 27: 1-10.

3 Sierra-Honigmann A, Krause PR: Live oral poliovirus vaccines do not contain detectable simian virus 40 (SV40) DNA. Biologicals 2000; 28: 1-4.

4 Girardi AJ, Sweet BH, Slotnick VB, Hilleman MR: Development of tumors in hamsters inoculated in the neonatal period with vacuolating virus, SV-40. Proc Soc Exp Biol Med 1962; 109: 649-660.

5 Strickler HD, Rosenberg PS, Devesa SS, Hertel J, Fraumeni JF, Jr., Goedert JJ: Contamination of poliovirus vaccines with simian virus 40 (1955-1963) and subsequent cancer rates. JAMA 1998; 279: 292-295.

6 Bergsagel DJ, Finegold MJ, Butel JS, Kupsky WJ, Garcea RL: DNA sequences similar to those of simian virus 40 in ependymomas and choroid plexus tumors of childhood. N Engl J Med 1992; 326: 988-993.

7 Carbone M, Pass HI, Rizzo P, Marinetti M, Di Muzio M, Mew DJ, Levine AS, Procopio A: Simian virus 40-like DNA sequences in human pleural mesothelioma. Oncogene 1994; 9: 1781-1790.

8 Carbone M, Rizzo P, Procopio A, Giuliano M, Pass HI, Gebhardt MC, Mangham C, Hansen M, Malkin DF, Bushart G, Pompetti F, Picci P, Levine AS, Bergsagel JD, Garcea RL: SV40-like sequences in human bone tumors. Oncogene 1996; 13: 527-535.

9 Lednicky JA, Garcea RL, Bergsagel DJ, Butel JS: Natural simian virus 40 strains are present in human choroid plexus and ependymoma tumors. Virology 1995; 212: 710-717.

10 Pepper C, Jasani B, Navabi H, Wynford-Thomas D, Gibbs AR: Simian virus 40 large T antigen (SV40LTAg) primer specific DNA amplification in human pleural mesothelioma tissue. Thorax 1996; 51: 1074-1076.

11 Shivapurkar N, Harada K, Reddy J, Scheuermann RH, Xu Y, McKenna RW, Milchgrub S, Kroft SH, Feng Z, Gazdar AF: Presence of simian virus 40 DNA sequences in human lymphomas. Lancet 2002; 359: 851-852.

12 Vilchez RA, Madden CR, Kozinetz CA, Halvorson SJ, White ZS, Jorgensen JL, Finch CJ, Butel JS: Association between simian virus 40 and non-Hodgkin lymphoma. Lancet 2002; 359: 817-823.

13 Shah KV: Evidence for an SV40-related papovavirus infection of man. Am J Epidemiol 1972; 95: 199-206.

14 Minor P, Pipkin P, Jarzebek Z, Knowles W: Studies of neutralising antibodies to SV40 in human sera. J Med Virol 2003; 70: 490-495.

15 Minor P, Pipkin PA, Cutler K, Dunn G: Natural infection and transmission of SV40. Virology 2003; 314: 403-409.

16 Brown P, Tsai T, Gajdusek DC: Seroepidemiology of human papovaviruses. Discovery of virgin populations and some unusual patterns of antibody prevalence among remote peoples of the world. Am J Epidemiol 1975; 102: 331-340.

17 Carter JJ, Madeleine MM, Wipf GC, Garcea RL, Pipkin PA, Minor PD, Galloway DA: Lack of serologic evidence for prevalent simian virus 40 infection in humans. J Natl Cancer Inst 2003; 95: 1522-1530.

Dr. P. Minor, N.I.B.S.C., Blanche Lane, South Mimms, Potters Bar, Herts EN6 3QG, U.K.
E-mail: pminor@nibsc.ac.uk

Petricciani J, Sheets R (eds): Vaccine Cell Substrates 2004.
Dev Biol (Basel). Basel, Karger, 2006, vol 123, pp 109-117.

Polio Vaccines, SV40 and Human Tumours, an Update on False Positive and False Negative Results

A.G. Elmishad[1], M. Bocchetta[1], H.I. Pass[2], M. Carbone[1]

[1] Loyola University Medical Center, Cardinal Bernandin Cancer Center, Room 205, Thoracic Oncology Program, Department of Pathology, Maywood, IL, USA

[2] Karmanos Cancer Center, Wayne State University School of Medicine, Detroit, MI, USA

Abstract: Simian virus 40 (SV40) has been detected in different human tumours in numerous laboratories. The detection of SV40 in human tumours has been linked to the administration of SV40-contaminated polio vaccines from 1954 until 1963. Many of these reports linked SV40 to human mesothelioma. Some studies have failed to detect SV40 in human tumours and this has caused a controversy. Here we review the current literature. Moreover, we present evidence showing how differences in the sensitivities of methodologies can lead to a very different interpretation of the same study. The same 20 mesothelioma specimens all tested negative, 2/20 tested positive or 7/20 tested positive for SV40 Tag by simply changing the detection method on the same immuno-precipitation/western blot membranes. These results provide a simple explanation for some of the apparent discordant results reported in the literature.

SV40, THE VIRUS

Simian Virus 40 (SV40) is a double-stranded circular DNA virus that was discovered as a contaminant of both the Salk and Sabin polio vaccines. The virus has proved to be highly oncogenic [1]. Injection of the virus in hamsters resulted in the development of sarcoma and ependymoma [2] while intra-pleural injection in hamsters resulted in up to 100 % of the animals developing mesothelioma [3]. The virus encodes two transforming proteins, the large T antigen (Tag) and the small t antigen (tag) plus three capsid proteins (VP1-3). The large T antigen is one of the most potent and studied oncogenic viral proteins. Its ability to transform rodent and human cells has already been established [1,4,5]. When the large T antigen is expressed inside the cell, it binds a number of intracellular proteins, including p53 and pRb [5,6]. These proteins are involved in the regulation of the cell cycle. Binding of large T antigen to these proteins results in inactivation of the protein, which in turn, disrupts numerous cellular pathways in the normal cell cycle.

Mesothelioma

Mesothelioma is a rare but aggressive malignancy of the pleural, pericardial and peritoneal lining. Mesothelioma was a very rare cancer in the first half of the last century but the incidence has risen significantly over the past five decades. Currently, there are 2000 to 3000 newly diagnosed cases every year in the United States alone [7], where mesothelioma accounts for about 2500 deaths per year. The increase in incidence has been attributed to exposure to asbestos. However, about 20 % of mesothelioma cases do not have a history of asbestos exposure and only 5 % of exposed individuals develop the tumour. These facts suggest that in addition to asbestos, other factors contribute to mesothelioma [8]. SV40 appears to be one of these factors [9,10] and SV40 and asbestos have been shown to be cocarcinogens in vitro [11].

SV40 and Mesothelioma

The evidence that points to a causative relationship between SV40 and mesothelioma was investigated during the last decade. Numerous studies support this hypothesis. For example, using PCR analysis and DNA sequencing in microdissected tumour cells, SV40 sequences were detected in 57 % of pleural and peritoneal epithelial mesothelioma tumour cells. On the other hand, these sequences were not detected in matched- adjacent microdissected lung tissue [12]. In another study, using primed in situ (PRINS) method and immuno-histochemistry, SV40 sequence was detected in 60 % of mesothelioma biopsies but not in adjacent normal stromal cells [13]. In addition, SV40 sequences were rarely detected in the normal mesothelial cells (one in 75 biopsies) or in any of the pleural carcinoma metastases biopsy samples that were included in these two studies [12,13].

SV40 and Human Tumours

SV40 sequences have been detected in different types of human tumours such as mesothelioma [9,14] bone tumours [15,16] and brain tumours [17-20]. SV40 sequences were also detected in non-Hodgkin's lymphoma [21-23]. The detection of SV40 in these tumours was often accomplished with the help of the polymerase chain reaction (PCR). PCR is a highly sensitive technique that, when appropriately designed, is able to detect viral sequences in the tumour biopsies. When using PCR as the sole method of detection, it is important to include appropriate controls to detect any possible plasmid or viral contamination. When controls are included, PCR is a sensitive and reliable methodology for detecting SV40 in human tumour biopsies. Overall, more than 50 different laboratories have detected SV40 in human tumours. The fact that so many laboratories have reported this finding establishes the reliability of this association. At the same time, not all these laboratories appeared to have taken the rigorous steps required to rule out PCR contamination and it is possible that some of these positive results were caused, at least in part, by laboratory artifacts. Other techniques have also been used to verify the presence of

SV40 [20]. Southern blot, sequence analysis, in situ hybridization and immunohistochemistry are some of the additional techniques that were used to confirm the presence of SV40 in human biopsies [13,20,24]. Recently, quantitative real-time PCR has been proposed as the method of choice for detecting SV40 in human biopsies [25].

Negative Findings

Some studies reported negative findings for SV40 in human tumours [10,24]. Some of these negative results were probably caused by geographical differences linked to the distribution of contaminated vaccines in some countries but not in others. This interpretation is supported by the fact that the same laboratories that did not detect SV40 in mesotheliomas from Finland, Turkey and Austria, countries that did not receive SV40 contaminated vaccines, detected SV40 in specimens from the U.S.A. or Italy, countries that received contaminated vaccines. This topic is discussed in detail elsewhere [10,24]. Some negative findings, however, are probably the results of methodologies that are not sufficiently sensitive to detect SV40 in human tumours. This is exemplified in Figure 1. Here we precipitated the SV40 Tag from 20 mesothelioma biopsies. The final reaction was developed with three different methodologies (Panels A, B and C). The methodology used in Figure 1 (Panel A), detects SV40 Tag only in the positive control and all the biopsies appear negative. In Panel B, we used an ultrasensitive methodology and biopsy 65 is clearly Tag positive and biopsy 76 is weakly positive. In Panel C, the same membrane shown in B was stripped and the reaction developed using a methodology that has an intermediate sensitivity between A and B. Seven of the 20 biopsies are Tag positive. These results underscore the importance of optimizing the procedures to detect low amounts of Tag in human biopsies. Importantly, the ultrasensitive technique used in Panel B detects only two positive samples, because the high sensitivity causes a high background noise that requires to limit to a few seconds the exposure time before the film becomes completely black. Instead, a less sensitive technique (Panel C) is more efficient in detecting SV40 Tag. Therefore, the most sensitive techniques are not necessarily the most reliable for detecting low amounts of Tag in human biopsies.

The importance of optimizing the methodology to detect SV40 reliably was recently underscored by the research team of U. Stadt. These investigators demonstrated how important are apparently minor technical details, such as the use of a silica gel-based DNA extraction methodology, to detect low copy numbers of SV40 reliably in real time quantitative PCR [25]. Our results (Fig. 1), the results of Heinsohn et al [25], and those of many other investigators, indicate that low amounts of SV40 and SV40 proteins are sometimes present in human tumour biopsies. It should be noted that low amounts of Tag expression by tumour cells are associated with a more oncogenic tumour phenotype in animals compared to clones expressing a high level of Tag, because the cells can evade immune detection [26]. Moreover, the research team of D. Shrump at the NCI has elegantly demonstrated that low amounts of Tag are sufficient to drive malignant growth and prevent apoptosis in human mesothelioma cell lines derived from SV40-positive tumours [27].

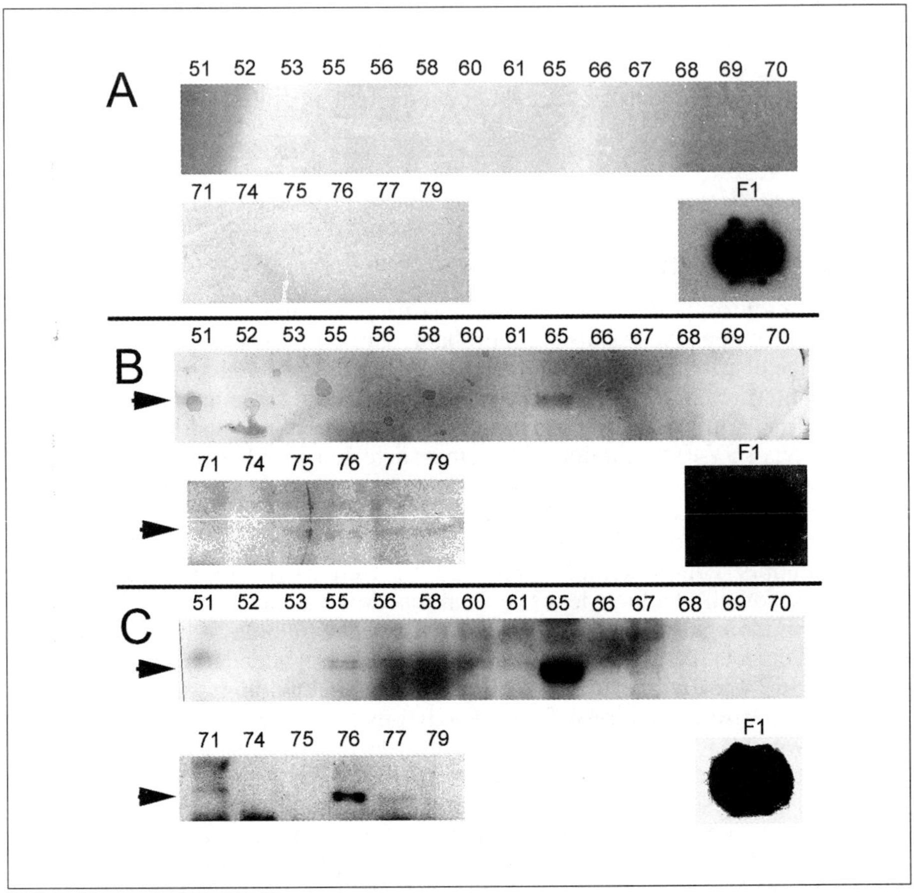

Fig. 1: **Immunoprecipitation of the SV40 large T antigen from 20 human mesothelioma frozen biopsies visualized using three diverse methods. Changes in detection methodologies lead to different results because of variation in sensitivity.** F1, positive control, SV40-transformed human mesothelial cell line. The arrow points to the ~90 kDa Tag band.

In **Panel A**, an anti-mouse goat polyclonal serum conjugated to horseradish peroxydase (HRP) was used as the secondary antibody. In **Panel B**, an anti-mouse, biotinylated chicken polyclonal antibody was used as the secondary antibody. The filters were then incubated with the ABC (Vectastain) reagent (which consists of a multimeric complex of streptavidin and biotinylated HRP. In **Panel C**, The membranes in B were stripped of antibodies at 50°C for 30' in 100 mM 2-mercaptoethanol, 2 % SDS, 62.5 Tris-HCl pH6.7. After blocking and exposure to the primary antibody, an anti-mouse, biotinylated chicken polyclonal antibody was used as the secondary antibody. The filters were then incubated with HRP-conjugated streptavidin. Bands were visualized by enhanced chemio-luminescence using the SuperSignal chemio-luminescence substrate (Pierce) in all cases.

Note that using the procedure of Panel A, all mesothelioma specimens were negative. Using the ultrasensitive ABC reagent (Panel B), the low signal-to-noise ratio allows identifying as Tag positive only two samples (65 and 76, though weakly positive). By using the intermediate sensitivity method of Panel C it was possible to identify as Tag positive samples 55, 56, 58, 60, 65, 76 and 77. Three samples were of uncertain status (51, 67 and 71). More specifically, samples 51 and 67 seemed to display a double band at the right molecular weight, but the signal was too low to make a definitive conclusion. Sample 71 showed a clear band but of slightly higher molecular weight. This band may either represent a mutated Tag or an experimental artifact. Reprinted with permission from Cancer Research (Carbone, et al, "SV40 detection in human tumor specimens" November 1st).

Methodology

500 µg of total protein lysates were immunoprecipitated with agarose-conjugated antibody mouse monoclonal PAb 101 (specific for the SV40 Tag; arrow-points). After three washes, agarose pellets were suspended in 50 µl of SDS-page loading buffer. 25 µl of this buffer were run on a 10 % SDS-page and electroblotted on to a nitrocellulose membrane. To avoid the positive control capturing most or all the antibody specific for Tag, the strip of membrane containing the positive control (F1) was cut out and hybridized separately and in parallel to those containing tumour samples. Membranes were subsequently blocked either in PBS-T containing 5 % non-fat milk (Panel A) or in Blocker casein in PBS (Pierce; Panels B and C). Western blot hybridization was carried out by using mouse monoclonal Pab 419 (specific for the SV40 Tag) as the primary antibody. After four washes, the membranes were handled differently (Panels A, B and C).

SV40 Infection of Different Cell Types

SV40 infection has different effects in different cell types. When SV40 infects rodent cells, the virus cannot replicate. These cells are described as non-permissive. In non-permissive cells, the SV40 genome is expressed and causes morphological transformation and cell division. Since SV40 cannot replicate, the cell progeny does not contain SV40 and these non-permissive cells revert to the normal phenotype following only a few passages in cell culture. This is what is called an abortive transformation. Malignant transformation in non-permissive cells can occur only if SV40 becomes integrated into the host genome. This rare event occurs in fewer than $1/10^7$ infected cells, ensuring that the SV40 genome is propagated and expressed in the cell progeny, thus maintaining the malignant phenotype [24]. Exposure of human fibroblasts and monkey cells to SV40 leads to different effects. Only about 20 % of the human fibroblasts will have productive SV40 infection, while monkey cell exposure to SV40 results in a 100 % infection rate. The two cell types are similar in that once productive infection is established, the virus replicates very fast and causes cell lysis. The inability of the human fibroblasts to survive the infection had led to the belief that SV40 could not cause malignant transformation of human cells. Cellular transformation of human fibroblasts was considered a very rare event that only took place when the virus became integrated into the human cell genome. The integration often causes deletions of the SV40 genome that prevent viral particle formation and therefore protects the cell from lysis. However, SV40 was detected in human mesotheliomas, not in fibrosarcomas. Mesotheliomas are tumours of the Mesothelial cells, the last remnant of mesoderm, and are thus very undifferentiated cells. Normal human mesothelial cells express high levels of p53. We found that when T antigen is expressed in these cells, it binds and inactivates p53 [11]. At the same time, p53 binding impairs Tag ability to promote SV40 replication. Therefore, fewer SV40 viral particles are produced in infected human mesothelial cells compared to infected human fibroblasts. This phenomenon is associated with a reduced cell lysis rate. About 80 % of human mesothelial cells survive infection and are capable of carrying episomal SV40 [22]. This is in contrast to the other human cell types in which episomal SV40 would lead to rapid viral replication, which in turn would cause cell lysis [24]. In conclusion, human mesothelial cells are very susceptible to SV40 infection and transformation, and in SV40-transformed mesothelial cells, SV40 can be found integrated or more often as an episome. SV40 infection in human mesothelial cells causes extensive genetic damage. For example, in one study, infected human mesothelial cell had 85 chromosomes and most of the cells showed multiple aberrations and chromosomal rearrangement [24].

SV40, a Cancer-Causing Virus

There are numerous studies that point to SV40 as a cancer-causing agent. SV40 induces telomerase activity in infected cells, which in turn renders the cell immortal [28]. The disruption of the intracellular pathways regulated by SV40 large T and small t antigens and telomerase activity is considered sufficient to create a human tumour cell [29]. Strong supporting evidence for SV40 oncogenicity was established in a 1964 study, in which SV40-transformed human cells grew subcutaneous tumours when injected into human volunteers [30]. It should be underlined that viruses are seldom complete carcinogens and most of the people exposed to SV40 infection are not expected to develop mesothelioma or other cancers. A competent immune system will recognize the transformed cells and destroy them. Therefore, SV40 ability to induce tumour formation in humans should be studied alongside the environmental and genetic factors associated with that particular tumour type (for example, asbestos exposure for mesothelioma).

Sources of SV40 Infection, the Polio Vaccine Link

There are known possible sources such as contaminated polio vaccines; direct contact with monkeys as well as direct exposure in laboratory personnel working with the virus. There is also the hypothesis that human to human transmission is possible. In 1960, it was discovered that SV40 contaminated 10-30 % of the poliovirus vaccines in the United States [31]. Several batches of polio vaccine prepared in the U.S. and in other countries during the years between 1955-1963 were contaminated with SV40. The source of SV40 was both the original seed vaccines prepared by Dr. Salk, Dr. Sabin and others and the monkeys used to propagate this vaccine in subsequent years. During that period, about 90 % of children and 60 % of adults (more than 98 million people in the U.S.A.) were vaccinated for polio and many of them were exposed to SV40 contaminated vaccines [32]. Epidemiological studies that investigated the relationship between SV40 infection and cancer risks have yielded mixed results. In a review by Dang-Tan et al [33], the authors have concluded that the studies can be grouped into three categories based on their exposure definition of SV40 infection: (i) the use of vaccination or birth cohorts as proxy variables for infection; (ii) follow-up of children of pregnant women who received polio vaccines and (iii) direct molecular detection of the virus or serological detection of anti-SV40 antibody responses. These authors indicated that a meta-analysis of five published studies did not support the hypothesis that SV40 exposure increases the overall risk of cancer incidence or cancer mortality. They also concluded that there are substantial problems that most studies have had in reliably defining exposure, defining latency effects, or dealing with confounding and other biases.

Conclusion

Following review of the published literature, the Vaccine Safety Committee of the Institute of Medicine (IOM), National Academy of Sciences, U.S.A., conclusions about polio vaccines and human cancer were: (i) "the evidence is inadequate to accept or reject a causal relation between SV40-containing polio vaccines and cancer" because the "epidemiological studies are sufficiently flawed" and (ii) "the biological evidence is of moderate strength that SV40 exposure from the polio vaccines is related to SV40 infection in humans". The epidemiological studies were considered flawed because it was not possible to distinguish reliably between infected and non-infected cohorts. As for SV40, the IOM concluded that (i) "the evidence is strong that SV40 is a transforming virus' and (ii) 'the evidence is of moderate strength that SV40 exposure could lead to cancer in humans under natural conditions" [34].

Two more panels have reached similar conclusions. First at an International consensus meeting on SV40 and human tumours held at the University of Chicago [35] and later in a workshop organized by the Biological Carcinogenesis Branch of the National Cancer Institute, Bethesda, MD [36].

The unavailability of epidemiological evidence results in our inability to establish a precise percent of tumours linked to SV40 and to establish the effects that the virus may have in exposed cohorts. Since only 10-30 percent of polio vaccines were SV40 contaminated, it is impossible to determine among individuals vaccinated from 1954 to 1963 who received contaminated vaccines and who did not. Moreover, it is expected that great variations in SV40 detection could be detected from one country, or even county to the next, depending on whether a contaminated or non-contaminated batch of vaccine was distributed in that region. Therefore, currently the contribution of SV40 to human cancer can be studied only in individual tumour biopsies. The overall effect of SV40 in carcinogenesis cannot be evaluated until reliable epidemiological studies among exposed and non-exposed cohorts can be conducted, given the uncertainties in clearly distinguishing between these two cohorts.

Acknowledgement

This work was supported by the following grant NIH, RO1 CA092657 to Michele Carbone

References

1 Butel JS, Lednicky JA: Cell and molecular biology of Simian virus 40: Implications for human infections and disease. J Natl Cancer Inst 1999; 91: 119-133.

2 Eddy BE, Borman GS, Grubbs GE, Young RD: Identification of the oncogenic substance in rhesus monkey kidney cell cultures as Simian virus 40. Virology 1962; 17: 65-75.

3 Cicala C, Pompetti F, Carbone M: SV40 induces mesothelioms in hamsters. Am J Pathol 1993; 142: 1524-1533.

4 Kirschtein RL, Gerber P: Ependymomas produced after intracerebral inoculation of SV40 into new-born hamsters. Nature 1962; 195: 299-300.

5 Carbone M, Rizzo P, Grimley PM, Procopio A, Mew DJ, Shridhar V, de Bartolomeis A, Esposito V, Giuliano MT, Steinberg SM, Levine AS, Giordano A, Pass HI: Simian virus-40 large-T antigen binds p53 in human mesotheliomas Nat Med 1997a; 8: 908-912.

6 Carbone M, Rizzo P, Pass H: Simian virus 40: the link with human malignant mesothelioma is well established. Anticancer Res 2000; 20 (2A): 875-877.

7 Testa JR, Pass HI, Carbone M: Molecular Biology of Mesothelioma, in De Vita V, Hellman S, Rosenberg S, Williams & Wilkinis (eds): Principles and Practice of Oncology. Philadelphia, PA, Lippincott, 2001, pp 1937-1943.

8 Carbone M, Kratzke RA, Testa JR: The pathogenesis of mesothelioma. Seminars in Onclogy 2002; 29 (1): 2-17.

9 Carbone M, Pass HI, Rizzo P, Marinetti M, Di Muzio MA, Mew DJ, Levine AS, Procopio A: Simian virus 40-like DNA sequences in human pleural mesothelioma. Oncogene 1994; 9: 1781-1790.

10 Gazdar AF, Butel JS, Carbone M: SV40 and human tumours: myth, association or causality? Nat Rev Cancer 2002; 2: 957-964.

11 Bocchetta M, Di Resta I, Powers A, Fresco R, Tosolini A, Testa JR, Pass HI, Rizzo P, Carbone M: Human mesothelial cells are usually susceptible to simian virus 40-mediated transformation and asbestos cocarcinogenicity. Proc Natl Acad Sci USA, 2000; 97: 10214-10219.

12 Shivapurkar N, Wiethege T, Wistuba II, Salmon E, Milchgrub S, Muller KM, Churg A, Pass H, Gazdar AF: Presence of simian virus 40 sequences in malignant mesotheliomas and Mesothelial cell proliferatios. J of Cellular Biochemistry 1999; 76: 181-188.

13 Ramael M, Nagels J, Heylen H, De Schepper S, Paulussen J, De Maeyer M, Van Haesendonck C: Detection of SV40 like viral DNA and viral antigens in malignant pleural mesothelioma. Eur Respir J: 1999; 14: 1381-1386.

14 Cristaudo A, Vivaldi A., Sensales G, Guglielmi G, Ciancia E, Elisei R., Ottenga F: Molecular biology studies on mesothelioma tumour samples: preliminary data on H-ras, p21, and SV40. J Environ Pathol. Toxicol Oncol 1995; 14:29-34.

15 Carbone M, Rizzo P, Procopio A, Guiliano M, Pass HI, Gebhardt MC, Mangham C, Hansen M, Malkin DF, Bushart G, Pompetti P, Picci P, Levine AS, Bergsagel JD, Garcea RL: SV40- like sequences in human bone tumours. Oncogene 1996; 13: 527-535.

16 Yamamoto H, Nakayama T, Murakami H, Hosaka T, Nakamata T, Tsuboyama T, Oka M, Nakamura T, Toguchida J: High incidence of SV40-like sequence detection in tumour and peripheral blood cells of Japanese osteosarcoma patients. Br J Cancer 2000; 82: 1677-1681.

17 Weiss AF, Portmann R, Fischer H, Simon J, Zang KD: Simian virus 40-related antigens in three human meningiomas with defined chromosome loss. Proc Natl Acad Sci U S A 1975; 72 (2): 609-613.

18 Lednicky JA, Garcea RL, Bergsagel DJ, Butel JS: Natural simian virus 40 strains are present in human choroids plexus and ependymoma tumours. Virology 1995; 212: 710-717.

19 Martini F, Iaccheri L, Lazzarin L, Carinci P, Corallini A, Gerosa M, Iuzzolino P, Barbanti-Brodano G, Tognon M: SV40 early region and large T antigen in human brain tumours, peripheral blood cells and sperm fluids from healthy individuals. Cancer Res 1996; 56: 4820-4825.

20 Carbone M, Bocchetta M, Cristaudo A, Emri S, Gazdar A, Jasani B, Lednicky J, Miele L, Mutti L, Pass HJ, Ramel, M, Rizzo P, Testa JR: SV40 and human brain tumours. Int J Cancer 2003a; 106: 140-142.

21 David H, Mendoza S, Konishi T, Miller CW: Simian virus 40 is present in human lymphomas and normal blood. Cancer Lett. 2001; 162: 57-64.

22 Shivapukar N, Harada K, Reddy J, Scheuermann RH, Xu Y, McKenna RW, Milchgrub S., Kroft SH, Feng Z, Gazdar AF: Presence of simian virus 40 DNA sequences in human lymphomas. Lancet 2002; 359: 851-852.

23 Vilchez RA, Madden CR, Kozinetz CA, Halvorson SJ, White ZS, Jorgensen JL, Finch CJ, Butel JS: Association between simian virus 40 and non – Hodgkin lymphoma. Lancet 2002; 359: 817-823.

24 Carbone M, Pass HI, Miele L, Bocchetta M: New developments about the association of SV40 with human mesothelioma. Oncogene 2003b; 22: 5173-5180.

25 Heinsohn S, Golta S, Kabisch H, Stadt UZ: Standardized detection of simian virus 40 by real-time quantitative polymerase chain reaction in pediatric malignancies. Hematologica/the hematology journal 2005:90 (1): 95-100.

26 Salewski H, Bayer TA, Eidhoff U, Preuss U, Weggen S, Scheidtmann KH: Increased oncogenicity of subclones of SV40 large T-induced neuroectodermal tumour cell lines after loss of large T expression and concomitant mutation in p53. Cancer Research 1999; 59: 1980-1986.

27 Waheed I, Guo ZS, Chen GA, Weiser TS, Nguyen DM, Schrump DS: Antisense to SV40 early gene region induces growth arrest and apoptosis in T-antigen-positive human pleural mesothelioma cells. Cancer Research 1999; 59 (24): 6068-6073.

28 Foddis R, De Rienzo A, Broccoli D, Bocchetta M, Stekala E, Rizzo P, Tosolini A, Jhanwar SC, Pass HI, Testa JR, Carbone M: SV40 infection induces telomerase activity in human mesothelial cells. Oncogene 2002; 21: 1434-1442.

29 Hahn WC, Dessain SK, Brooks MW, King JE, Elenbaas B, Sabatini DM, DeCaprio JA, Weinberg RA: Enumeration of the simian virus 40 early region elements necessary for human cell transformation. Mol Cell Biol 2002; 22 (7): 2111-2123. Erratum, in Mol Cell Biol 2002; 22 (10): 3562.

30 Jensen F, Koprowski H, Pagano JS, Ponten J, Ravdin RG: J Natl Cancer Inst 1964; 32: 917-937.

31 Sweet BH, Hilleman MR: The vaculating virus SV40. Proc Soc Exp Biol Med 1960; 105: 420-427.

32 Shah KV, Nathanson N: Human exposure to SV40. Review and comment. Am J Epidomiol 1976; 103: 1-12.

33 Dang-Tan T, Mahmud S, Puntoni R, Franco E: Polio vaccines, Simian Virus 40, and human cancer: the epidemiologic evidence for a causal association. Oncogene 2004:23: 6535-6540.

34 IOM Report: Immunization safety review. SV40 contamination of poliovaccine and cancer, in Stratton K, Almario DA, McCormick M (eds): The National Academy of Sciences. Washington, DC, The National Academies Press, 2003, www.nep.edu.

35 Klein G, Powers A, Croce CM: Association of SV40 with human tumours. Oncogene 2002; 21: 1141-1149.

36 Wong M, Pagano JS, Schiller JT, Tevethia SS, Raab-Traub N, Gruber J: New associations of human papillomavirus, Simian virus 40, and Epstein-Barr virus with human cancer. J Natl Cancer Inst 2002; 94 (24): 1832-1836.

Dr. M. Carbone, Loyola University Medical Center, Department of Pathology, Bldg 112, Room 205, 2160 South First Avenue, Maywood, IL 60153, USA.
E-mail: mcarbon@lumc.edu

Petricciani J, Sheets R (eds): Vaccine Cell Substrates 2004.
Dev Biol (Basel). Basel, Karger, 2006, vol 123, pp 119-132.

PANEL II DISCUSSION

Potential Oncogenicity of Viruses Latent in Cells

Panel Members: *David Onions (Panel Chair), Ruth Jarrett, John Coffin, Audrey Chang, Phil Minor, Michele Carbone, Phil Krause, Alan Shaw*

PANEL QUESTIONS

Question 1:
Do we need to implement more general screening assays for oncogenic viruses?
- If so, which viruses? e.g. polyomaviruses, herpesviruses and retroviruses?
- If so, what type of assays?
  - PERT for retroviruses?
  - RDA, redundant PCR?
  - In vivo assays in newborn mice, rats, hamsters?

Question 2:
Should we look for novel viral interactions
- (e.g., pseudotype formation) in novel cell substrates?
- Potential complementation of defective oncogenic viruses by vaccine viruses?

Question 3:
Do the observations of SV40 sequences in human tumours have implications for the continued use of primary cells?

Question 4:
Is the potential for infectious virus to be transmitted in DNA form relevant to the safety of vaccines?

Dr. Onions: If SV40 is the initiating agent in mesotheliomas, then what one would expect from other tumour virus models is that that virus would be clonally involved in those tumours. At a crude level you would expect that there would be at least one genome copy per cell and, if that was integrated, you should also find a clonal integration. Do data support the contention that in these tumours that are positive there is a clonal pattern of SV40 infection?

Dr. Carbone: Except for one case that I am studying in which SV40 appears to be clonal and integrated into the mesothelioma cell line that we have derived from an SV40 positive mesothelioma biopsy, I do not think any have been done. I do not know whether SV40 is necessarily involved in the initiation of mesothelioma. It could be that SV40 acts as a carcinogen with other factors that could be the initiators, such as asbestos. I do not know if SV40, in other words, is the initiator or if it is the promoter.

The integration of SV40 is not frequent. When it is integrated it is easy to find it if it is clonal. In the other cases it is less easy. We and others have found that sometimes cell lines that you derive from human mesotheliomas that are SV40 positive contain a minority of cells that are SV40 positive and a majority of cells that are SV40 negative. I do not know what it means. I have found that by passaging these cells in nude mice you increase the number of positive SV40 cells.

The data that we found in that cell line in which we have a minority of cells expressing T antigen and you select for those cells when you put it in nude mice, which may support this hypothesis.

Dr. Peden: I was very struck with what you said about 776 appearing in mesotheliomas. Do you know whether other strains have been found in mesotheliomas apart from 776?

Dr. Carbone: Yes. I have found strains of wild type SV40. I used 776 as a reference for wild type, meaning virus that contains 272 base pairs in the regulatory region. We found about 30 percent of samples that contained various deletions that are often present in T antigen. I do not know exactly how we want to call those strains. There has been a report of one archetypal SV40 isolated from a human mesothelioma that I remember, and that was published in Cancer Research in 1999. There are differences among 776-like strains in the carboxyl terminus but the general message is that except for that single paper showing archetypal SV40, most reports about SV40 mesothelioma concern a virus that has 272 base pairs in the regulatory region similar to wild type 776.

Dr. Peden: Why are 272 base pairs better for mesotheliomas?

Dr. Carbone: We want to prove it, and to do so, I need human primary or animal cells and have been unable to do so until now. Wild type SV40 will lyse most cells but will not lyse mesothelial cells. I think that the only place where it can be found is in mesothelial cells because they can survive it. Instead, it would not be easy to find it in other cell types because it would kill them. Lymphocytes are killed by SV40 wild type.

By wild type, even if this is the wrong terminology because now they call it differently, I mean the virus with 272 base pairs in the regulatory region, which is also a virus that usually replicates much faster and is more aggressive than other viruses in terms of viral replication.

Participant: Phil (Minor), did you test your cynomolgus sera with BKV (human BK virus) and JCV (human JC virus)?

Dr. Minor: We did not, but Denise Galloway did the things that we sent her. As I recollect, they were reasonably specific.

Dr. Cichutek: I would suggest that we conclude that if SV40 is transferred by vaccines to humans, you would find it in normal cells and in tumour cells. I would not be convinced about a causative role of SV40 unless you do the right experiments. In papillomaviruses, it has been shown that the oncogenes derived from the papillomaviruses are present in the tumour cells. They are active in the tumour cells, and if you silence the gene they change the phenotype and are no longer malignant. Have you done any of these experiments with mesothelioma cells? I would like to see data of active large T. If you take the genome away by inactivation, which can readily be done, do the cells change their phenotype and not grow in nude mice?

Dr. Carbone: I showed a couple of slides of a paper that we published in Nature in 1997, showing that in human mesothelioma biopsies SV40 T antigen was bound to cellular p53.

A paper was published by David Strand in Cancer Research in 1999 that was designed to address the issue you raised. He derived some cell lines from his mesothelioma patients and found that there was some T antigen expression but that the amount of T antigen expression was extremely low. He was convinced that T antigen was not biologically important. He did the same type of experiment that was done to demonstrate that HPV was the causative agent of cervical carcinoma.

To his surprise, he found that T antigen caused growth arrest and apoptosis only of the SV40 positive mesothelioma cell culture and not of SV40 in negative mesothelioma cell culture, which I think is an indication that the virus, at least in that mesothelioma, was playing a role in the pathogenesis of the mesothelioma.

Cancer is always a multifactorial phenomenon that is caused by different things, so I would not consider SV40 a sufficient cause, but I am not aware of any. The only necessary cause that I know of is HBV and even that is insufficient. There are a number of other things that need to go wrong for a cancer to develop.

We have very strong evidence that SV40 and asbestos are co-carcinogens.

Briefly, we found first in tissue culture that SV40 and asbestos were co-carcinogens in transforming human mesothelial cells. Now we have found that in animals there is an incredible co-carcinogenicity between SV40 and asbestos in causing mesothelioma. That is the first time that mineral fibre and virus are shown to cooperate in causing cancer in animals.

Participant: As I understood it, you mentioned one individual who had been born after the U.S. distribution of SV40 containing polio vaccine had stopped, and then went on and talked about the Russian experience. But I wondered if you knew anything in particular about that individual and whether there was any possibility that he or she might have received a Russian vaccine.

Dr. Carbone: No, unfortunately, I know nothing about that case. There is another interesting coincidence about the detection of SV40, which comes from one of the so-called negative papers that was published in 1999 by Wagon et al. in Germany. If I am correct, they analysed 256 and of these 256, 254 were negative for SV40. Only two were positive and these happened to be the only two samples that were coming from the United States. The other 254 were from West Germany. There seems to be some clustering in SV40 positivity, but whether this clustering is caused by the polio vaccines is not known.

Dr. Onions: Were the DNAs prepared in the same laboratory? Were the DNAs prepared from Germany samples in one laboratory and the DNA samples prepared from the American set in another laboratory? This refers to issues not just of geographical origin but preparation of materials.

Dr. Carbone: Yes, they were prepared in the same place and were used on two American soldiers who had been operated on at the same hospital.

Participant: I am very impressed with growing the cell lines from the human mesotheliomas and testing for SV40 T antigen, integration, etc. In the meeting in 1997, the issue came up with the possibility of human-to-human transmission of SV40 before the introduction of the polio vaccine. There are several tumour banks around that exist with samples of material that was collected before polio vaccine was introduced in 1954. Do you know of any work that has been done on those that suggests that there may or may not have been SV40 DNA in any of those samples, early samples for example taken in the 1930s, 1940s or 1950s?

Dr. Carbone: I know only of the study that I am doing on samples that I have received pre-1954 and after 1954 on bone tumours. Unfortunately, I still do not have enough data to be able to talk about it.

Participant: Dr. Audrey Chang. Can you do PERT assays on cell supernatants where you have lytic infection? Clearly, this is more complex because of DNA polymerases. Are there methodologies that are being used and validated to show that you can reduce that effect?

Dr. Chang: The recommendation is to try to obtain the best suitable sample for the PERT assay, which is why I suggested the cell supernatant. You are correct; there are some instances where that is not possible. There are two things that can be done. One is to add the activated nicked DNA, and it serves as a preferential template for the DNA polymerase. In that case I would recommend that assay because you know that your sample does have DNA polymerase. In situations where you can avoid that, the addition of activated DNA can sometimes reduce the assay sensitivity, but in this case you have at least an assay which is better than nothing.

Dr. Peden: We have specific primers that if you do the conventional, non-quantitative PERT assay, you can eliminate cell lysate activity, but we have been unable to adapt that to the quantitative PERT assay. In our experience the oligonucleotides and activated calf thymus DNA do absolutely nothing.

Dr. Onions: We have a slight difference. We feel that activated calf thymus DNA, used correctly, is effective, but it does reduce sensitivity a little.

Dr. Peden: Well, there was a little hyperbole when I said it does not do anything. In cell lysates it did not do very much, but in cell supernatants, with a little activity you are right, it does.

Dr. Cichutek: I think we all agree that transferring an infectious agent through vaccines is bad and if they carry oncogenes, that is worse. If we come up with tumorigenic cell lines that carry these oncogenes and we have identified them, I suggest that the recommendation should be to reduce the amount of oncogenic DNA, as specified already, in residual DNA as much as we can even if the DNA is less than 10 ng per dose. Would the panel agree with that or would you think that when you have a 10 ng per dose level you would not have to look at the specific levels of oncogenes?

Dr. Coffin: There are very few cell lines I can think of that have that property now. I would think for cells that had identifiable oncogenes such as HeLa cells, one might be more stringent in the absence of additional knowledge to the contrary, e.g. a specific assay that at least told you something, but it is not something to which I have given serious thought.

Dr. Onions: This is a very important question for the industry because actually we are now moving towards cell substrates that have defined oncogenic sequences. Per.C-6 is a very good example where we have a well-defined E1-AB region. We know exactly what that construct is. We can evaluate by quantitative PCR whether or not there is an intact element at whatever level you would like to define in residual DNA. I am not sure that you need to push it below 10 ng. I think that would be a slight concern to several people in this audience. If that were true, I would like to know a strong reason why you would want to do that.

Dr. Coffin: One could certainly quantify the amount of that specific oncogene quite accurately in these preparations.

Dr. Cichutek: The theoretical argument would be that oncogenes in cells in culture can transform them. We also know that we do not really have a good in vivo model because nothing has been found in vivo, which does not necessarily argue that this does not occur. It may tell us the test is not so good. So, at least for theoretical arguments, that may lead to a recommendation to specify a certain level, look at whether these are intact genes or whether they are fragmented pieces of them and, if they are intact, reduce them as much as you can so that may lead to a specification below 10 ng in such cases.

Dr. Onions: What I am trying to say is that the absolute level of DNA may be less relevant than the cogent point that was made this morning, i.e. that the level of shearing of that DNA may be a much more important factor, given that some methodologies only give a modal value and that certainly using gels is not a sufficient criterion for assessing the degree of shearing; it has to be slightly more sophisticated than that.

Dr. Minor: I am not quite sure why we pick on viral oncogenes in particular. If you consider something like a wildly oncogenic MDCK cell, which I believe exist, I am presuming that has some kind of oncogene in there or constellation of oncogenes which do just as much from MDCKs as E-6, E-7 or whatever it is does for HeLa. Why pick on E-6, E-7?

Dr. Onions: As Dr. Carbone very elegantly showed in his first slide, if you look at something such as SV40, SV40 crams a whole set of functions to contribute to transformation into that transforming domain and another region adjacent to it. It has an RB binding component, p53 binding component plus other components that are involved in transformation. Shearing becomes more important in this case.

I think one of the values of having defined genes that we know contribute to the oncogenicity of the cell line is actually a plus because we can look at the consequences. Do we need to implement more general screening assays for oncogenic viruses? If so, which viruses should we be looking for and what types of assays should we use?

We have heard about the PERT assay which is already implemented effectively as a broad screening assay for retroviruses. Ruth has talked about both RDA and redundant PCR. Although we have not heard about it, Andy Lewis, for instance, has made a very cogent case for the use of multiple animal models to screen for polyomaviruses and other transforming viruses. I would like to try and take you through those questions in turn.

Dr. Minor: The SV40 incident really took place the best part of 50 years ago, and has come back to bite us. I do not believe that we have got to the bottom of it. I do not know whether SV40 is circulating in humans or not. From a product point of view, I think this is not a situation you want to be in. If it is possible to implement more general screening assays for oncogenic viruses, I think they should be done. That begs the question of whether it is possible to do so. But I think you do it if you possibly can.

Dr. Onions: This has implications. For instance, we always talk about screening cell lines for SV40. Until very recently people did not bother to screen Vero cells for lymphotropic polyomavirus, a virus which is actually a virus of African green monkeys. We know the sequence of this virus so it is fairly straightforward. Until recently there were a number of vaccines in use based on BHK cells or primary hamster cells which can be infected by a transforming polyomavirus. It causes both lyphomas and skin tumours. We can screen for it because we have PCR assays for that virus. But as Ruth Jarrett pointed out, there are probably a large number of polyomaviruses that we simply do not know about; we have sequences for relatively few. So, it would not be a surprise to find in some immortalized lines sequences of SV40-like viruses and polyomaviruses that we know nothing about and that really does induce a need to look for these viruses using some form of assay.

Dr. Carbone: I certainly agree that it is important to screen the vaccines since we are giving them to our children and therefore it is a very sensitive thing not only for us and for the public as a whole. In fact, I was a little surprised that SV40 PCR testing was not done as suggested by Phil in a paper that he published in 1999, by adding PCR testing in addition to tissue culture testing to test for SV40.

The problem with PCR is that you can only test for what you know, and the problem is that SV40 has already happened so it is hoped it is not going to happen again. It is not really SV40 that we have to test for. We have to test for what we do not know, and a primer cannot be designed for that. The suggestion of using an animal model could be an excellent idea. The most sensitive technology among all those I have used was the lipofection of the DNA into the cells. SV40 grew very well in all of them. With PCR, for example, when the amount of DNA is not sufficient, it will give a positive result on one run and a negative result on the other. If it is at the limit of detection, that is what you get. People have the wrong idea that with PCR if there is one molecule which has to be found but that is not true and everyone who does PCR knows this.

The issue is to do a tumour test if it is reasonably economical to do. Based on my experience, lipofecting the DNA into different cell lines and if something grows out or causes damage, it can be a relatively simple way to detect agents that we do not know and against which we cannot design a primer.

Dr. Lewis: I think our emphasis on the animal assay has essentially been driven by the fact that it is the most general type of screening procedure we can use. It has been successfully used to detect oncogenic agents in tissue culture, especially in tissue cultures that may be used to manufacture vaccines. In our search of the literature, at least the viruses that were known six or eight years ago, using animals could detect most of them.

The problem with these assays is that they are extremely insensitive. The best I am aware of for detecting either SV40 or adenovirus in any of these models is about 1,000 to 10,000 infectious units. That is a lot of virus/ml to expect to be present in a cell substrate. I think the possibility that concerns me is that oncogenic agents exist that are very similar to hepatitis C which occur in very low concentrations. They are hard to find because of the concentration, which can change over time. In spite of the sensitivity of the PCR and the PERT assay, it is very easy to conceive of a situation, especially where you are using a primary cell for example, and you had e.g. 10-20 T-150 flasks full of cells, you might have only five foci of infectivity in those 20 or 30 flasks, and you would never see it by PCR. Only if it amplifies itself by replication over time would you stand any chance at all of detecting it either by PCR, animal inoculation, tissue culture, or anything else. If you happen to have a situation where it is very slow growing, it would be almost impossible to detect by any current technique.

That is the real threat, and I think it represents a threat we really do not know how to address. But it is also the argument for not using primary cells. If you have continuous cell lines and cell lines which you can study over a longer period in a much greater sample, you would be more confident that something like that does not exist.

Dr. Onions: I think the last point is perhaps one of the most pertinent ones, and I think it was also borne out by Ruth's comments that clearly, although we may have concerns about using immortalized and tumorigenic cell lines, in many ways the capacity to characterize those cell lines very thoroughly outweighs the disadvantages of using them.

Dr. Lewis: With regard to the animal models, we are attempting to look at other types of animals, especially some of the more genetically modified strains, such as the CD epsilon mice which are more immunocompromised than the nude mouse, to see if they may be more sensitive. I am really not enthusiastic that we are going to find differences of major orders of magnitude in the sensitivity of hardly any of these things. I think we are left with that.

Dr. Jarrett: Sensitivity is an issue with primary cells. The RDA would not be a good technique. The sensitivity would be fine when we are dealing with clonal cell substrates. Although I said it was technically demanding and time-consuming, I think where you have an important cell substrate you want to make sure it does not have something exogenous in it. There, perhaps, would be a rule for testing with RDA.

Dr. Krause: May I add to what Andy said because we have some data on that general question. When we looked at simian cytomegalovirus in primary monkey kidney cells, we checked them by a PCR that is much more sensitive than the redundant PCRs you are talking about because we were using highly specific primers that we knew would detect the strains we were looking for.

These cells were all negative by PCR when we received them. Subsequently, over time, they would ultimately come down with simian cytomegalovirus cytopathic effect, and would also be PCR positive a little before.

The other comment that I would make on RDA is that there are a couple of other disadvantages. One of them is the need to find something to suppress. If you have a cell line that comes from a tumour, you need to find something that is genetically as identical as possible to that to subtract, to try to find the virus. If you have a sample from the same individual, as did Chang and Moore when they discovered HHV-8, then you have an identical genetic background. But if you do not have such a sample, then you are dramatically increasing the number of those annoying bands you were talking about which have to be identified.

Dr. Jarrett: I did not have enough time to go into this, but the slide I showed had a reciprocal subtraction on it to make that very point because we were subtracting a cell line from a central sample and the reciprocal subtraction had some fragments which turned out to be HLA.

Dr. Krause: Although I know that some RNA viruses such as GBV (GB virus) have been discovered by RDA, in general this is a technique which is designed to be most effective for finding DNA viruses because you can obtain a representation of what is in the genome. Cell-to-cell variability, even within the same individual, in terms of the RNA that is expressed can be greater, which also increases the number of bands that you need to chase down and decreases your likelihood of finding virus if you are looking for an RNA virus. I am intrigued by the use of RDA as well, but I also see significant issues with sensitivity. Many laboratories have tried looking for viruses by RDA and have found nothing.

Dr. Jarrett: In fairness, every time we have looked at a sample which we have known contains a virus, we have detected it. But I think one of the problems is that if you do a number of RDA experiments, the result is many bad bands and you have to decide what to pursue, and that is the horror because cloning and sequencing can be done forever.

Dr. Onions: Should we look for novel virus interactions, particularly the retroviruses? For instance, is pseudotype formation perhaps important in novel cell substrates? Should the more theoretical issue of potential of complementation of detecting oncogenic viruses by vaccine viruses be something that we should be thinking of? Is this just too much of an academic theoretical point of view to be of consequence? The kinds of things I was thinking of are, for instance, a number of flaviviruses being used on novel cell substrates. What is the possibility of flavivirus pseudotyping? Let us assume a virus that is defective does not have an envelope gene of its own, but the flavivirus puts on an envelope for it and allows that retrovirus, at least in a one-hit kinetics, to get into the target human cell?

The second question is that some of the new cell substrates being studied are screened and are known to be PERT-negative. They do not release retroviruses. On the other hand, for instance, Vero cells contain the MAC family of retroviruses within them and, as far as I am aware, they have not been checked for expression of envelope gene or envelope protein. If that occurs would it be of any significance?

Dr. Coffin: These kinds of things can happen and there are many instances in the gene therapy and retroviral vector literature where you find unexpected recombinants in general between vector construct and packaging constructs. An example is the case where macaques were inoculated with a globulin containing vector that had a fair amount of contamination of recombinant virus. Mal Martin analysed them and found a great number of tumours with many recombinants. They had picked up various endogenous proviruses from the original mouse cell line. Although recombination with the added helper virus contributed to this, it was very clear that the endogenous viruses could also contribute in the context of where one had the opportunity for recombination to occur with the vector construct.

I think in the case where vector constructs have been put into cells to generate the lines or to provide the product you are interested in, one has to look for these things very carefully. Very sensitive vector rescue tests have to be developed to ensure there is nothing replicating.

Dr. Onions: One of the related issues goes back to the PERT assay which is used now as a broad screening assay to determine whether a cell substrate is negative. We have already heard there are issues in infections where there is a degree of difficulty in interpretation because you can find DNA polymerase contamination. The danger is that a retrovirus activated under stress of the vaccine virus infection might be masked.

Dr. Coffin: It is very important to know your cells. We know what happens with certain kinds of mouse cells very well, for example, NIH 3T3 cells compared to cells from other strains of mice, and the same thing can be true for other cells.

Dr. Onions: Would that suggest, John, that in the old days we used to do a number of induction assays on cell lines with IDU or methylation inhibitors, but those seem to have gone out of general cell substrate screening. Should we be incorporating those kinds of induction assay systems back into new cell substrates such as Vero and MDCK and other cell substrates?

Dr. Coffin: I think particularly for primate cell lines it is not a bad idea to do that because we do not have very good characterization of what is in these, particularly for

D-type proviruses. The other thing probably would be co-cultivation assays, particularly with a variety of human cell lines, to see if anything infectious would come up.

Dr. Cichutek: In the gene transfer world, we have a European requirement that you look for replication of recombinant retrovirus or adenovirus. You actually use cells as much as you can to amplify the virus contaminant and then use specific detection methods to specify what you have. If you find any, the requirement would be to have less than one particle of adenovirus or RCR in the dose.

This is difficult in specific instances, for example with the HIV vectors we are currently using where there is no good positive control. People have used synthetically cloned HIV-1 vectors that have VSV envelope genes.

A specific question I would like to ask about is the human endogenous retrovirus and the VL-30 element. Is that a case we should rest because nobody has found anything attributable to endogenous human retroviruses as contaminants? What about the VL-30 cases? Is that something we should still consider? Has anybody found any risk associated with those specifically?

Dr. Coffin: There has clearly not yet been an example where someone has been able to isolate an infectious version of the human endogenous retrovirus or, for that matter, a recombinant with some other virus that is infectious and has acquired HER sequences. I do not think that means we should drop our guard completely on this issue. It is still possible that there is perhaps in a minority of people, some polymorphism. They are not, as we once thought, completely monomorphic in the human population. I know of at least four proviruses that are polymorphic. None of them has been recently inserted but they may still be there, but it would require a huge census. There could be humans around who have some infectious version or could easily yield some. I think we have to keep on looking and asking, but I do not think it is a major area of concern because we have not seen anything yet, and we have considerable experience.

Dr. Minor: You mentioned the issue about induction of retroviruses which was fashionable about 15 years ago. The difficulty was what to do with the data? If you induce a murine cell to produce something, does that mean you do not use the cell anymore? I think those tests fell out of use because nobody knew what to do with the results.

Dr. Coffin: In mice, we know the players in the inbred strains of mice. So, in some cases you may induce particles; you may induce ectropic virus that will spread through the culture, although that often happens spontaneously before you do the induction. In some cases you can induce the xenotropic provirus from BXV-1. That is something to be avoided.

I would suggest that mouse cell lines should be specifically screened to ensure they do not contain that particular provirus. That is the one mouse provirus from the endogenous viruses that could conceivably cause a problem. It should be avoided, both by screening the cells and also by screening inocula to make sure that they do not contain it. That can be done very sensibly by PCR.

Dr. Onions: I think there is a real difference between the kinds of considerations we had then, when we were trying to characterize processes particularly for monoclonal antibodies with a great deal of downstream purification. There is a huge difference between a monoclonal antibody product and, say, a vaccine product from an MDCK cell line that has little purification.

Dr. Lewis: I may be mistaken, but I think there is a huge range in the concentration of inducing agents that needs to be used. Did not Dusty Miller induce a retrovirus in mouse cells using about 10 or 100 times the concentration of IUDR than that other people generally used?

Dr. Coffin: He induced such a provirus. I do not remember what the concentrations were.

Dr. Lewis: In murine systems it is reasonably straightforward because there is considerable experience, but in primate systems (Vero and other types of cells) it is hard to know exactly what to recommend.

Dr. Coffin: The apparent exclusion of the foreign envelope glycoprotein probably occurs because different groups of viruses have slightly different mechanisms for incorporating glycoproteins. HIV envelope glycoprotein has a very long tail. It probably interacts fairly specifically with matrix glycoprotein, but that does not seem to be true in ALV and probably is not true in gamma retroviruses either. We need to test that specifically with different viruses.

Dr. Onions: It should be noted that, for instance, people are using BHK cells in different parts of the World. The BHK cell will actually produce a retrovirus that would infect human cells at a low level.

Dr. Khan: Induction strategies can identify the presence of a potential, inducible infectious virus in novel cell substrates. It should be part of the characterization of your starting material. If you can induce a retrovirus from the cell substrate, initially identified by the PERT assay, you can follow it up to see whether it is infectious and what its properties are so that you can create specific detection assays for it. That does not preclude the possible use of the cell line if you can develop a procedure to monitor the production of the virus and follow it through your production scheme.

Dr. Krause: My impression is that most of the cases in nature where you have this kind of pseudotyping or recombination between viruses, occur in circumstances where there is some kind of selection pressure, or selection advantage for the recombinant. I think that is generally true also in vaccine production. It does not mean that you do not have to worry just because these kinds of events may be rare. If you have a system in which you are trying to grow one thing optimally which you know will probably have a selective advantage in that system, it is going to be much less likely that these kinds of events occur. If they do, it is less likely that they will propagate through the culture relative to whatever else is going on in that culture. On the other hand, if you are worried about it, that makes it that much harder to detect.

I also wanted to echo a comment that in the end everything depends on what your final product is and what you are doing with it. If you are inactivating your final product, then it could be argued that many of these concerns about viruses are less serious. On the other hand, you could also argue that if you are inactivating it, unless you know it was there to start with, you do not know the degree to which you need to demonstrate inactivation. I think you have both sides of that coin also.

One should not only be worried about infectious viruses and recombination but also nucleic acids in these viruses. I think that was one of the questions in the first panel discussion which basically stated that, assuming that there are no viral genomes in these cells, how do we deal with this? Of course, the question is how do you generate the data that make you sufficiently confident that you agree to that assumption? If one cannot generate those data, then one does have to worry about potential viral infectivity, especially in cells with unusual properties that could be caused by viruses, latent or otherwise. For instance, if there is a transformed cell and you do not know why it is transformed, perhaps one would be more worried that there was some virus that had been associated with that event.

Participant: Regarding the induction assays, do they cause the induction of the quiescent retrovirus, for example, or are they potentially causing a recombination event between endogenous retroviruses and something that normally would not happen with those cells without induction?

Dr. Coffin: Recombination is a function of replication. An agent that induces replication could induce recombination. But if it does so, it is something that you should know. All you are doing with the inducing agent is enhancing the frequency of an event which you expect to occur naturally at some lower frequency. I do not think you are creating any circumstance that would not exist naturally.

Dr. Onions: The potential of transmitting viruses through infectious DNA is probably less of a risk than we think, partly because in certain situations these viruses are integrated and 'screwed up', and the rates of efficiency of some of these transmissions are very low.

Dr. Krause: The most sensitive scheme for finding SV40 is to extract DNA in speciality cells, implying that there is infectious nucleic acid in these. Perhaps someone contemplated making vaccines in such cells 20 years ago before he did all this work showing the association of SV40 here?

All these new kinds of assays are intriguing, and they all have potential to detect new things that we are not currently detecting. A key element is assay validation and understanding what it is that they detect and what they do not detect. Very often if one suggested that an assay be added, somebody will ask what assay can be replaced. If one is adding assays one cannot necessarily replace existing assays because they are testing new things. Then one might ask to what type of cells are these assays most relevant. Unless we keep a very close eye on these kinds of new developments and new ways of protecting matters, and seriously consider using new assays, especially as novel cell substrates are introduced, we are doing the public a disservice.

Dr. Shaw: It becomes a rather daunting challenge to account for all the "if's" and "what if's", and if you do this, then what happens. I would hope that we could come to some sort of conclusion about what is a reasonable series of questions to ask in a way that would generate a credible answer within a reasonable time frame that we can apply them to various cell types we would like to use for future vaccines. Otherwise, you are not going to have any new vaccines.

Dr. Carbone: Screening requires different techniques because these work in different circumstances. Relying on one technique means you are going to miss something.

The qualifications of the people who are performing the assays is also very important.

Dr. Sheets: Given that you really cannot take something like polyoma DNA which would contaminate an animal facility to do these kinds of in vivo tests, how can you address this issue of taking essentially an unvalidated system, something that historically has been used to detect viruses but has not necessarily been validated in a testing vendor's facility, and not use a positive control? How can we rely on that to make regulatory decisions or decisions about the acceptability of cell substrates? What kind of positive controls can you use that are not infectious and would not propagate through any animal facility to address this issue of viral oncogenesis?

Dr. Peden: You just do what you can do, and historically newborn rats, newborn hamsters and newborn mice have been the animals that have been sensitive to these agents. The positive control for oncogenicity is a little different because that is not an amplifiable thing so I do not think that is the problem for the testing facilities.

Dr. Coffin: In the case of viruses, I am having trouble thinking of where a broad, non-specific screen for oncogenicity would help us. It certainly would not help us for retroviruses. There is almost no retrovirus I can think of that might be a serious contaminant that would show positive in an oncogenicity test, particularly in mice. HIV would not be apparent, nor would HTLV or Xenotropic MLV. Intact viruses probably would not show up in a mouse. Maybe there are other virus systems where unknown viruses of that group might show up in such a test where they could not be easily detected any other way, but I am having difficulty with the value of an animal test in these cases.

Summary Statement

New techniques should be implemented routinely only if they are robust and can be validated to ICH criteria and have the real potential to add to our safety information. But where novel cell substrates are being used, investigational techniques may well be of value in providing additional background information.

As an example, redundant PCR techniques, particularly for herpes viruses and polyomaviruses, may well be robust enough for use in that context.

RDA could possibly have a place under certain circumstances, but it is not a technique that should be routinely applied because there are a number of technical problems with the routine implementation of that technology.

PERT has effectively become established as a useful broad screening test for retroviruses in cell assays, but we must also remember some of its limitations when screening master virus seeds because of the issue of cell lysis and DNA polymerases, and there is some controversy about the extent to which that issue can be overcome.

Small viruses might be transmitted as a DNA element, for example with polyomaviruses. But that problem should be overcome either by suitable shearing of the DNA or simple reduction of the level of the DNA.

The issue of whether transforming sequences of viruses are more dangerous than an activated oncogene (e.g. ras) in a cell was discussed. Small DNA transforming viruses often have transforming functions within a short genome area. Therefore, shearing to lower than 600 base pairs may be required to inactivate those sequences.

Retrovirus elements are ubiquitous in eukaryotes. They have been identified in yeast cells, and some endogenous viruses are replication-competent and may infect cells of other species.

Infectivity assays are preferred wherever possible as the gold standard. But often for retroviruses, we do not know what the appropriate target cell substrates might be.

Work done at FDA has shown that in a lytic virus situation, pseudotyping between lentiviruses and vaccine virus is probably not an issue. But this might require further study for alphaviruses and gamma retroviruses where pseudotype formation might occur more readily.

Induction may be an important part of the characterization of new cell substrates, followed by screening with broad infectivity assays and by PERT.

SV40 sequences have certainly been detected by a number of laboratories with some regularity in mesotheliomas, brain tumours, and osteosarcomas.

However, some reports provide negative data for these tumour types. In the case of lymphomas, for example some laboratories have reported a high frequency of positivity, but other laboratories have reported uniformly negative results.

There appears to be a constant low level of seropositivity to SV40 across age cohorts. ELISA adsorption assays suggest that this may be associated with cross-reactive antibody to BK virus, but there are some anomalies that might not be fully explained by that observation.

The role of SV40 in human populations and its role in tumours clearly requires further study.

It also emphasizes the value of using well-characterized cell substrates.

Thank you.

Petricciani J, Sheets R (eds): Vaccine Cell Substrates 2004.
Dev Biol (Basel). Basel, Karger, 2006, vol 123, p 133.

SESSION III

Routine Adventitious Agent Tests – Can in Vivo Methods be Replaced?

Petricciani J, Sheets R (eds): Vaccine Cell Substrates 2004.
Dev Biol (Basel). Basel, Karger, 2006, vol 123, pp 135-145.

Adventitious Agent Test Methods

R.L. Sheets

Commander, U.S. Public Health Service
U.S. National Institutes of Health, National Institute of Allergy and Infectious Diseases, Bethesda, Maryland, USA

Each manufacturer of viral vaccines characterizes the cell substrate used in production. This characterization involves defining the history of isolation and banking, growth characteristics, karyology, identity (e.g. iso-enzyme analysis), tumorigenicity, and freedom from adventitious agents [1-3]. Adventitious agent tests include bacterial and fungal sterility (as promulgated in 21 CFR 610.12), mycoplasma (cultivatable and non-cultivatable; and, in the case of insect cells, spiroplasma), mycobacteria (animals and/or culture), viruses (by in vitro and in vivo methods) including acute (lytic, haemadsorbing, haemagglutinating) and latent (e.g. retroviruses or other oncogenic viruses), specific tests for retroviruses, other agent-specific tests (primarily polymerase chain reaction-based), and tests for bovine and porcine agents (as promulgated in 9 CFR 113). In addition, bovine-derived materials used in production should be sourced from countries in which BSE is neither known to exist nor which represent an undue risk for BSE because their import requirements are less restrictive than would be acceptable for import into the U.S. (as listed in 9 CFR 94.18). If the passage history of the cell substrate did not carefully document the sourcing of bovine-derived materials used in culturing or if sourcing was from countries that have subsequently been identified to have a risk of BSE, risk assessment may be necessary to address this theoretical potential risk. It is important to note that not all tests are necessary for all substrates and a testing regimen should be selected considering the donor history and tissue source of the substrate, its passage history (exposure to animal-derived or human-derived materials), banking strategy, and intended use. So, for example, a master cell bank of a human or simian cell substrate should be tested for bovine or porcine viruses if, during its passage history, animal-derived (bovine or porcine) materials were used which were not adequately controlled for adventitious agents or not documented appropriately for such control. However, careful quality control of raw materials used during manufacturing might preclude inclusion of bovine or porcine virus tests on a lot-by-lot basis once the cell bank has been demonstrated to be free of such agents.

Adventitious virus test methods are found in various FDA guidance documents or revoked regulations. The basis for revocation of the regulations in 21 CFR 630 was that these regulations duplicated the general standards for vaccines and with existing vaccine licences and were over-restrictive on the development of new vaccines using more modern technology. For example, if primary monkey kidney cell cultures were used to manufacture poliovirus vaccines, as was intended in the original regulations, then tests lot-by-lot for mycobacteria were prudent, but if well-characterized banks of Vero cells or MRC-5 cells were used – as is currently done – such testing would not be necessary lot-by-lot and having regulations requiring it would be over-restrictive. Nevertheless, the test methods for the adventitious agent tests, performed when appropriate for the product (i.e. not required to be performed if inappropriate), that were promulgated in those revoked regulations are still recommended by FDA.

Insight into the purpose of many of the tests can be obtained by reviewing historical documents that evolved into the 21 CFR 630 regulations. For example, on March 16 and 17, 1961, a hearing was held before a House subcommittee to the committee on interstate and foreign commerce [4]. During this hearing, regulations were proposed for the control of oral poliovirus vaccines. The record of this hearing indicates that poliovirus harvests were to be tested for "adventitious and other infectious agents including polioviruses of other types or strains, simian agents, Mycobacterium tuberculosis, pox viruses, lymphocytic choriomeningitis virus, Echoviruses, Coxsackieviruses, and B-virus." The test methods required to do so included inoculation into rabbits, adult mice, suckling mice, guinea pigs, monkey kidney tissue cultures, human tissue cultures, and rabbit kidney tissue cultures. Each test was intended to detect one or more of the specific agents listed. They were not originally proposed as a package of broad general screening tests as they are now used, but rather as tests for specific agents known or suspected in the simian cell substrates used to propagate oral poliovirus vaccines.

In Vivo Tests

Adult mice

The test in adult mice was described in 21 CFR 630.35 [5], which was revoked after 1996, although the test method is still recommended. The text of that regulation reads as follows:

"Each of at least 20 adult mice each weighing 15-20 grams shall be inoculated intraperitoneally with 0.5 mL and intracerebrally with 0.03 mL amounts of each virus pool to be tested. The mice shall be observed for 21 days. Each mouse that dies after the first 24 hours of the test, or is sacrificed because of illness, shall be necropsied and examined for evidence of viral infection by direct observation and subinoculation of appropriate tissue into at least five additional mice, which shall be observed for 21 days. The virus pool may be used only if at least 80 percent of the original group of mice remain healthy and survive the observation period and if none of the mice shows evidence of a transmissible agent or other viral infection, other than measles virus, attributable to the vaccine."

For purposes of use of this test for cell substrates, viral seeds (other than measles virus seeds), and production harvests, the phrase "other than measles virus" should **not** be applied. For purposes of use of this test for cell substrates, the phrases "each virus pool" and "the virus pool" should **not** be applied. Intact cells (10^7 cells per mL of culture fluid) should be inoculated instead. Alternatively, cell lysates of an equivalent number of cells may be used.

By way of clarification regarding the statement that "at least 80 percent of the original group of mice remain healthy and survive the observation period", if 20 mice are used in the test, 16 must survive to the end of the 21-day observation period. If 25 mice are used in the test, 20 mice must survive to the end of the 21-day observation period, etc. This phrase has been interpreted to mean 80 % of the mice that survived past the first 24 hours, so that if half the mice (10/20) died in the first 24 hours, then they would expect 8/10 to survive to day 21 and consider the test as having passed or been successful. This would be a misinterpretation. If more than 20 percent of the mice die in the first 24 hours, the test is either invalid or the test material has failed.

Although the original purpose of this test was to detect lymphocytic choriomeningitis virus (LCMV) [4]; currently, this test is performed to detect adventitious viruses including but not necessarily limited to LCMV, Coxsackieviruses, flaviviruses, and rabies virus [6,7].

Suckling mice

The test in suckling mice was described in 21 CFR 630.35 [5], which was revoked after 1996, although the test method is still recommended. The text of that regulation reads as follows:

> "Each of at least 20 suckling mice less than 24 hours old shall be inoculated intracerebrally with 0.01 mL and intraperitoneally with 0.1 mL of the virus pool to be tested. The mice shall be observed daily for at least 14 days. Each mouse that dies after the first 24 hours of the test, or is sacrificed because of illness, shall be necropsied and examined for evidence of viral infection. Such examination shall include subinoculation of appropriate tissue suspensions into an additional group of at least five suckling mice by intracerebral and intraperitoneal routes and observed daily for 14 days. In addition, a blind passage shall be made of a single pool of the emulsified tissue (minus skin and viscera) of all mice surviving the original 14-day test. The virus pool is satisfactory for measles virus vaccine only if at least 80 percent of the original inoculated mice remain healthy and survive the entire observation period, and if none of the mice used in the test shows evidence of a transmissible agent or other viral infection, other than measles virus, attributable to the vaccine."

For purposes of use of this test for cell substrates, viral seeds (other than measles virus seeds), and production harvests, the phrases "measles virus vaccine" and "other than measles virus" should **not** be applied. For purposes of use of this test for cell substrates, the phrase "the virus pool" should **not** be applied. Intact cells (10^7 cells per mL of culture fluid) should be inoculated instead. Alternatively, cell lysates of an equivalent number of cells may be used.

The clarification regarding the meaning of the 80 % survival is as described in the test for adult mice.

Although the original purpose of this test was to detect Coxsackieviruses [4], currently this test is performed to detect adventitious viruses including: human viruses: Coxsackie viruses types A and B (type B detectable in cell culture), and other picornaviruses such as poliovirus and Echovirus, alphaviruses, bunyaviruses, phleboviruses, nairoviruses, arenaviruses, flaviviruses, rabies, and herpesviruses (herpes simplex virus). This test can also detect many murine agents [6,7].

Lymphocytic Choriomeningitis Virus (LCMV)

A specific in vivo test for LCMV is recommended for characterizing rodent cell substrates. It involves inoculating mice used in MAP testing (see below) with live virulent LCMV to determine viral interference. Animals infected during the MAP test with non-virulent strains will survive this challenge because of the protective immune response they developed.

Guinea pigs

The test in guinea pigs was described in 21 CFR 630.65 [5], which was revoked after 1996, although the test method is still recommended. The text of that regulation reads as follows:

> "Each of at least five guinea pigs, each weighing 350-450 grams, shall be inoculated intracerebrally with 0.1 mL and intraperitoneally with 5 mL of the undiluted virus pool. The animals shall be observed for at least 42 days. Each animal that dies after the first 24 hours of the test or is sacrificed because of illness, shall be necropsied. All remaining animals shall be sacrificed and necropsied at the end of the observation period. The virus pool is satisfactory only if at least 80 percent of all animals remain healthy and survive the observation period and if all of the animals used in the test fail to show evidence of infection with *M. tuberculosis* or any viral infection."

For purposes of use of this test for cell substrates, the phrase "the virus pool" should **not** be applied. Intact cells (10^7 cells per mL of culture fluids) should be inoculated instead. Alternatively, cell lysates of an equivalent number of cells may be used.

By way of clarification regarding the statement "at least 80 percent of all animals remain healthy and survive the observation period," if five guinea pigs are used in the test, four must survive to the end of the 42-day observation period. If 10 animals are used in the test, eight animals must survive to the end of the 42-day observation period, etc. If more than 20 percent of the guinea pigs die in the first 24 hours, the test is either invalid or the test material has failed.

Although originally this test was intended to detect LCMV and *Mycobacterium tuberculosis* [4], currently this test is performed to detect *Mycobacterium tuberculosis* and adventitious viruses including, but not necessarily limited to: paramyxoviruses (including Sendai virus), reoviruses, and filoviruses [6,7].

Rabbits

The test in rabbits was described in 21 CFR 630.65 [5], which was revoked after 1996, although the test is still recommended. The text of that regulation reads as follows:

"A minimum of 15 mL of each virus pool shall be tested by inoculation into at least five healthy rabbits, each weighing 1500-2500 grams. Each rabbit shall be injected intradermally in multiple sites with a total of 1.0 mL and subcutaneously with 2.0 mL of the virus pool, and the animals observed for at least 30 days. Each rabbit that dies after the first 24 hours of the test or is sacrificed because of illness shall be necropsied and the brain and organs removed and examined. The virus pool is satisfactory only if at least 80 percent of the rabbits remain healthy and survive the entire period and if all the rabbits used in the test fail to show lesions of any kind at the sites of inoculation and fail to show evidence of any viral infection."

For purposes of use of this test for cell substrates, the phrase "the virus pool" should **not** be applied. Intact cells (10^7 cells per mL of culture fluids) should be inoculated instead. Alternatively, cell lysates of an equivalent number of cells may be used.

By way of clarification regarding the statement "at least 80 percent of the rabbits remain healthy and survive the entire period", if five rabbits are used in the test, four must survive to the end of the 30-day observation period. If 10 animals were to be used in the test, eight animals must survive to the end of the 30-day observation period, etc. If more than 20 percent of the rabbits die in the first 24 hours, the test is either invalid or the test material has failed.

The original purpose of this test was to detect adventitious viruses, specifically simian herpes B virus [4].

Embryonated chicken eggs

The test in embryonated hens' eggs was described in 21 CFR 630.35 [5], which was revoked after 1996, although the test methods are still recommended. The text of that regulation reads as follows:

"Embryonated eggs, 10 to 11 days old, shall be inoculated by the allantoic route using 0.5 milliliter per egg. Following incubation at 35° C for 72 hours, the allantoic fluids shall be harvested, pooled, and subpassed by the same route into fresh, embryonated eggs, 10 to 11 days old, using 0.5 milliliter per egg and incubated at 35° C for 72 hours. Both the initial pool and the subpassage harvest shall be tested for the presence of haemagglutinin. The virus pool is satisfactory if the embryos appear normal and there is no evidence of haemagglutinating agents."

"Embryonated eggs, 6 to 7 days old, shall be inoculated by the yolk sac route using 0.5 milliliter per egg. Following incubation at 35°C for at least 9 days, the yolk sacs shall be harvested and pooled. A 10 percent suspension of yolk sacs shall be sub-passed by the same route into fresh embryonated eggs, 6 to 7 days old, using 0.5 milliliter of inoculum per egg and incubated at 35°C for at least 9 days. The virus pool is satisfactory if the embryos in both the initial test and the sub-passage appear normal."

For purposes of use of this test for cell substrates, the phrase "the virus pool" should **not** be applied. Intact cells (10^7 cells per mL of culture fluids) should be inoculated instead. Alternatively, cell lysates of an equivalent number of cells may be used. For purposes of use of this test for viral seeds and viral harvests, if vaccine virus needs to be neutralized to perform a valid test, then 21 CFR 630.35 indicates that antisera used for neutralization should **not** be of human, simian, or avian origin.

The number of eggs to be used in each test is not described in the revoked 21 CFR 630.35. However, it is recommended that at least 10 eggs per each route of inoculation be used.

This test is performed to detect adventitious viruses [6,7] including, but not necessarily limited to:

By the allantoic route: orthomyxoviruses (influenza virus) and paramyxoviruses (mumps, measles, parainfluenza viruses), alphaviruses, and vesiculoviruses.

By the yolk sac route: herpes viruses (HSV), poxviruses, rhabdoviruses, as well as rickettsiae, mycoplasmas, and bacteria.

HAP, RAP, MAP

Antibody production tests are performed in the species-appropriate model (hamster, rat, mouse). The test article is inoculated into specific-pathogen-free (SPF) animals that are subsequently tested for antibodies to specific agents. The hamster agents include: Hantaan virus, Kilham rat virus (KRV), LCMV, mouse poliovirus (GD VII), minute virus of mice, pneumonia virus of mice, reovirus type 3, Sendai virus, simian virus 5 (parainfluenza), and Toolans H-1 virus. The rat agents include: Hantaan virus, KRV, LCMV, mouse adenovirus, mouse poliovirus (GD VII), pneumonia virus of mice, rat rotavirus, reovirus type 3, sialodacryoadenitis virus (SDA/RCV), Sendai virus, and Toolans H-1 virus. The mouse agents include: Ectromelia virus, EDIM (murine rotavirus), Hantaan virus, LCMV (including challenge to detect interference from non-lethal strains), LDH-elevating virus, minute virus of mice, mouse adenovirus, mouse encephalomyelitis (GD VII virus), mouse hepatitis virus, mouse salivary gland virus (murine CMV), pneumonia virus of mice, polyoma virus, reovirus type 3, Sendai virus (murine parainfluenza virus), and thymic virus. K virus (mouse pneumonitis virus) is not included in this list, because it is not believed to be present in the U.S. [8,9].

Further guidance on these tests can be obtained from CBER "Points to Consider in the Manufacture and Testing of Monoclonal Antibody Products for Human Use" [9].

In Vitro Tests

Tissue Culture (TC) safety test

The tissue culture safety test was described in the 1993 "Points to Consider in the Characterization of Cell Lines Used to Produce Biological Products" (CBER) [1] as follows:

"An appropriate volume should be inoculated on to monolayer cultures of at least three cell types:

1. monolayer cultures of the same species and tissue type as that used for production;
2. monolayer cultures of a human diploid cell culture; and
3. monolayer cultures of a monkey kidney cell culture.

The sample to be tested should be diluted as little as possible. The cell cultures should be observed for at least two weeks. If the production cell culture is known to be capable of supporting the growth of human cytomegalovirus, the human diploid cell cultures should be observed for at least four weeks. The cultures should be tested for haemadsorption at the end of the observation period."

In addition, if the production cell culture is known to be capable of supporting the growth of simian cytomegalovirus, the human cell cultures should be observed for at least four weeks.

A subculture on to fresh cells is advisable to distinguish between non-specific CPE and virus-induced CPE, as any toxic effects of the initial specimen or length in culture will be diluted, whereas virus-induced CPE should be amplified.

It is not stated in the 1993 PTC what the appropriate volume should contain (i.e. cells or fluid) nor what volume it should be. Either cells (co-cultivation), cell lysate, or culture fluids may be used as an inoculum to the cell cultures. If a viral seed is being tested, the vaccine virus may be neutralized with appropriate anti-serum and inoculated. Control cell cultures, run in parallel with the production of the viral seed, may also be tested to determine whether a contaminant was introduced during the production of the viral seed.

The volume to be tested should be assessed based on the total culture or batch size. Generally, it is recommended that 10^7 cells or lysate equivalent to that number of cells be used when testing the cell substrate. When testing viral seeds or vaccine harvests, 21 CFR 630.4 (f) required that at least 500 doses of monovalent or trivalent inactivated poliovirus vaccine be tested in monkey kidney cell cultures for SV40 contamination, and 630.4 (a) (2) required that 5 mL from a poliovirus harvest or pool should be tested for SV40 contamination after antibody neutralization of the poliovirus. Similarly, 630.18 (a) (5) and (6) required that at least 500 doses or 50 mL (whichever was a greater volume of virus) of monovalent poliovirus vaccine be tested on monkey kidney cell cultures and on human cell cultures after antibody neutralization of the poliovirus, and 630.18 (a) (7) required that 500 mL of virus pool (harvest or monovalent bulk) be tested on rabbit kidney cell cultures. Similar volumes (500 doses or 50 mL) were specified in the revoked 630s for the testing of other vaccines (e.g. measles, mumps, rubella vaccines).

Haemadsorption is generally performed using guinea pig, chicken, and human type O red blood cells (RBCs), which are each added to portions of the cell cultures at 2-8°C for $^1/_2$ hour, rinsed, and observed for haemadsorption/haemagglutination. This is followed by an incubation with RBCs for $^1/_2$ hour at room temperature, rinsing, and observation for haemadsorption/haemagglutination. Sometimes, Rhesus monkey RBCs are used.

These cultures are performed to detect adventitious viruses, as follows: production cell or cell line of same species and tissue: Production harvest may be taken at an earlier time point or vaccine virus propagation may result in a lytic infection. If so, then by allowing the production cell substrate or a cell line of similar virus susceptibilities to grow for longer observation periods, this may permit the detection of viruses capable of propagation in the production cell substrate.

Human diploid cells may detect (but are not necessarily limited to) the following [6,7]:

Human viruses: picornaviruses (medium sensitivity for enteroviruses, low sensitivity for polio, Coxsackie B, echovirus, high sensitivity for rhinoviruses; low sensitivity for HAV), orthomyxoviruses (low sensitivity), paramyxoviruses (low sensitivity, including measles), adenoviruses (medium sensitivity), herpesviruses (HSV – medium sensitivity, VZV and CMV – high sensitivity, EBV – negative), coronaviruses; reoviruses; alphaviruses; rubella; flaviviruses; and rabies.

Simian viruses: CMV

Monkey kidney cells:

Human viruses: picornaviruses (high sensitivity for enteroviruses, medium sensitivity for polio, medium/high sensitivity for Coxsackie B and Echoviruses; HAV can be grown in African green or rhesus monkey kidney cells); rhinoviruses (low sensitivity), orthomyxoviruses (high sensitivity), paramyxoviruses (medium to high sensitivity including parainfluenza, mumps, and measles), adenovirus (low sensitivity), herpesviruses (HSV & VZV – low sensitivity, CMV & EBV – negative); reoviruses; poxviruses (vaccinia); SV40, rotavirus; alphaviruses; rubella; flaviviruses; rabies; vesiculoviruses; filoviruses; influenza; bunyaviruses; phleboviruses; nairoviruses; arenaviruses, rotaviruses; reoviruses; and polyomaviruses.

Simian viruses: herpes B virus

Not detected in these routine cell cultures or in vivo tests described above: retroviruses, various hepatitis viruses, Norwalk viruses and other human caliciviruses, some togaviruses, papillomaviruses, EBV, HHV-6, HHV-7, and HHV-8 [6,7].

Not detected in these routine cell cultures but detected by in vivo tests: Coxsackie A viruses [6,7].

Transmission Electron Microscopy (TEM)

TEM is a test that can detect viral particles present in a cell substrate, including those from an expressed endogenous retrovirus. While this test does not have a particularly low limit of detection, it is a generic assay that can detect adventitious agents of many types. TEM can also be used as a method for estimating the viral load of endogenous retroviruses to support viral validation studies. Methods include negative staining and thin section. These methods are described in Bierley et al, Dev Biol Stand, 1996 [10].

Reverse Transcriptase (RT) tests

PCR-based assays to detect reverse transcriptase (RT) activity from retroviruses (exogenous and endogenous) are currently recommended for the assessment of vaccine cell banks and viral seeds. It is acknowledged that avian cells and vaccine (or viral seeds) propagated in them will have a positive signal in such PCR-based RT assays. It is advisable to consult Office of Vaccines Research and Review/CBER (OVRR) review staff in such circumstances regarding what, if any, further testing may be recommended. This decision would be based on the source and experience with the avian cells (e.g. whether or not they come from well-characterized, genetically-closed, specific-pathogen-free flocks). Assay methodologies and modifications to minimize positive signals from polymerases other than RT can be obtained [11-17].

Infectivity tests for retroviruses

Rodent cell substrates and viral seeds that are potentially contaminated with murine retroviruses should be tested for infectivity by co-cultivation with *Mus dunni* cells or other highly susceptible cells and/or S + L- assays. The following excerpt is taken from the 1993 PTC [1] in this regard:

"For murine retroviruses, amplification of low-level contaminants may be achieved by co-cultivation of cells with a highly susceptible cell line, e.g. *Mus dunni* cells [18]. The latter cells are susceptible to infection by all tested murine leukaemia viruses (MuLVs) except Moloney MuLV, in which case another susceptible cell line, e.g. SC-1 [19], should be used. Fluid from the co-cultures should be further passaged on *Mus dunni* cells and subsequently assayed for MuLV. A variety of other assays may be useful, depending on the circumstances. Some examples of such assays include viable cell immunofluorescence (IFA) on the infected *Mus dunni* cells using a broadly reactive monoclonal antibody (e.g. HY95) for the detection of ecotropic, xenotropic, mink cell focus-forming, and amphotropic viruses; feline S + L- assay using PG4 cells [20] for detection of amphotropic viruses; mink S + L- assay for detection of xenotropic viruses [21]; and mouse S + L- assay using D56 [22] cells for detection of ecotropic viruses."

For non-murine retroviruses, infectivity testing on appropriate indicator cells may be recommended and such testing strategies would be determined case-by-case. It would be advisable to discuss such testing strategies with OVRR before implementation.

Specific PCR or other specific in vitro tests

The polymerase chain reaction (PCR) has important benefits and significant limitations for detecting viral genomes. Due to the amplification process, small quantities of contaminating viruses may be detectable. However, only small amounts of sample can be tested in a single reaction, and defective genomes will be detected. This is in contrast to in vitro cultures, upon which litres of material could potentially be assayed. Due to the specificity of PCR, multiple PCR assays may need to be performed to detect the full range of viruses detectable in a single general biological assay. In cases where viruses cannot be readily grown in culture (e.g. Hepatitis C Virus, human papillomaviruses), PCR may be the most effective tool to assess a cell substrate for contamination with such viruses.

Other in vitro tests may include Southern hybridization analysis of DNA extracted from the cell substrate or tissue culture tests using cell lines known to be susceptible for specific pathogens.

Testing of human cell lines should take into consideration the tissue source and medical history of the donor from which the cell line was derived. Specific tests to be considered include those for: Hepatitis A, B, and C Viruses; human papillomaviruses; human adenoviruses; Epstein-Barr virus; human cytomegalovirus; and human herpes viruses-6, -7, and -8.

Infectivity test for bovine viruses

The test for bovine adventitious viruses is described in 9 CFR 113 [23]. Agents of concern from bovine sources listed in 9 CFR 113.47 (b) (1) and (2) are: bovine viral diarrhoea virus (BVDV), reovirus, rabies, bluetongue virus, bovine adenoviruses, bovine parvovirus, and bovine respiratory syncytial virus. Additional agents of concern include: infectious bovine rhinotracheitis virus (IBR, bovine herpesvirus) and bovine parainfluenza virus type 3 (BPIV-3) [9 CFR 113.53, 113.46 (a) and (b)]. The sensitivities of the 9 CFR tests for the various bovine viruses should range from 1-100 infectious units (depending on the virus).

In addition to the agents listed above, bunyaviruses, circoviruses, bovine polyoma virus, bovine immunodeficiency virus, and bovine leukaemia virus may be included in a testing strategy depending upon exposure history and species of origin of a cell substrate.

Furthermore, the agent that causes bovine spongiform encephalopathy is of particular concern. The current tests for this agent are of inadequate sensitivity or incompletely validated for use as a test for the purity and safety of biologicals. Over time, this technology is expected to improve and be validated, and specific testing may, at that time, become required. At present, however, use of bovine materials derived from animals from the countries on the USDA list is proscribed. The latest guidance on issues regarding BSE may be obtained from http://www.fda.gov/cber/bse/bse.htm. Additionally, the FDA's TSE Advisory Committee has convened to discuss use of various bovine-derived materials in the production of vaccines and transcripts of those meetings may be obtained from http://www.fda.gov/cber/advisory/tse/tsemain.htm.

Infectivity test for porcine viruses

9 CFR 113.47 (b) (1) and (6) list the following agents that porcine material should be tested for: bovine viral diarrhoea virus (BVDV), reovirus, rabies, porcine adenovirus, porcine parvovirus, transmissible gastroenteritis virus (TGE) and porcine haemagglutinating encephalitis virus. Porcine trypsin should be tested for porcine parvovirus.

Porcine agents that may be included in a testing strategy in addition to those listed in 9 CFR 113 (and above) are: circoviruses, enteroviruses, porcine respiratory and reproductive syndrome virus (PRRS), porcine cytomegalovirus, porcine influenza viruses, pseudorabies, swine pox, swine fever (hog cholera virus, African), and vesicular stomatitis virus (VSV) [24].

Conclusion

Adventitious agent tests should be considered in characterizing a cell substrate based on the donor history and tissue origin, the passage history of cells (and exposure to animal-derived or human-derived materials), the banking strategy used, and the intended use (minimally purified product versus highly purified product, live virus or inactivated virus, and so on). A checklist approach to adventitious agent testing does not make scientific sense and thus testing strategies should be developed on the familiar "case-by-case" basis using sound scientific justifications for the choices made.

Acknowledgements

The author would like to thank Valerie Randolph for historical insights into the original purposes of various test methods.

References

1 CBER's Points to Consider on the Characterization of Cell Lines Used to Produce Biologicals. CBER Guidance Documents, 1993.

2 Q5D: Derivation and Characterisation of Cell Substrates Used for Production of Biotechnological/Biological Products. International Conference on Harmonisation, 1998.

3 Q5A: Viral Safety Evaluation fo Biotechnology Products Derived from Cell Lines of Human or Animal Origin. International Conference on Harmonisation, 1998.

4 Hearings before a Subcommittee of the Committee on Interstate and Foreign Commerce House of Representatives Eighty-Seventh Congress First Session on Developments with Respect to the Manufacture of Live Virus Polio Vaccine and Results of Utilization of Killed Virus Polio Vaccine. Washington, U.S. Government Printing Office, 1961, pp 349-356.

5 Title 21 Code of Federal Regulations part 630.35 and part 630.65. U.S. Government Printing Office, Washington, 1996, pp 103-105 and p 113.

6 Knipe DM, Howley PM: Fields Virology, 4th Ed. Philadelphia, Lippincott Williams & Wilkins, 2001.

7 Specter and Lancz: Clinical Virology Manual, New York, Elsevier, 1992.

8 Facklam TJ, Geyer S: The preparation and validation of stock cultures of mammalian cells. Bioprocess Technol 1991, 13:54-85.

9 CBER's Points to Consider in the Manufacture and Testing of Monoclonal Antibody Products for Human Use. CBER Guidance Documents, 1997.

10 Bierley ST, Raineri R, Poiley JA, Morgan EM: A comparison of methods for the estimation of retroviral burden, in Brown F, Lubiniecki AS (eds): Viral Safety and Evaluation of Viral Clearance from Biopharmaceutical Products. Dev Biol Stand. Basel, Karger, 1996, vol 88, pp 163-165.

11 Silver J, Maudru T, Fujita K, Repaske R: An RT-PCR assay for the enzyme activity of reverse transcriptase capable of detecting single virions. Nucleic Acids Res 1993; 21 (15): 3593-3594.

12 Pyra H, Boni J, Schupbach J: Ultrasensitive retrovirus detection by a reverse transcriptase assay based on product enhancement. PNAS 1994; 91 (4): 1544-1548.

13 Heneine W, Yamamoto S, Switzer WM, Spira TJ, Folks TM: Detection of reverse transcriptase by a highly sensitive assay in sera from persons infected with human immunodeficiency virus type 1. J Infect Dis 1995; 171 (5): 1210-1216.

14 Maudru T, Peden K: Elimination of background signals in a modified polymerase chain reaction-based reverse transcriptase assay. J Virol Methods 1997; 66 (2): 247-261.

15 Chang A, Ostrove JM, Bird RE: Development of an improved product enhanced reverse transcriptase assay. J Virol Methods 1997; 65 (1): 45-54

16 Lugert R, Konig H, Kurth R, Tonjes RR: Specific suppression of false-positive signals in the product-enhanced reverse transciptase assay. Biotechniques 1996; 20 (2): 210-217.

17 Maudru T, Peden KW: Adaptation of the fluorogenic 5'-nuclease chemistry to a PCR-based reverse transcriptase assay. Biotechniques 1998:25 (6) 972-975.

18 Lander MR, Chattopadhyay SK: A *Mus dunni* cell line that lacks sequences closely related to endogeous murine leukemia viruses and can be infected by ecotropic, amphotropic, xenotropic, and mink cell focus-forming viruses. J Virol 1984; 52 (2): 695-698.

19 Hartley JW, Rowe WP: Clonal cell lines from a feral mouse embryo which lack host-range restrictions for murine leukemia viruses. Virology 1975; 65 (1): 128-134.

20 Bassin RH, Ruscetti S, Ali I, Haapala DK, Rein A: Normal DBA/2 mouse cells synthesize a glycoprotein which interferes with MCF virus infection. Virology 1982; 123 (1): 139-151.

21 Peebles PT: An in vitro focus-induction assay for xenotropic murine leukemia virus, feline leukemia virus C, and the feline-primate viruses RD-114/CCC/M-7. Virology 1975; 67 (1): 288-291.

22 Bassin RH, Tuttle N, Fischinger PJ: Rapid cell culture assay technic [sic] for murine leukaemia viruses. 1971; 229 (5286): 564-566.

23 Title 9 Code of Federal Regulations Part 113. U.S. Government Printing Office, 2003.

24 Diseases of the Swine 7th Edition, in Leman AD, Straw BE, Mengeling WL, D'Allaire S, Taylor DJ (eds): Ames, Iowa State Press, 1992.

Dr. R.L. Sheets, NIH/NIAID, 6700B Rockledge Dr. Rm. 5145, Bethesda, MD 20892, USA.
E-mail: rsheets@nih.gov

Petricciani J, Sheets R (eds): Vaccine Cell Substrates 2004.
Dev Biol (Basel). Basel, Karger, 2006, vol 123, pp 147-152.

Scope and Practicality of in Vivo Testing for Adventitious Agents

M.D. Whiteman

BioReliance Corporation, Invitrogen Bioservices, Rockville, MD, USA

Key words: Biologicals testing, adventitious viral contaminants, in vivo virus screen.

Abstract: An overview of the in vivo assays currently performed to test for the presence of adventitious contaminants is available. Retrospective validation data have been reviewed. Assay problems specifically related to the testing of vaccine viral stocks and cell substrates may be discussed.

INTRODUCTION

A biological product is defined in The Code of Federal Regulations, Title 21, Section 600.3 as "any virus, therapeutic serum, toxin, antitoxin or analogous product for prevention, treatment, or cure of human diseases or injuries to man" [1]. For more than 30 years, the concept of using transformed cells in the production of human pharmaceuticals had been rejected. The use of animal cell cultures in the production of biologicals was not new. Vaccines for polio and rubella products used this technology. However, until 1987, regulatory authorities required that the manufacturers of human biologicals use cells with normal morphology, growth patterns and genetic stability. In 1987, a major change in FDA policy amended the regulations to allow the use of transformed cells and in doing so ensured the survival of much of the new biotechnology industry. As with the earlier products prepared in normal primary or diploid cell cultures, the same fear of contamination with deleterious agents exists. Potential dangers of biologicals include oncogenes, the potential to stimulate oncogenes, bacterial, fungal and mycoplasmal contamination, and viruses of both human and rodent origin. The "portals of entry" of contaminants into manufacturing processes, include the cell substrate, raw materials, equipment, facilities and the operator. Cell substrate contaminants may be endogenous or exogenous. Safety testing of early biological products was based on vaccine safety testing described in 21 CFR [2] as well as on a publication in the "Journal of Biological Standards" in 1981 entitled "Guidelines for the Acceptability Management and Testing of Serially Propagated Human Diploid Cells Used for the Production of Live Virus Vaccines for Use in Man" [3]. Currently there are several key regulatory guidelines and documents which describe testing recommended or required to evaluate the viral safety of biological products. In the U.S. these documents include the 21CFR parts 58, 200-299, and 600-680, "Points to Consider" FDA/CBER, for both cell lines (1997) and monoclonal antibodies (1993) [4-6]. The European Union CPMP issued the "Notes for Guidance" (1998) and the International Conference on Harmonization issued the ICH Viral Safety Document in 1997 [7-9]. There is reason for concern as there are historical examples of viral contamination of biologicals. These include: yellow fever vaccines contaminated with avian leucosis virus and hepatitis B virus, poliovirus vaccine contaminated with SV40, poliovirus vaccine contaminated with live poliovirus, adenovirus vaccine contaminated with SV40, and blood products contaminated with HIV.

Products which require viral safety assessment include: biopharmaceutical products such as monoclonal antibodies and proteins from genetically engineered mammalian cell lines, gene therapy products (viral vectors), human blood derived products, biological pharmaceuticals (heparin, collagen), vaccines, xenotransplants, as well as supplements/raw materials used in pharmaceuticals (bovine serum, sheep blood, peptones/amino acids, bovine insulin/transferin). A three pronged approach is utilized to control the potential contamination of biotechnology products: (i) Selection and testing of source materials for the absence of infectious virus; (ii) testing is performed at appropriate stages of production for the absence of virus and (iii) the capacity of the production process to remove or inactivate viruses is tested. This type of approach is necessary and important because no single test can demonstrate the absence/presence of all known viruses and all known viruses require a minimum level of viral contamination to record a positive. If the sample contains a low concentration of virus and only a fraction of the sample is assayed, the fraction may test negative due to random and unequal distribution throughout the test sample. Therefore, no single approach can establish the safety of a product [10]. To screen for the absence of adventitious viral contaminants both in vivo and in vitro tests are recommended. Viruses that do not cause cytopathic or other noticeable effects in a cell culture system may be detectable in an animal system. This paper discusses the methods and pitfalls of the in vivo virus screening assays.

In Vivo Virus Screening Assay Methods

The animal systems generally recommended for screening for adventitious viral contaminants are suckling mice, adult mice, guinea pigs, and embryonated hen's eggs. For cell lines, this testing is most often performed once on the master cell bank and once on end of production cells (cells at the limit of in vitro cell age used for production). The testing is also performed on viral seed stocks and viral vaccines. Cell line samples are submitted as lysates at a concentration of 1 x 10^7 cells per ml suspended in conditioned medium. What follows are brief descriptions of the methodologies for each of the systems mentioned and an indication of the viruses most likely to be detected in the specific systems. In each system, after injection, the health of the animals is monitored and any abnormality is investigated to establish the cause of the illness.

Suckling Mice

Two litters of suckling mice are injected using three routes of injection, intraperitoneal, intracranial, and per os. The suckling mice are observed for 14 days and on day 14 a single pool of emulsified tissue of all surviving mice is prepared and passaged. These animals are then observed for clinical signs of illness for another 14 days. This blind passage serves as an enhancement and acclimation step designed to optimize the sensitivity of the assay. Suckling mice are considered an efficient isolation system for many viruses including, arboviruses, coxsackie A and B viruses, herpes simplex virus (types 1 and 2), rhabdoviruses (including rabies), toga viruses (LDV), junin, and herpes B.

Adult Mice

Ten adult mice are injected using four routes, intraperitoneal, intracranial, per os and intranasal. The animals are observed for 28 days for clinical signs of viral infection. Adult mice are a sensitive isolation system for certain viruses. These include: rhabdoviruses (including rabies), toga viruses (LDV) and LCM.

Guinea Pigs

In some instances guinea pigs are recommended as a test system. Each guinea pig is injected intraperitoneally and intracranially. The injection sites are observed weekly for lesions. The animals are observed for clinical signs for at least 28 days. For cell cultures used in vaccine production or vaccine viral seeds the guinea pigs are observed for 42 days. For some test materials, necropsy and gross examination for tubercular lesions may be appropriate. The guinea pig test system is useful in the isolation of rhabdoviruses, LCM, Lassa, Junin, Marburg, ebola and vaccinia viruses.

Embryonated Hen's Eggs-Allantoic Route

Embryonated hen's eggs are a long used and sensitive system for the isolation of viruses. Eggs are injected via the allantoic route and allantoic fluids are harvested three days post-injection. The allantoic fluids are tested for hemagglutinins at room temperature and 4 degrees centigrade using guinea pig, human O and chicken erythrocytes. The fluids are also pooled and passaged into another set of eggs and the fluids from these subpassage eggs are tested for hemagglutinins.

Embryonated Hen's Eggs-Yolk Sac Route

This route of injection insures that the embryo is directly exposed to the test material. Embryos six to seven days old are injected and incubated for at least 9 days. They are examined for viability and the yolk sac material is pooled, homogenized and injected into a second set of 6-7 day old eggs which are then examined for viability at least 9 days post-passage injection. Embryonated hen's eggs are a very efficient system for the isolation of many viruses. Viral contaminants detectable using the yolk sac route of injection include: herpes simplex, rhabdoviruses, herpes B virus, mumps, influenza, parainfluenza (types 1, 2 and 3,) and vaccinia.

Assay Pitfalls

For a test material to be considered not contaminated with adventitious viral contaminants the following criteria must be met: 80 % of the animals/eggs must remain healthy, survive the observation period and show no evidence of viral infection. This evaluation can be complicated and compromised by problems both related and unrelated to the test materials, but not indicative of adventitious viral contamination of the test material. In the case of assay problems unrelated to test material, the two most vulnerable test systems are the suckling mouse and the embryonated hen's eggs portions of the assay. In suckling mice, deaths and/or clinical signs can be the result of natural causes, runting, inadequate maternal care, failure to thrive, cannibalization, and injection trauma. If the number of animals which die constitutes more than 80 % of the test article group, a repeat assay is necessary. Unexpected deaths in guinea pigs and adult mice can be the result of natural causes or an injection related trauma sometimes associated with the intracranial route of injection, but intercurrent deaths in these two test systems are less frequent (see Table 1). The embryonated hen's egg portion of the assay tends to be more problematic. Prior to injection the "health" of the egg is determined by candling. This technique demonstrates egg viability at the time of candling, but is not a good indicator of egg hardiness. The eggs may appear viable at 24 and 48 hours post-injection but may not be sufficiently hardy to survive the observation period.

Table 1: Trend analysis.

Test system	Number of assays examined	Number of system suitability problems	System suitability
Suckling mice	98	7	93 %
Adult mice	98	1	99 %
Guinea pigs	88	2	98 %
Allantoic route	100	3	97 %
Yolk sac route	100	5	95 %

Data from assays performed at BioReliance, Invitrogen Bioservices, Rockville, MD from March 28, 2003 through August 31, 2003.

Investigating assay problems related to test materials is more complex, but as a result of testing experience, it has been possible to categorize these issues and relate them to the type of material being tested, thus avoiding concerns that an adventitious contaminant is present. Toxicity can be an issue when gene therapy viral vectors are tested. Even if the virus is not replication competent, toxicity of unprocessed bulk samples may occur. At certain virus particle concentrations, and this level can vary depending on the vector, animal death may occur within the first 10 days post-injection. Other vector related issues include infection of the animal in the case of replication competent viruses and hemagglutination. If the viral vector itself hemagglutinates erythrocytes, the allantoic route of injection may yield false positive results. Media components of bulk harvest materials can also cause problems in animals and eggs. Media components may be toxic to chick embryos. It was observed in the BioReliance facility in Scotland that a common media component used as an emulsifier seemed to induce the growth of bacteria in the yolk sac portion of the assay. Vaccine viral stocks and or substrates present a unique set of problems for in vivo assays. These can include a bioburden in the substrate, the ability of the virus to infect one or more test systems, the hemagglutination potential of the virus, viral neurovirulence when the virus is injected intracranially and toxicity of the antiserum used to neutralize the vaccine virus. Evaluation of the potential problem areas prior to initiation of the GLP assay is essential. At BioReliance, feasibility and/or qualification studies are performed and used to gather information. These are small studies which help us determine how the test material will react in the various test systems.

Solutions to the issues described above include dilution, neutralization of the virus in the test material, the addition of antibiotics, the adjustment of antiserum concentration, the use of antiserum control animals, and the elimination of one or more test systems or routes of injection. The feasibility and qualification studies assist in the selection of the solution which has the least impact on the sensitivity of the assay being performed.

Discussion

In vivo safety testing of biotechnology products and vaccines in particular can be challenging, but in most cases the assays can be successfully completed. Table 1 demonstrates that when appropriate preliminary testing is performed the effect of the test material on system suitability can be minimized. The risk of viral contamination exits for all biotechnology products derived from cells of human or animal origin and undetected could have serious clinical consequences. In vivo assays are an important part of a viral testing program designed to assure the safety of biotechnology products.

References

1 Code of Federal Regulations: Biological Products: General. Title 21, Part 600.3 (US Government Printing Office, Washington, DC), 2003.

2 Code of Federal Regulations: Animals and Animal Products. Title 9, Part 113.53 (US Government Printing Office, Washington, DC), 2003.

3 Jacobs JP, Magrath DI, Garrett AJ, Schild: Guidelines for the acceptability, management and testing of serially propagated human diploid cells for the production of live virus vaccines for use in man. J Biol Stand 1981; 9: 331-342.

4 FDA/CBER: Points to Consider in the Characterization of Cell Lines Used to Produce Biologicals (FDA/CBER, 1993) Rockville, MD: Food and Drug Administration, Center for Biologics Evaluation and Research, 1993.

5 FDA/CBER: Points to Consider in the Manufacture and Testing of Monoclonal Antibody Products for Human Use (FDA/CBER, 1997) Rockville, MD: Food and Drug Administration, Center for Biologics Evaluation and Research, 1997.

6 FDA/CBER: Guidance for Industry: Guidance for Human Somatic Cell Therapy and Gene Therapy (FDA/CBER, 1998) Rockville, MD: Food and Drug Administration, Center for Biologics Evaluation and Research, 1998.

7 International Conference on Harmonization: Guideline on Viral Safety Evaluation of Biotechnology Products Derived from Cell Lines of Human or Animal Origin (ICH 5A, 1997) Geneva, Switzerland: International Conference on Harmonization, 1997.

8 International Conference on Harmonization: Guideline on Quality of Biotechnological/Biological Products: Derivation and Characterization of Cell Substrates Used in the Production of Biotechnological/Biological Products (ICH Q5D, 1997) Geneva, Switzerland: International Conference on Harmonization, 1997.

9 EMEA/CPMP (European Agency for the Evaluation of Medicinal Products. Committee for Proprietary Medicinal Products): Note for Guidance on the use of Bovine serum in the manufacture of human biological medicinal products. London, 2003, pp 1-8.

10 Fenno J, Luczak J, Moore A, Poiley J, Raineri R, Whiteman M: The characterization of transformed cell substrates. BioPharm 1988; September: 36-42.

Dr. M. Whiteman, BioReliance, Invitrogen Bioservices, 14920 Broschart,, Rockville, MD20850, USA.
E-mail: mary.whiteman@invitrogen.com

Petricciani J, Sheets R (eds): Vaccine Cell Substrates 2004.
Dev Biol (Basel). Basel, Karger, 2006, vol 123, pp 153-164.

Detection of Adventitious Viruses in Biologicals – A Rare Occurrence

R.W. Nims

Adventitious Virus Testing, BioReliance, Invitrogen Bioservices, Rockville, MD, USA

Key words: Biologicals testing, adventitious viral contaminants, in vitro virus screen

Astract: Adventitious virus assays are performed as part of raw materials testing, cell-line characterization, and lot-release testing of biologicals such as monoclonal antibodies, gene therapy vectors, recombinant proteins, and vaccines. The testing methods follow guidance provided in the 9 CFR (bovine and porcine raw materials testing, and certain vaccine products) or Points to Consider documents (cell line characterization and evaluation of the majority of biologicals). The methodologies used and the types of adventitious viruses detected during testing of the various types of samples are discussed in this paper. The detection of adventitious viruses is quite rare, especially during evaluation of cell banks and biologicals produced in human, mouse, or insect cell substrates. The most common detection scenarios include bovine viral diarrhoea virus in foetal bovine serum samples, porcine parvovirus in porcine substrates, and murine minute virus, REO virus, and Cache Valley virus in Chinese hamster cell-derived bulk harvests. The two last-named viral entities are believed to be introduced via bovine serum used during the manufacturing process (during scale-up or during the entire process). Knowledge of the types of agents being detected is useful in designing viral clearance methodologies for purification processes and in engineering manufacturing processes and facilities.

INTRODUCTION

The viral safety of biologicals (therapeutic and diagnostic modalities manufactured using processes involving biological systems such as cells, bacteria, plants, or animals) is ensured through several complementary approaches. These include: (i) the implementation of Good Manufacturing Practices (cGMP); (ii) the thorough safety testing of raw materials used in the process; (iii) viral safety evaluation of unprocessed bulk harvests using in vivo and/or in vitro virus screening assays and (iv) validation of the viral clearance capacity of the purification processes used to generate the final product. Adventitious viral impurities (contaminants) may find their way into a biological either through the use of a contaminated raw material (cell substrate, trypsin, bovine serum, or other medium components of animal origin) [1] or as a result of inadvertent introduction during the manufacturing campaign for a given lot. Due to the possibility of the latter mode of contamination, biologicals are evaluated for viral safety on a lot-by-lot basis.

The detection of adventitious viral contaminants in biologicals is a relatively rare occurrence. Certain classes of biologicals appear to be more susceptible to viral contamination than others. The results discussed in this paper have been generated over the course of nearly a decade of evaluation of cell banks and biologicals, including cellular vaccines, gene therapy vectors (adenovirus, adeno-associated virus, baculovirus, retrovirus, lentivirus, Sendai, and REO virus), monoclonal antibodies, and recombinant proteins. The types of adventitious viral contaminants that have been detected in these classes of biologicals are discussed, with special emphasis on the results obtained through use of the in vitro virus screening assay. An additional paper in the series discusses in more detail the use of the in vivo adventitious virus screening assay [2].

Background and Regulatory Basis

Viral safety testing has been mandated to ensure the safety of patients undergoing therapy with biologicals. Viral detection assays were already in place when the first biological products began to be formulated in the 1980s. The underlying principle upon which all viral infectivity detection methods have been based is that viruses are not able to replicate in the absence of host cells. Viruses lack the energy sources and machinery required to produce their own genomic material and structural proteins, and therefore must enter and subordinate a host cell for this purpose [3]. Infectivity-based assays provide such host cells, which are then available to rescue viral contaminants from test samples (typically conditioned medium, cell lysates, or raw materials).

The infectivity-based assays may be placed in three broad classes on the basis of the types of viruses to be detected: (i) retroviral assays; (ii) virus-specific assays (e.g. bovine and porcine virus assays) and (iii) viral screening assays. The regulatory guidance underlying the various viral safety tests depends on the nature of the test samples. Raw materials testing typically follows methods outlined in the 9 CFR (U.S. Code of Federal Regulations) 113.53 [4] and ICH (International Conference on Harmonization) Q5A [5], Q5D [6]. Cell line characterization is driven by the Points to Consider in the Characterization of Cell Lines Used to Produce Biologicals [7], while bulk harvest (lot-release) testing is driven in the U.S. by a variety of Points to Consider documents (e.g. Points to Consider in the Manufacture and Testing of Monoclonal Antibody Products for Human Use [8], Guidance for Industry: Guidance for Human Somatic Cell Therapy and Gene Therapy [9]) and in the EU by ICH Q5A [5], Q5D [6]. This paper emphasizes the in vitro adventitious virus screening assays. Additional virological methods, including detection of retroviruses [1,10], and polymerase chain reaction (PCR)-based methodologies for viral detection which are capable of detecting viruses from each class [11,12] are discussed in the cited papers.

In vitro Virus Screening Methodologies

The in vitro viral screening methodologies are not optimized for any given viral entity. The screening aspect of the assays is accomplished in part by providing multiple host cell types. These typically include, as a minimum, the following three indicator cells: (i) a human diploid cell such as MRC-5; (ii) a primate monkey kidney cell such as Vero and (iii) a third indicator cell which should be a monolayer cell of the same or similar species to the cell substrate used to manufacture the biological. It was common in the 1980s and early 1990s for the actual production cell substrate to be used as the third indicator cell line. There were disadvantages to this, as the production cell substrate was often a suspension cell and these tended to be poor indicator cells due to the difficulties encountered in assessing morphological changes. In addition, the use of positive controls was hindered under these circumstances, as the testing laboratory typically had little experience with the substrate cells, especially with respect to potential positive controls. The practice was discontinued later in the 1990s, as the 1993 Points to Consider in the Characterization of Cell Lines Used to Produce Biologicals [7] specified the use of monolayer cultures.

Additional indicator cell lines may be added to enhance the screening aspect of these assays further, the intention being that the various cell lines included may have complementary virus susceptibility profiles. Selection of additional cell lines may also be based on concern for the potential introduction of viral contaminants from raw materials. If, for example, there is concern about the status of a bovine component in the medium used in the process, a bovine indicator cell might be added to the viral screen used for the unprocessed bulk testing.

In addition to the use of multiple indicator cell lines, the screening aspect of these assays is also conferred through use of multiple assay endpoints. These include: (i) light microscopic observation of monolayers for viral cytopathic effect (virus-induced changes in cell morphology, confluency, or subcellular organization) and (ii) haemadsorption and haemagglutination of multiple species of erythrocytes at one or more temperatures. The response profiles of various viruses capable of being detected in the screening assay may differ with respect to these three endpoints. By including multiple endpoints, the assay is not only able to detect a wider range of viruses, but also provides a greater amount of diagnostic information in the rare case that a viral contaminant is detected.

In vitro virus screening is performed on the following types of test samples: (i) certain types of raw materials; (ii) cell banks (master, working) and cellular vaccines; (iii) viral seed stocks and vaccines; (iv) unprocessed bulk harvest and (v) end of production cells (cells at the limit of in vitro age). Cell-based samples are normally evaluated as cell suspensions (1 x 10^7 cells/ml, in conditioned medium), while unprocessed bulk samples typically represent samples harvested from the manufacturing process without further manipulation. Viral vaccines and viral seed stocks may need to be neutralized with specific antisera before evaluation. The use of selection media containing certain antibiotics, such as hygromicin or methotrexate, for cell banks or during the manufacturing process, can sometimes lead to problems in the in vitro virus detection assays. These agents, as well as any other cytotoxic agents that may be present in the test sample, may lead to cell death in the indicator cultures used in the screening assay, confounding interpretation of the results. For the same reason, the test samples must be isotonic, as hypotonic or hypertonic samples may lead to rapid indicator cell death.

The in vitro virus screening method may be described, briefly, as follows: The test sample is administered to the various indicator cell monolayers for ~one hour (replacing the normal medium), after which the sample is removed and replaced with maintenance medium appropriate to the cells. The cultures are incubated for a period of 14 days, during which time they are observed for cytopathic effect. On day 14, the haemagglutination endpoint is performed on conditioned medium from the cultures, while the cultures themselves are used for the haemadsorption endpoint. Typically, erythrocytes from several animal species are used (e.g. guinea pig, chick, and Rhesus monkey or human type O). At this point the test is either terminated (14-day method) or conditioned medium from the test cultures is used to inoculate fresh indicator cell cultures which are then incubated for an additional 14 days (28-day method). The cultures continue to be observed and at the end of the second 14-day incubation period, the haemagglutination and haemadsorption endpoints are performed.

Assays for Specific Viral Contaminants

Virus-specific in vitro methodologies have been optimized for the detection of certain viruses of concern. For instance, assays designed to detect a panel of bovine or porcine viruses are used, primarily for the evaluation of raw materials (trypsin, serum, collagen, thrombin, etc.), but also in some cases for cell line characterization and unprocessed bulk testing. The methodologies used in these assays are covered in reference [1]. The primary difference in methodology, compared to the screening assay, is that the virus-specific assays are optimized for detection of certain viruses. This optimization is accomplished through use of susceptible indicator cell lines (bovine turbinate, porcine testicle, etc.) and the addition of an immunofluorescent staining endpoint using specific antisera. The types of test samples used in these assays are similar to those mentioned above for the screening assays. However, the mode of exposure to the indicator cells may be unique to the test sample being evaluated (especially in the case of raw materials). For example, a bovine serum sample may be exposed to the indicator cells by incorporation at 15 % (v/v) into the growth medium for the duration of the assay. Other medium components may also be evaluated by being incorporated into the growth medium at nominal (or more typically – three times nominal) concentration (nominal being the concentration of the component as used in the manufacturing process).

Detection of Viral Contaminants

The typical scenario for detection of a viral contaminant in a cell-based infectivity assay is as follows: a focal area of cytopathic effect may be observed in the monolayers for one or more of the indicator cells used in the screening or virus-specific assay. One of the characteristics of a viral infection, which helps differentiate this from a purely cytotoxic effect, is that a viral infection typically takes longer to show up and should progress. The latency of the infection (duration of time from inoculation until the effect is observed) may depend both upon the specific virus involved and the amount of virus initially inoculated into the culture. A small focal area of infection will usually progress to involve more and more of the monolayer. On the other hand, a cytotoxic effect of the test sample typically occurs within a day or two of inoculation and, providing that at least some of the indicator cells survive the insult, recovery in the form of increasing confluency may occur over time.

Progressive deterioration of one or more of the indicator cultures inoculated with the test sample may or may not, depending on the viral contaminant involved, be associated with a positive haemagglutination and/or haemadsorption endpoint. In a 28-day screening assay, an apparent cytopathic effect occurring before day 14 may be amplified during the passage on to fresh indicator cells that is performed on day 14. The passage process not only facilitates amplification of any virus present (leading therefore to increased sensitivity of the assay), but also provides an opportunity to confirm the initial cytopathic effect. This is useful as an additional means of differentiating viral cytopathic effect from cytotoxicity. The bovine and porcine virus assays include sub-culturing at one or more intervals which again facilitates amplification and confirmation of initial effects. In the virus-specific assays, the use of appropriate antisera in the immunostaining endpoint allows for direct identification of the contaminating virus. In contrast, a positive endpoint in a screening assay does not lead immediately to an identification. The totality of the information provided by the screening assay can, however, assist in the identification process. The latency information, the host range results (which indicator cells were affected and which were not), the presence or absence of haemagglutination and/or haemadsorption of one or more of the erythrocytes used, as well as the characteristics of the cytopathic effect, all provide clues that may be used to rule out certain candidate viruses.

Isolation of the viral contaminant involves harvesting conditioned medium and/or cells from the affected indicator cell monolayers. This material may then be used to inoculate fresh indicator cells for the purpose of generating samples to be used specifically for the purpose of identifying the contaminant. On the basis of the known information (described above) about the contaminant, a limited set of focused assays may be performed. These may include (i) transmission electron microscopy of infected cells; (ii) immunofluorescent staining of infected cells; (iii) detergent inactivation studies or (iv) PCR testing of infected cells or conditioned medium for specific viral genomic sequences. The electron microscopy can provide cellular localization, size, structural detail, and often, whether the virus possesses a lipid envelope. If the latter feature is not clearly established from the electron microscopy, the detergent inactivation study can provide definitive evidence of this. Immunostaining results may provide identity to the genus or serogroup level, while PCR testing (with eventual genomic sequencing of amplicons and comparison with GenBank sequence database information) has the potential to provide definitive identification at the individual virus level.

Adventitious Viruses Detected in Raw Materials

Raw material testing, conducted primarily using the virus-specific assays, occasionally demonstrates the presence of viral contaminants and this information may be used to reject such lots of reagents before they are used in manufacturing processes. From the standpoint of bovine virus testing, the most common findings are bovine viral diarrhoea virus (BVDV) and bovine polyoma virus (BPyV) in bovine serum samples. The BVDV, especially the non-cytopathic variant, is a common enough contaminant that the requirements for serum testing in the E.U. [13] include evaluation of neutralizing antibodies to BVDV in the serum. Such antibodies can have adverse effects on assay sensitivity, effectively masking the presence of the contaminant in bovine serum samples. Bovine polyoma virus is another bovine virus of concern, especially in the EU. It is believed, on the basis of PCR testing, that BPyV is present in a high percentage (~70 %) of bovine serum lots [14,15]. It is not clear whether these PCR results are indicative of the presence of infectious virus. The existing infectivity assays for this virus require the use of an indicator cell line such as MDBK (bovine kidney) for amplification, with a PCR endpoint for detection. Unfortunately, insufficient numbers of serum lots have been evaluated with the infectivity assays to establish whether infectious BPyV is present in the PCR-positive lots. With porcine-derived raw materials, such as trypsin, the most common finding in the viral detection assays is porcine parvovirus (PPV). Certain other common porcine viral contaminants of raw materials (eg., porcine circovirus, swine hepatitis) typically require alternative methodologies with reverse transcriptase or PCR-endpoints for detection [16]. For additional details on viruses that should be detectable using various methodologies, see reference [1].

Adventitious Virus Detection in Cell Line Characterization

Cell banks (Master and Working) produced in the biologics industry in accordance with the various controls stipulated in the Good Manufacturing Practice requirements have not been found by the author to harbour viral contaminants. Cell banks that have been used in basic research laboratories and later adapted for use in biologics manufacturing should be considered at risk, however. As a matter of fact, viral contaminants have been isolated from such cell bank materials (derived from Chinese hamster cells). The practice of submitting pre-bank samples for viral screening before the generation of Master Cell Banks has been instituted by certain of the biologicals manufacturers and this is considered to be prudent.

Adventitious Viruses in Unprocessed Bulk Harvest Lots

Manufactured biologicals are expected, according to the Points to Consider guidelines, to be tested for viral safety in a lot-by-lot manner at the bulk harvest (unpurified bulk) level. Testing after purification is not appropriate, for two reasons: (i) purification may change the osmolality of the sample, leading to hypertonic or hypotonic materials that would rapidly kill the indicator cells used in the detection assays and (ii) the purification process may intentionally or unintentionally eliminate or inactivate any viral contaminants present in the bulk harvest, thereby masking the presence of the contaminants. Adventitious virus testing is done for each lot, since opportunities for the introduction of viral contaminants may occur anew during each campaign, from scale-up through final harvest.

The detection of viral contaminants in bulk harvests is rare, with different degrees of risk appearing to be associated with different types of biologicals. Cellular vaccines, retroviral vectors, and monoclonal antibody products have rarely been found to contain viral contaminants. On the other hand, adenoviral vectors, which are E1-deleted and therefore expected to be replication-defective in most of the indicator cells used for testing, may contain small numbers of particles which, through recombination, are rendered replication-competent. These particles are therefore capable of productively infecting these cells, producing cytopathic effects characteristic of adenoviral infection. While this recombination phenomenon is a natural process and not a case of introduction of an adventitious virus during the manufacturing cycle, the outcome is the same, as the recombinant adenovirus is an unwanted contaminant and its detection in an adventitious virus evaluation may mask the presence of other potential viruses.

The highest frequency of detection of adventitious viral contaminants in bulk harvests has, by far, been associated with the production of recombinant proteins in processes using Chinese hamster cell substrates. It is not clear whether this increased risk reflects a higher potential (i.e. greater susceptibility) of these cells to serve as viral hosts, relative to the substrates used for other types of manufacture, or to the relatively common use of this cell substrate in manufacturing processes. The viral contaminants that have been detected with some frequency in Chinese hamster cell processes in the past two decades have included murine minute virus (MMV), REO virus (REO), and Cache Valley virus (CVV).

Documentation of MMV in a Chinese hamster cell process first appeared in the literature in 1996 [17]. As a result of a number of incidents involving the detection of this murine parvovirus in Chinese hamster cell processes, certain of the Points to Consider documents [8] now mandate that this virus be assayed for in bulk harvests from such processes. MMV represents a special challenge, in that it is relatively small (~20 nm) and therefore difficult to remove by filtration and to inactivate by gamma irradiation. It is non-enveloped and thus resistant to chemical and physical inactivation strategies. Like most parvoviruses, MMV is relatively hardy and the potential for survival of the virus outside cell cultures therefore exists [18]. In fact, the most likely route of introduction of this virus into the manufacturing processes involved was believed to be contamination of environmental surfaces used for raw materials processing and packaging [17]. Another documented isolation of MMV occurred in a process involving a Syrian hamster embryo cell line, BHK [19,20]. In screening in vitro virus assays, MMV may be detected using Chinese hamster indicator cells such as CHO-K1 and certain murine indicator cells such as NIH/3T3. These fibroblast cell lines are susceptible to the fibrotropic strain of MMV (prototype strain), but not the lymphotropic strain of MMV. An unusually susceptible host cell is the SV-40-transformed human newborn kidney fibroblast (324K) cell line described by Tattersall [21]. This cell line is not only especially sensitive to the prototype strain of MMV but also, surprisingly, to the lymphotropic strain of MMV as well. As a result of these factors, the 324K indicator cell line is commonly used in optimized infectivity assays for MMV.

REOvirus (family Reoviridae, genus *Orthoreovirus*) represents another contaminant that has been isolated from Chinese hamster cell cultures on a number of occasions. This virus may propagate slowly in the cell substrate, causing minimal effects on oxygen or base demand, or on the yield of recombinant protein from the process. For this reason, it is not uncommon for the presence of this virus to be totally non-apparent to the manufacturer until the contaminant is detected during evaluation of the bulk harvest in a screening in vitro virus assay. The virus particles are spherical, 60-80 nm in diameter, and lack a lipid envelope. In in vitro viral screening assays, it causes cytopathic effect in CHO-K1 and Vero indicator cells. The human 324K cell line has also shown susceptibility to this virus (Fig. 1). It is difficult to determine the animal species of origin of the various REO viruses (types 1, 2, and 3), confounding investigation of the route cause of the contamination in manufacturing processes.

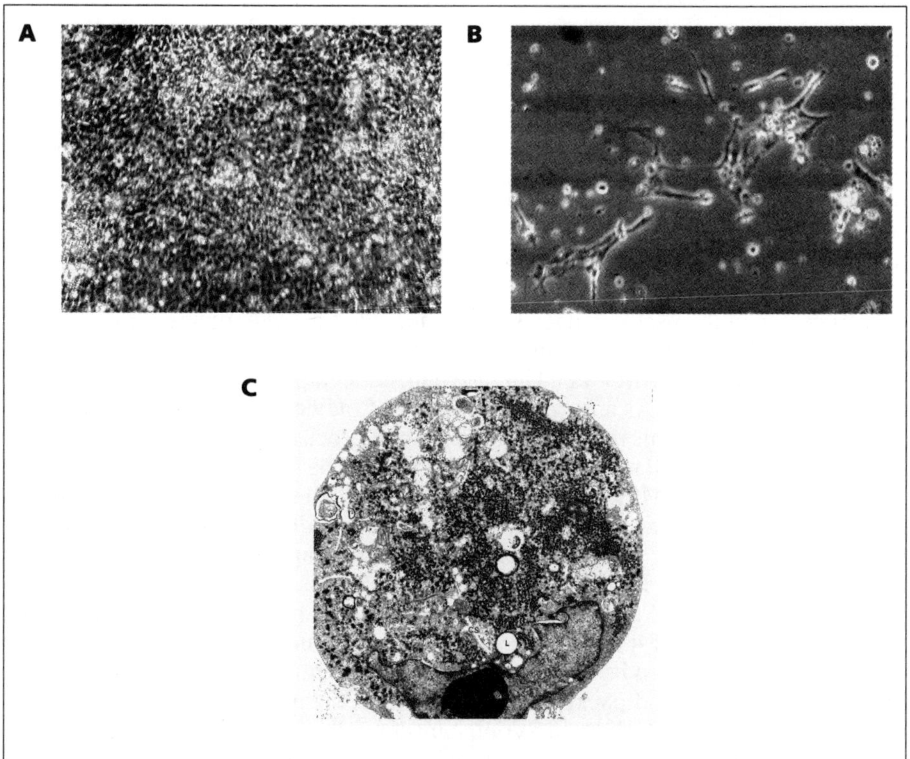

Fig. 1: REO virus infection in 324K cells (contaminant originally isolated from a Chinese hamster cell process). A) uninfected 324K cells ≅ 100 x magnification; B) REO virus-infected 324K cells ≅ 100 x magnification; C) a REO virus-infected 324K cell ≅ 15000 x magnification.

The third viral contaminant that has been detected with some frequency in bulk harvests from Chinese hamster cell processes is Cache Valley virus, a member of the family Bunyaviridae, genus *Bunyavirus.* Cache Valley virus is spherical, 80-120 nm in diameter with a lipid envelope containing glycoprotein spikes. In contrast to the case for REO virus, amplification of Cache Valley virus during manufacturing campaigns usually leads to detectable changes in substrate viability, and associated changes in oxygen and base demand. The virus is readily detected in in vitro viral screens, as marked cytopathic effects are observed with short latency in a variety of host cells, including CHO-K1 and Vero (Fig. 2), and to a lesser extent, human cell lines such as MRC-5 and 324K.

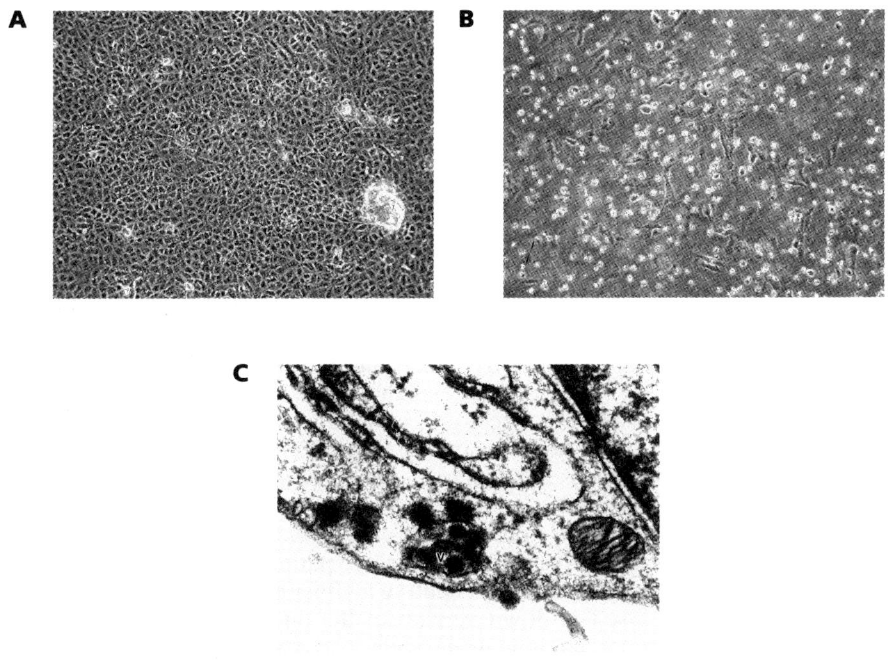

Fig. 2: Cache Valley virus infection in Vero cells (contaminant originally isolated from a Chinese hamster cell process). A) uninfected Vero cells ≅ 100 x magnification; B) Cache Valley virus-infected Vero cells ≅ 100 x magnification; C) Cache Valley virus-infected Vero cells ≅ 100000 x magnification.

The presence of REO or Cache Valley virus in bulk harvest samples is usually ascribed to non-homogeneously contaminated bovine serum used during the manufacturing process. Cache Valley virus is known to infect livestock, being transmitted through insect vectors. As with the REO viruses, a low-level contamination of bovine serum lots with this virus may lead to an inability to detect the agent during quality control testing of the serum. During raw materials testing of foetal bovine serum, it is typical for one bottle from a lot to be evaluated, and the amount of serum actually administered to the test system may be of the order of 100-200 ml. The large amounts of serum incorporated into the culture medium during the manufacturing process, coupled with the relatively long culture times used, may allow these viruses to propagate to the point that the virus is detectable in the bulk harvests.

The three adventitious viral contaminants discussed above are emphasized since the author is aware of more than one instance of detection of each virus in unprocessed bulk harvest samples collected from Chinese hamster cell manufacturing processes. At least two additional findings of adventitious viral contaminants in Chinese hamster cell systems have been reported. Epizootic haemorrhagic disease virus (family Reoviridae, genus *Orbivirus*) was isolated from a genetically-engineered Chinese hamster cell substrate used for biologic production [22], and a calicivirus, family Caliciviridae, genus *Calicivirus*, was isolated from a Chinese hamster cell exhibiting cytopathology [23].

Discussion

Detection of an adventitious viral contaminant in a manufactured biological is a rare occurrence, with the risk appearing to be greater for certain types of **products** (Table 1). The Points to Consider and ICH documents governing adventitious viral testing of bulk harvests do not specify the volumes of harvest material to be evaluated. This fact has important implications. In a typical screening assay, each cell line may be inoculated with 3 ml of bulk harvest. For an assay using three cell lines, this means a total of 9 ml of bulk harvest is applied to the system. This is true regardless of the volume of bulk harvest generated in the course of the manufacturing process, which may range from 2 L to 2000 L or more. Due to limitations in the sensitivity of the various viral detection assays, and to the fact that only a relatively small volume of each bulk harvest is evaluated during testing, a negative result in a viral safety assay does not necessarily mean that adventitious viral contaminants are not present in the harvest lot being evaluated. These screening assays would appear to have been intended to detect gross contamination of the manufacturing process, and a negative outcome from such testing means that no adventitious viral entities were detected, but does not guarantee that there are absolutely none present in the lot. In recognition of this, viral clearance studies are performed for each purification process in addition to subjecting bulk harvest samples on a lot-by-lot basis to viral screening assays, as a means to ensure patient safety. This purification process validation testing is done in a **proactive** manner by spiking known amounts of model viruses into a scaled-down version of the purification process that has been validated to represent manufacturing scale. Clearance studies involving specific viral entities may be performed in response to the detection, during routine lot release testing, of a viral contaminant.

Table 1: Adventitious viruses isolated by product category.

Biological	**Viruses detected**
Cell banks (Master, Working)	None[a]
Gene therapy vectors	
• retrovirus	None
• baculovirus	None
• lentivirus	None
• adenovirus	Replication-competent adenovirus
• adeno-associated virus	None
• Sendai	None
• REO	None
Monoclonal antibodies	None
Recombinant proteins	
• non-CHO cell process	None
• CHO cell process	MMV, REO, Cache Valley virus

[a] Viruses have been isolated from research cell lines. This information is beyond the scope of this paper.

SUMMARY

Adventitious viral contaminants are rarely detected in biologicals. This reflects adherence to Good Manufacturing Practices in the control of the manufacturing environment and adequate viral safety testing of raw materials, including Master and Working cell banks. Lot-by-lot evaluation of bulk harvests from the manufacturing campaigns for biologicals, in combination with viral clearance information obtained through purification process validation studies, provides assurance of patient safety. Knowledge of the types of viral agents being detected in viral screening of bulk harvests is useful in assessing the appropriateness of viral clearance methodologies for purification processes and the engineering manufacturing processes and facilities.

REFERENCES

1 Dusing SK: Cell line characterization, in Sofer G, Zabriskie DW (eds): Biopharmaceutical Process Validation. New York, Marcel Dekker, 2000, pp 77-99.

2 Whiteman MD: Scope and practicality of in vivo testing for adventitious agents, in Petricciani J, Sheets R (eds): Vaccine Cell Substrates 2004. Dev Biol (Basel). Basel, Karger, 2006, vol 123, pp 147-152.

3 Knipe DM: Virus-Host-Cell Interactions, in Fields BN, Knipe DM, Chanock RM, Hirsch MS, Melnick JL, Monath TP, Roizman B (eds): Fundamental Virology, 2nd edition, New York, Raven Press, Ltd, 1991, pp 267-289.

4 Code of Federal Regulations: Animals and Animal Products. Title 9, Part 113.53 (US Government Printing Office, Washington, DC), 2003.

5 International Conference on Harmonization: Guideline on Viral Safety Evaluation of Biotechnology Products Derived from Cell Lines of Human or Animal Origin (ICH 5A, 1997). Geneva, Switzerland, International Conference on Harmonization, 1997.

6 International Conference on Harmonization: Guideline on Quality of Biotechnological/Biological Products: Derivation and Characterization of Cell Substrates Used in the Production of Biotechnological/Biological Products (ICH Q5D, 1997). Geneva, Switzerland, International Conference on Harmonization, 1997.

7 FDA/CBER: Points to Consider in the Characterization of Cell Lines Used to Produce Biologicals (FDA/CBER, 1993) Rockville, MD: Food and Drug Administration, Center for Biologics Evaluation and Research, 1993.

8 FDA/CBER: Points to Consider in the Manufacture and Testing of Monoclonal Antibody Products for Human Use (FDA/CBER, 1997) Rockville, MD: Food and Drug Administration, Center for Biologics Evaluation and Research, 1997.

9 FDA/CBER: Guidance for Industry: Guidance for Human Somatic Cell Therapy and Gene Therapy (FDA/CBER, 1998) Rockville, MD: Food and Drug Administration, Center for Biologics Evaluation and Research, 1998.

10 Coffin J: Significance of retrovirus contaminants of cell substrates (Abstract), in Petricciani J, Sheets R (eds): Vaccine Cell Substrates 2004. Dev Biol (Basel). Basel, Karger, 2006, vol 123, p 89.

11 Jarret R, Johnson D, Wilson K, Gallagher A: Molecular Methods for Virus Discovery, in Petricciani J, Sheets R (eds): Vaccine Cell Substrates 2004. Dev Biol (Basel). Basel, Karger, 2006, vol 123, pp 77-88.

12 Chang A, Dusing S: Practical assay issues with PERT/PBRT assay: A highly sensitive reverse transcriptase assay, in Petricciani J, Sheets R (eds): Vaccine Cell Substrates 2004. Dev Biol (Basel). Basel, Karger, 2006, vol 123, pp 91-87.

13 EMEA/CPMP (European Agency for the Evaluation of Medicinal Products. Committee for Proprietary Medicinal Products): Note for Guidance on the use of Bovine serum in the manufacture of human biological medicinal products. London, 2003, pp 1-8.

14 Schuurman R, van Steenis B, van Strien A, van der Noordaa J, Sol C: Frequent detection of bovine polyomavirus in commercial batches of calf serum by using the polymerase chain reaction. J Gen Virol 1991; 72: 2739-2745.

15 Kappeler A, Lutz-Wallace C, Sapp T, Sidhu M: Detection of bovine polyomavirus contamination in fetal bovine sera and modified live viral vaccines using polymerase chain reaction. Biologicals 1996; 24 (2): 131-135.

16 Potts BJ, Kivens WJ: Detecting and removing porcine viruses to prevent contamination of biopharmaceutical raw materials. BioPharm 2000; 26-31.

17 Garnick RL: Experience with viral contamination in cell culture, in Brown F, Lubiniecki AS (eds): Viral Safety and Evaluation of Viral Clearance from Biopharmaceutical Products. Dev Biol Stand. Basel, Karger, 1996 vol 88, pp 49-56.

18 Jacoby RO, Ball-Goodrich LJ, Besselsen DG, McKisic MD, Riley LK, Smith AL: Rodent parvovirus infections. Lab Anim Sci 1996; 46: 370-380.

19 Nettleton PF, Rweyemamu MM: The association of calf serum with the contamination of BHK21 clone 13 suspension cells by a parvovirus serologically related to the Minute Virus of Mice (MVM). Arch Virol 1980; 64: 359-374.

20 Zoletto R: Parvovirus serologically related to the minute virus of mice (MVM) as contaminant of BHK 21 cl. 13 suspension cells, in Spier R, Griffiths B, Hennessen W (eds): Production and Exploitation of Existing and New Animal Cell Substrates. Dev Biol Stand. Basel, Karger, 1985, vol 60, pp 179-183.

21 Tattersall P, Bratton J: Reciprocal productive and restrictive virus-cell interactions of immunosuppressive and prototype strains of minute virus of mice. J Virol 1983; 46 (3): 944-955.

22 Rabenau H, Ohlinger V, Anderson J, Selb B, Cinatl J, Wolf W, Frost J, Mellor P, Doerr HW: Contamination of genetically engineered CHO-cells by epizootic haemorrhagic disease virus (EHDV). Biologicals 1993; 21: 207-214.

23 Oehmig A, Buttner M, Weiland F, Werz W, Bergemann K, Pfaff P: Identification of a calicivirus isolate of unknown origin. J Gen Virol 2003; 84 2837-2845.

Dr. R.W. Nims, BioReliance, 9900 Blackwell Road, Rockville, MD 20850, USA.
E-mail: rnims@bioreliance.com

Petricciani J, Sheets R (eds): Vaccine Cell Substrates 2004.
Dev Biol (Basel). Basel, Karger, 2006, vol 123, pp 165-176.

Approaches to the Release of a Master Cell Bank of PER.C6 Cells; A Novel Cell Substrate for the Manufacture of Human Vaccines

J.A. Lewis[1], *E.L. Brown*[2], *P.A. Duncan*[2]

[1] Department of Live Virus Vectors

[2] Department of BioProcess and BioAnalytical Research
Merck Research Laboratories, Merck and Co., West Point, PA, USA

Abstract: At Merck and Co. we have developed a recombinant E1 deficient adenovirus type 5 vaccine vector for HIV-1 and have adopted the PER.C6 cell line as a cell substrate for the manufacture of this vector for Phase I and II clinical trials. The PER.C6 cell line was developed at Crucell by the transfection of human primary embryonic retinoblasts with a transgene of E1 constructed with a minimum of E1 coding sequences to preclude homologous recombination generating replication-competent adenovirus, between E1 sequences in PER.C6 and adenovirus vectors with E1 deletions of the same molecular coordinates.

We have developed a Master Cell Bank (MCB) of PER.C6 cells under serum-free conditions of suspension culture from a vial of PER.C6 cells obtained from Crucell. This MCB has been released according to an extensive panel of testing for the detection of adventitious viral agents, including assays for sterility and mycoplasma, in vivo and in vitro assays for the detection of viruses of human, bovine and porcine origin, replication competent adenoviruses, sensitive PERT assays for the detection of RT in supernatants of co-cultivations, electron microscopy and a panel of PCR-based assays for specific human and animal viruses. This MCB has been used for the manufacture of vaccine vector supporting a number of IND submissions for Phase I clinical trials over a three-year period during which the panel of PCR testing applied to the MCB has been judiciously expanded. Advances in QPCR technology, liquid handling systems, and more recently mass spectrometry offer the possibility that very broad panels of primers and probes capable of the detection of all known human viruses can be applied routinely to support the release of biologicals for human clinical trials. The impact of this breadth of testing on the continued reliance of classical in vivo and in vitro assays for adventitious viruses is clearly an emerging issue worthy of serious debate.

Introduction

At Merck and Co. we have developed adenovirus type 5 (Ad5) E1 deficient vaccine vectors for HIV-1 and have adopted the PER. C6 cell line as a cell substrate for the manufacture of vaccine for Phase I and II clinical trials [1]. The PER. C6 cell line was developed at Crucell by the transfection of primary human embryonic retinoblasts (HER) with a transgene composed of 3052bp of the E1A/E1B coding sequences of Ad5 (bp459-3510) flanked at the 5' end by the human phosphoglycerate kinase gene promoter and at the 3' end by poly A termination sequences of hepatitis B [2]. The transfection of HER led to the outgrowth of a number of foci of transformed cells, one of which, designated PER. C6, was selected for its favorable properties of in vitro growth, E1A/E1B expression and its support of high-level replication of various E1 deleted adenovirus vectors. The E1 transgene was so constructed as to preclude the generation of replication competent adenoviruses (RCA) during the infection of PER. C6 cells with adenoviruses carrying E1 deletions spanning the same molecular coordinates. The use of PER. C6 cells for the manufacture of adenovirus vectors is expected to make routine the manufacture of vaccine which meets or exceeds the current specification of < 1 RCA per 10^{10} adenovirus particles (Biological Response Modifiers Advisory Meeting 5 April 2001) and thereby contributing an element of additional safety to vaccine products.

The PER. C6 cell line was subsequently expanded in larger scale culture at Crucell and at P29 was used to prepare a Research Master Cell Bank (rMCB), from which Research Working Cell Banks (rWCB) at P33 and P35 have been prepared and been made available by Crucell to licensees for the manufacture of vaccines, monoclonal antibodies, and other biological products.

Manufacture of a cGMP MCB of PER.C6 at Merck and Co

Merck's adoption in this novel cell substrate as a manufacturing platform for an Ad5HIV vaccine was predicated on: (i) a confidence in the attentions paid by Crucell to cGMP aspects of operational, environmental, and documentary control; (ii) the sensitivity and redundancy of the in vivo and in vitro assays used for the detection of adventitious agents supporting the release of the Crucell rMCB and rWCB, and (iii) the demonstrated ability of the cell line to propagate appropriately developed Ad5 viral vectors without the development of RCAs. At Merck cryopreserved vials of the PER. C6 rMCB at P29 obtained from Crucell were expanded in suspension culture under serum free cGMP conditions to P41 at which level a MCB was cryopreserved for use in vaccine manufacture. This MCB has supported the manufacture of Ad5HIV vaccine vector over a three-year period and its release is the subject of the following sections.

In Vivo and in Vitro Assays for Adventitious Viruses Supporting the Release of the Merck MCB

This P41 cGMP bank was released as a cell substrate for manufacturing in 3Q/2000 according to an extensive panel of assays for the detection of adventitious microbial and viral agents presented in Figure 1. This panel of assays was developed with due consideration to (i) compendial precedents, guidances and literature precedents [3,4] and (ii) a judicious review of the tissue of origin, isolation of primary cells, transfection and development of the continuous cell line and reagents used in the course of all cell culture operations. These assays include (i) broth and agar cultures to assay for sterility and mycoplasma, as well as cultures specific for the cultivation of *M. tuberculosis;* (ii) embryonated eggs inoculated by the yolk-sac, allantoic, and amniotic routes; (iii) i.c., i.m. and i.p. inoculation of adult and suckling mice and i.m. inoculation of guinea pigs; (iv) monolayer cell cultures using MRC-5, 293, Vero, RK 13, and RD 114 cells for the detection of various human viruses; (v) continuous suspension cultures of H9 and Raji for the specific detection of retroviruses analysed for reverse transcriptase by PERT assay and (vi) bovine turbinate, Vero, and primary porcine kidney cells for the detection of bovine and porcine viruses specified by 9CFR 113.53 which may have been adventitiously introduced through the use of raw materials of animal origin.

In concert, this panel of in vivo and in vitro assays is capable of the sensitive detection of an especially broad spectrum of both human and animal viruses. This spectrum has been well-characterized. The sensitivity of embryonated eggs for the replication of a wide spectrum of viruses has been well-studied for many decades beginning with the seminal report of Beveridge and Burnet [5], as are the sensitivities of the primary human diploid fibroblast cell line MRC-5 and the continuous African green monkey kidney line Vero, especially the Vero E6 subclone [6-8]

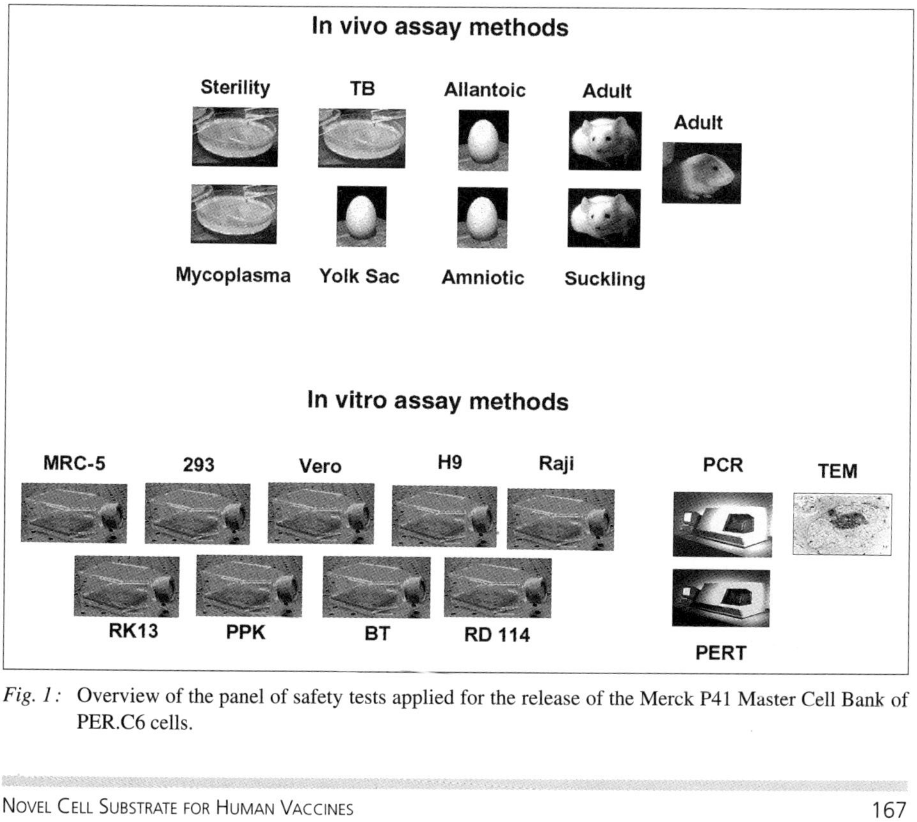

Fig. 1: Overview of the panel of safety tests applied for the release of the Merck P41 Master Cell Bank of PER.C6 cells.

whose broad spectrum of sensitivity to a wide variety of viruses has been demonstrated once again as a cell substrate for the isolation of the SARS virus from clinical specimens independently in two laboratories [9,10].

These in vivo and in vitro assays are in general inoculated with from 10^6-10^7 cell or cell lysate equivalents depending on compendial requirement [11] and in concert they provide numerous opportunities for the redundant detection of human viruses of numerous virus families. These opportunities for redundancy are summarized in Table 1 which uses the symbol (•) to indicate the use of the assay system for the detection of a given family of human viruses. Table 1 allows the generally useful conclusion to be drawn that viruses of six families of human RNA viruses including the *Togaviridae, Flaviviridae, Arenaviridae, Bunyaviridae, Rhabdoviridae* and

Table 1: Overview of the sensitivity of in vivo and in vitro assay systems for the detection of human viruses.

		In vivo		In vitro
Family	**R/D**[1]	**Eggs**	**Mice/G. pigs**	**Human/simian**
Togaviridae	R	•	•	•
Flaviviridae	R	•	•	•
Arenaviridae	R	•	•	•
Bunyaviridae	R	•	•	•
Rhabdoviridae	R	•	•	•
Paramyxoviridae	R	•	•	•
Orthomyxoviridae	R	•		•
Picornaviridae	R		•	•
Reoviridae	R		•	•
Filoviridae	R		•	•
Retroviridae	R			•
Astroviridae	R			
Caliciviridae	R			
Coronaviridae	R			
Poxviridae	D	•	•	•
Herpesviridae	D	•	•	•
Adenoviridae	D			•
Polyomaviridae	D			•
Papillomaviridae	D			
Hepadnaviridae	D			
Parvoviridae	D			

[1] RNA or DNA genome.

Paramyxoviridae are likely to be detected with a three-fold redundancy, using in vivo assays of embryonated eggs, rodents and in vitro cultures, whereas viruses of the *Orthomyxoviridae, Picornaviridae, Reoviridae,* and *Filoviridae* are likely to be detected redundantly through a combination of in vivo egg or rodent assays in conjunction with in vitro cell culture assays. Viruses of the *Retroviridae* are likely to be detected only with in vitro co-cultivation assays, with essentially no opportunities for redundancy with in vivo assays. By contrast, viruses of the *Astroviridae, Caliciviridae* and *Coronaviridae* of RNA viruses have in general proved to be refractory to replication in both the in vivo or in vitro assay systems described above, although they may be detectable in specialized cell culture systems following periods of adaptation. Clearly, exceptions to these generalities can be identified including the replication of SARS virus in Vero cells which have not previously been reported as a generally useful host for the replication of *coronaviridae* [9,10], and the failure of hepatitis C virus, a member of the pestivirus genus of *flaviviridae* to be cultured in vitro despite numerous experimental attempts with a wide variety of cell culture systems.

Table 1 provides a perspective of the opportunities for redundant detection of adventitious human DNA viruses as well. Of the seven families of human DNA viruses, only viruses of the *Poxviridae* and HSV1, HSV2, and CMV of the *Herpesviridae* are likely to be redundantly detected with in vivo and in vitro systems. Viruses of the *Adenoviridae* and BKV of the Polyoma genus *Papapovaviridae* are likely to be detected only with the in vitro cell culture assays with no redundancy, whereas viruses of the papillomavirus genus of *Papovaviridae, Hepadnaviridae* and *Parvoviridae* have yet to be cultured in any assay system used routinely for the detection of adventitious viruses.

PCR Assays for Adventitious Viruses Supporting the Release of the Merck P41 MCB of PER.C6

The MRL P41 MCB was released for the manufacture of vaccine vector, for which an Investigational New Drug application (IND) was submitted to the Center for Biologics Evaluation and Research (CBER) for Phase I clinical trials in 3Q/2000. This MCB release was supported with the in vivo and in vitro assays described above as well as a panel of PCR-based assays for the specific detection of adventitious viruses. This PCR panel for the 3Q/2000 release encompassed 18 specific viral agents, 15 with a preferred human tropism and 3 with an animal tropism. The viruses for which specific PCR testing are presented in Tables 2 and 3 under the column IND 1. This panel of PCR-based assays was reviewed extensively in pre-IND discussions with scientists from the Center for Biologics Evaluation and Research (CBER) in June 1999 and at the time subsumed nearly all the specific human and animal PCR assays then available by contract testing laboratories.

The PCR assays were performed by contract laboratory (Q-One BioTech Glasgow, Scotland). 13 of 15 assays were conducted using Southern blot (SB) hybridization technology for the detection of reaction product amplicons of the appropriate molecular size. The human BKV and JC viruses of the *papovaviridae* were assayed using QPCR technology [12] The SB method, though well established, is laborious, time-consuming, and creates issues of waste management with the use of radioactivity as a means of detection, although non-radioactive detection methods of similar sensitivity have recently been generally available. The sensitivity of the various PCR assays shown in Tables 2 and 3 are most commonly reported to be 10-1000 genome copies per reaction with most reported as 100. The levels of sensitivity of the SB or QPCR assays are essentially indistinguishable. Assuming conservatively a physical-to-infectious particle ratio of 10 for both DNA and RNA viruses, PCR assays of this reported sensitivity can detect 10 infectious units. The sensitivity of the PCR assay is largely governed by the input of genomic DNA to the PCR reaction, with sensitivity declining as the mass of genomic DNA exceeds 1µg, a mass of DNA extractable from 10^5 cells. It follows then that the reported limit of detection of the assay expressed per dose of vaccine would be that number of doses produced in 10^5 of the cells used as substrate.

Of the 15 human viruses in the PCR panel for 3Q/2000 release, only 6, CMV, HIV1, 2, HTLV1, 2 and the polyoma virus BK might have been expected to be redundantly detected by the panel of in vivo and in vitro assays; CMV to be detectable in both in vivo and in vitro assays and the five others with in vitro cell culture testing and for these agents PCR offers a useful opportunity for redundant detection. Of the nine other human viruses for which PCR assays were applied, not one of these would be expected to be detectable in the testing systems used, and the use of PCR therefore remains the only opportunity for such detection. While it might be argued that the intractability of HBV, HCV, AAV B19 to propagation in any system might render an adventitious appearance in an MCB unlikely, it is important to recall, given the novelty of the PER. C6 cell strain, that the breadth of its susceptibility to viral agents was at the time unknown and its origin as a retinoblastoma cell line may have permitted the replication of viruses with unanticipated tropisms. More recently the sensitivity of PER. C6 cells for the replication of influenza [13] and West Nile Virus [14] has been reported.

For the filing of a second IND in 3Q/01, cell lysates of PER. C6 held in cryopreservation were assayed by QPCR for human papilloma virus and a broader generic assay for enteroviruses, a major cause of aseptic meningitis and febrile illness in children [15]. These assays were encouraged by interaction with CBER developed under contract with Q-One BioTech (Glasgow, Scotland) and were applied for PCR testing of the MCB and release for Phase I clinical trials in September 2001. The PCR panel was further expanded to include an assay for measles virus and a broadly sensitive assay for adenoviruses, as shown in Table 2 [16], bringing the panel of PCR agents for human viruses to a total of 19 human agents.

The panel of PCR assays for bovine and porcine viruses was similarly expanded in the course of Ad5HIV vaccine development to detect adventitious contaminants of reagents of animal cell origin, and these assays were applied as they became available by contract testing (Table 3). The use of these PCR assays for bovine and porcine agents is indisputable, resting in large measure on their ability to detect adventitious viruses heretofore uncharacterized and whose tropism in the cell substrates of the 9 CFR 113.53 have yet to be carefully documented. Indeed it might be argued that given the ease and rapidity of the PCR testing, that the assays of 9 CFR113.53 for which specific detection is requested, might be better served with PCR rather than the specific fluorescence testing described.

Table 2: PCR testing for human viruses applied to the release of the Merck P41 MCB through successive IND filings with opportunities for redundant expression.

	IND submission				In vivo		In vitro
Virus	**1**	**2**	**3**	**PCR**	**Eggs**	**Mice/G. pigs**	**Human simian**
Hepatitis B virus	•	•	•	SB[1]			
Hepatitis C virus	•	•	•	SB			
Cytomegalovirus	•	•	•	SB			•
Epstein-Barr virus	•	•	•	SB			
Human herpes virus 6	•	•	•	SB			
Human herpes virus 7	•	•	•	SB			
Human herpes virus 8	•	•	•	SB			
Human immunodeficiency virus 1	•	•	•	SB			•
Human immunodeficiency virus 2	•	•	•	SB			•
Human T-cell leukaemia virus 1	•	•	•	SB			•
Human T-cell leukaemia virus 2	•	•	•	SB			•
Adeno-associated virus	•	•	•	SB			
Human parvovirus B19	•	•	•	SB			
Human polyoma virus BK	•	•	•	QPCR[2]			
Human polyoma virus JC	•	•	•	QPCR			
Human papilloma virus[1]		•	•	QPCR			
Enterovirus		•	•	QPCR		•	
Adenovirus			•	QPCR			•
Measles			•	QPCR	•		•
	15	**17**	**19**				

1 Southern blot   2 Quantitative polymerase chain reaction

Table 3: PCR testing for simian, bovine and porcine viruses applied to the release of the Merck P41 MCB through successive IND filings with opportunities for redundant expression.

	IND submission				In vivo		In vitro
Virus	**1**	**2**	**3**	**PCR**	**Eggs**	**Mice/G. pigs**	**Human simian BT or PPK**
SV40	•	•	•	SB[1]			•
Simian foamy virus	•	•	•	SB			
Bovine polyoma virus	•	•	•	SB			
Bovine leukaemia virus		•	•	SB			
Bovine immunodeficiency virus		•	•	SB			
Porcine parvovirus		•	•	SB			•
Bovine circovirus		•	•	Q-PCR[2]			
Porcine circovirus		•	•	Q-PCR			
Bovine herpesvirus I and II			•	Q-PCR			•[3]
Bovine rotavirus (G6 and G10)			•	Q-PCR			
	3	**8**	**9**				

1 Southern blot.
2 Quantitative polymerase chain reaction.
3 Egyed [16].

Future Applications of Methods for Molecular Detection of Adventitious Viral Agents in the Release of Biologicals

Clearly we are enjoying, at the time of this Vaccine Cell Substrate Meeting in June 2004, an ever-expanding capability for the rapid and exquisitely sensitive molecular detection of the genomes of essentially all human viruses. This capability is predicated on numerous technical advances, including (i) the ease with which the nucleotide sequence of primary viral isolates can be generated; (ii) the ready identification by BLAST homology searches of highly conserved nucleotide sequences within the regulatory elements of viral genomes at the level of genus, family, genus, and serotypic subgroup; (iii) the expansion of QPCR testing capabilities from 96 to 384 well formats allowing the replication of test articles with concomitant increases in levels of sensitivity, (iv) developments in micro-array technology [17,18]; (v) developments in SNP detection capabilities and (vi) systems for high throughput mass spectrometry for nucleic assay identification [19] With these capabilities it is not unreasonable to envisage within a few years the availability of very large assemblies of PCR primers and probes, with a degeneracy sufficient to encompass even minor sequence variations in the most conserved genetic control elements of human viruses to be available in a QPCR or PCR/micro-array format as an opportunity for a reassuring redundant control on the detection of viral agents by classical in vivo and in vitro assays. The simplicity of this molecular testing will of course invite a continuation of the dialogue begun here, seeking to define when, if ever, confidence in the breadth of detection of known human agents made available by the redundancy of probes and formats will perhaps allow a gradual diminishing on the reliance of the classical in vivo and in vitro methods which have served so well to ensure the freedom of generations of vaccines and biologicals from adventitious viral agents.

Additional Considerations Related to PER.C6

RCA and HDEP viruses

As described above, the E1 sequences in the transgene used for the development of the PER.C6 cell line was constructed to preclude the generation of replication competent adenoviruses (RCA) during the infection of PER. C6 cells with adenoviruses carrying E1 deletions spanning the same molecular coordinates. Indeed, as experimental data accrue on the detection of RCA in adenovirus vectors prepared on a large scale by various manufacturers, it becomes apparent that the theoretical advantages of PER. C6 have been clearly realized in practical terms, allowing several manufacturers, Merck and Crucell among them, to report the manufacture of more than 10^{13} particles without the detection of RCA. More recently, Murakami et al [20] have described a recombinant adenovirus designated Helper-Dependent E1 containing Particles (HDEP) reported by an occasional PER. C6 licensee at a frequency of 1 HDEP per 10^{13} particles which contains E1 sequences derived from the transgene of PER. C6, although its viral genome is otherwise grossly defective and wholly dependent on the replication of E1 deficient vectors for its genome replication and packaging. Clearly the defectiveness of this HDEP and its dependence on other adenovirus replication is likely to diminish significantly any concerns about potential risk.

Tumorigenicity of the PER.C6 cell line

As might be expected of a cell line established with the adenoviral E1 genes of well-established immortalizing functions, the PER. C6 line is tumorigenic when assayed in nude mice at the level of 10^7 cells per inoculum. Insofar as the licensing of PER. C6 may establish a precedent as the first human tumorigenic cell line licensed for the manufacture of human vaccines, special attention has been paid therefore by Merck and Co., working in close collaborative interaction with scientists from CBER, to assay PER. C6 DNA for tumorigenicity in a variety of rodent models. The results of these studies conducted over a two-year period in MRL, which demonstrate persuasively the lack of tumorigenicity of PER. C6 DNA in a variety of rodent models, are reviewed in this volume by Dr. B.J. Ledwith and Dr. J.A. Lebron [21]. Similar studies demonstrating the lack of tumorigenicity of a variety of transforming DNAs are well established in the literature and their general acceptance is expected to lead to a broader consideration of the cell substrates appropriate for the manufacture of vaccines for use in humans. In this context it is of interest to note a recent report of the release of HeLa cells as a substrate for the manufacture of an AAV vector for an HIV vaccine for clinical evaluation [22].

Expression of PrP^c in PER.C6 Cells

Insofar as the development of the PER. C6 cell line was accomplished during an era (1985-2000) which witnessed the emergence of bovine spongiform encephalopathy (BSE) in Great Britain and on the continent of Europe, thoughtful consideration has been given to TSEs as adventitious contaminants of PER.C6 cells. PER. C6 cells do express PrP^c, as do many continuous cell lines of both neuronal and non-neuronal origin, including both MRC-5, and 293 which are used or are contemplated for use in the development of human vaccines (P. Minor personal communication and Lewis et al, unpublished.). Although the expression of PrP^c appears to be requisite for infection and propagation of PrP^{sc} in cell lines in vitro, a cell line of neuronal origin is clearly not prerequisite. Indeed, recent work has demonstrated the susceptibility of cell lines of both epithelial and mesenchymal lineages to exogenous infection with inocula from homologous species [23,24], rendering untenable the original notion, based on the susceptibility of the neuroblastoma cell line N2a, and PC12, and GT1 to murine-adapted scrapie, that neuronal cells, and perhaps PER. C6 cells as well, were uniquely susceptible to infection with exogenous TSE agents.

Concerns then of adventitious contamination of the PER. C6 remain an exceedingly remote possibility allayed by (i) due diligence reviews conducted by MRL scientists of the control of raw material sourcing by Crucell, and further strengthened by experimental data which demonstrate (ii) a lack of detectable levels of PrP^{sc} in bovine serum from cattle during pre-clinical and clinical phases of the disease when assayed in sensitive intracranial assays in cattle [25]; (iii) the requirement for inocula of very high titre to establish TSE infections in those cell lines demonstrated in vitro and [23]; (iv) a failure to report either evidence of BSE transmission in vitro or cross species (heterologous) infection of cell lines known to be susceptible to infection with inocula of homologous murine PrP^{sc} and (v) importantly the application of western blot based assays which demonstrated the absence of PrP^{sc} in cell lysates of PER. C6.

Insofar as PER. C6 cells express PrP^c, the coding sequences of the PRNP gene of PER. C6 were amplified by PCR, cloned and then sequenced at the P29 rMCB level by Crucell and by Merck and Co., in collaboration with Lark (Houston, Texas). Both analyses reported PER. C6 to be heterozygous at the PRNP locus, encoding both a methionine 129 and a valine 129 PrP^c allele, a genotype which has been shown to operate against the development of TSE disease in humans [26]. Both studies reported additionally these PRNP alleles to be free of any mutations predisposing to the development of aberrant endogenous PrP^{sc} [27].

Acknowledgements

The authors wish to thank Dr. Rebecca Sheets of the Vaccine Research Center of NIAID for her kind invitation to speak of the Merck experience with the use of PCR for the detection of adventitious agents in the release of live virus vaccines, as well as many colleagues at Merck, especially Dr. Douglas Martinez for his numerous contributions to the development of a judicious panel of PCR assays for release and Dr. Jose Lebron and Dr. Jayanthi Wolf for their collaborations on the development of approaches to BSE testing and risk assessments.

References

1 Shiver JW, Emini EA: Recent advances in the development of HIV-1 vaccines using replication-incompetent adenovirus vectors. Ann Rev Medicine 2004; 55: 355-372.

2 Fallaux F, Bout A, van der Velde I, van den Wollemberg D, Hehir KM, Keegan J, Auger C, Cramer S, Van Ormondt H, Van der Eb AJ, Valerio D, Hoeben RC: New helper cells and matched early region1-deleted adenovirus vectors prevent generation of replication competent adenoviruses. Human Gene Therapy 1998; 9: 1909-1917.

3 Robertson JS: Strategy for adventitious agent assays, in Brown F, Lubiniecki AS (eds): Viral Safety and Evaluation of Viral Clearance. Dev Biol Stand. Basel, Karger, 1996, vol 88, pp 25-29.

4 Minor PD: Mammalian cells and their contaminants in Brown F, Lubiniecki J (eds): Viral Safety and Evaluation of Viral Clearance. Dev Biol Stand. Basel, Karger, 1996, vol 88, pp 37-40.

5 Beveridge WIB, Burnet FM: The cultivation of viruses and rickettsiae in the chick embryo. Medical Research Council Special Report Series No. 256, 1946.

6 Jacobs JP, Jones CM, Baille JP: Characteristics of a human diploid cell designated MRC-5. Nature 1970; 227: 168-170.

7 Rhim JS, Schell K, Creasy B, Case W: Biological Characteristics of an African green monkey kidney cell line (Vero). P.S.E.B.M. 1969; 132: 670-677.

8 Emeny J, Morgan MJ: Regulation of the interferon system; evidence that Vero cells have a genetic defect in interferon production. J Gen Virology 1979; 43: 247-252.

9 Ksiak TG, Erdmann D, Goldsmith C, Zaki SR et al: A novel coronavirus associated with severe acute respiratory syndrome. N.E.J.M. 2003; 348: 1953-1966.

10 Fouchier RAM, Hartwig NG, Bestebroer TM, Niemeyer B, De Jong JC, Simon JH, Osterhaus AD: A previously undescribed coronavirus associatd with respiratory disease in humans P.N.A.S. 2004; 101: 6212-6216.

11 Jacobs JP, Mcgrath DI, Garrett AJ, Schid GG: Guidelines for the acceptability, management, and testing of serially propagated human diploid cells for the production of live virus vaccines for use in man. J Biol Stand 1981; 9: 331-342.

12 Heid CA, Stevens J, Livak KJ, Williams PM: Real-time quantitative PCR. Genome Research 1996; 6: 995-1001.

13 Pau MG, Ophorst C, Koldjik MH, Schouten G, Mehtali M, Uytedehaag F: The human cell line PER. C6 provides a new manufacturing system for the production of influenza vaccines Vaccine 2001; 19: 2716-2721.

14 http://www.crucell.com/.

15 Stellrecht KA, Harding I, Woron AM, Lepow ML, Venezia RA: The impact of enteroviral RT-PCR assay on the diagnosis of aseptic meningitis and patient management. J Clinical Virology 2002; 25: S19-26.

16 Egyed L: Replication of bovine herpes type 4 in human cells in vitro. J Clinical Microbiology 1998; 36: 2109-2111.

17 Vernet G: DNA-chip technology and infectious diseases. Virus Research 2002; 82: 65-71.

18 Clewley JP: A role for arrays in clinical virology: fact or fiction? J Clin Virology 2003; 29: 2-12.

19 Hofstadler SA, Samopath R, Blyn LB, Eshoo, Hall TA, Jiang Y, Drader D, Hannis JC, Sannes-Lowery KA, Cummins LL, Libby B, Walcott D, Schink A, Massire C, Ranken R, Guitierrez J, Manali S, Ivy C, Melton R, Levine H, Barrett-Wilt G, Li F, Zapp V, White N, Samant V, McNeil JA, Knize D, Robbins, Rudnick K, Desai A, Moradui E, Ecker DJ: TIGER: The Universal Biosensor International Journal of Mass Spectrometry 2005; 242: 23-41.

20 Murakami P, Havenga M, Fawaz F, Vogels R, Marzio G, Pungor E, Files J, Do L, Goudsmit J, McCaman M: Common structure of rare replication – deficient E1 positive particles in adenoviral vector batches. J Virology 2004; 78: 6200-6208.

21 Lebron JA et al: Adaption of the WHO guideline for residual DNA in parenteral vaccines produced on continuous cell lines to a limit for oral vaccines, in Petricciani J, Sheets R (eds): Vaccine Cell Substrates 2004. Dev Biol (Basel). Basel, Karger, 2006, vol 123, pp 35-44.

22 Tatalick L, Gerard C, Takeya R, Price DN, Thorne BA, Wyatt LM, Ankelesaria P: Safety characterization of HeLa based cell substrates used in the manufacture of a recombinant adeno-associated virus-HIV vaccine. Vaccine 2005; 23: 2628-2638.

23 Vorberg I, Raines A, Story B, Priola SA: Susceptibility of common fibroblast cell lines to transmissible spongiform encephalopathy agents. J Inf Disease 2004; 189: 431-439.

24 Vilette, D, Andreoletti O, Archer F, Madelaine MF, Vilotte JL, Lehmann S, Laude H: Ex vivo propagation of infectious sheep scrapie agent in heterologous epithelial cells expressing prion protein. P.N.A.S. 2001; 98: 4055-4059.

25 Wells GAH, Hawkins S, Green R, Austin A, Dexter I, Spencer I, Chaplin M, Stack M, Dawson M: Preliminary observations of experimental bovine spongiform encephalopathy (BSE) an update. Veterinary Report 1998; 142: 103-106.

26 Wadsworth JDF, Asante EA, Desbrusias M, Linehan JM, Joiner S, Gowland I, Welch J, Stone L, Lloyd SE, Hill AF, Brandner, Collinge J: Human prion protein valine 129 prevents expression of variant CJD phenotype. Science 2004; 306: 1793-1796.

27 Amexis G, Ridge J, Cervenakova L, Enterline JC, Chumakov KM, Asher DM: Stability of the prion protein encoding PRNP gene in HeLa cells, 2003.

Dr. J.A. Lewis, Retired from Department of Live Virus Vectors, Merck Research Laboratories, Merck and Co., West Point, PA 19486, USA.
E-mail: paul_duncan@merck.com

Petricciani J, Sheets R (eds): Vaccine Cell Substrates 2004.
Dev Biol (Basel). Basel, Karger, 2006, vol 123, pp 177-180.

Newer Molecular Approaches to Detect Known Viruses and Previously Unrecognized Viruses in Cell Substrates

P.F. Wright

Vanderbilt University Medical Center, Nashville, TN, USA

Abstract: The theme of this brief presentation is that there are tremendous advances not only in PCR-based detection but in other novel ways of detecting viruses and other agents. Much of the research is now being driven by the bio-defence agenda. Somehow those with manufacturing and regulatory responsibilities have to keep abreast of these developments and incorporate them as they become available into defining the safety of cell substrates. I am concerned that these techniques are and may remain academic research tools without validation or standardization, and that individual academic laboratories will not develop the capacity and special skills necessary to satisfy regulatory requirements, but that there is a tremendous research incentive in this area that needs to be linked with the development of safe prophylactic and therapeutic biologicals. With that, I will finish what I would consider as a report rather than an original talk.

INTRODUCTION

Newer molecular techniques may allow detection of known or previously unrecognized viruses in cell substrates. My interest in the applicability of micro-array and similar approaches arose from a conversation with Dr. Donald Ganem at the University of San Francisco. Our group at Vanderbilt had recently described the importance of a new virus, human meta-pneumovirus, in human respiratory disease using PCR amplification of RNA from archived specimens in young children [1]. Dr. Ganem enquired about the use of these specimens for detection of previously unrecognized viruses based on techniques that his group had recently published [2].

Dr. Ganem's motivation has been and continues to be to use these techniques for discovery of new agents that can be implicated in human disease. Alternative, but similar, molecular detection techniques were used by Dr. David Relman at Stanford to discover the agent of Whipple's disease and he is now interested in using the host cellular response as an indicator of perturbation by infectious agents, see below [3,4].

The biodefence initiative is leading to a further broadening of research in organism detection. As examples within my institution, investigations into cellular detection of organisms are being done by Dr. John Wikswo using "Massively Parallel,

Multi-Phasic Cellular Biological Activity Detector (MP^2-CBAD)". The goal of this project is to merge cellular biophysics, micro-circuits and micro-fluidics, plus information technology, to create sensing biosensors for chemical and biological agents. Also at Vanderbilt and other institutions, major initiatives in nano-medicine are aimed at organism detection.

When I discussed with Dr. Ganem the possibilities of using these techniques in the context of ensuring the safety of cell substrates being used for vaccine production, he indicated that it was a good idea but not a direction his laboratory had yet considered for their research.

Molecular Approaches

I will, as a novice, present the approach being used by the San Francisco group. They are using massively parallel ways of looking for unknown viruses. To do this they create a DNA micro-array in which there are 10,000 spots, roughly 10 spots for each virus out of 1,000, representing all the known sequences that they can find that are shared or have homology between the viruses. A small number of these are shown in Figure 1 (reprinted with permission of the Proceedings of the National Academy of Science and Drs. Wang and Ganem). These spots are displayed in a linear fashion and hybridized with tissue culture extracted or amplified viral RNA. The presence and intensity of hybridization when compared to uninfected cell substrate can be represented colorimetrically and patterns of fingerprints emerge for viruses (Fig. 1). Shown are examples of adenovirus, parainfluenza type 3, respiratory syncytial virus, poliovirus and rhinovirus, and then in panel C, Kaposi's sarcoma associated herpes virus (human herpes virus 8).

The presentations in Figure 1 were all done with tissue culture harvests of viruses, to show that RNA as a direct reaction requires a large amount of virus. To detect smaller amounts of virus, an amplification step was introduced using degenerate primers. With this amplification, the San Francisco group has shown that they can detect viral specific genetic material from nasal secretions from individuals with illness. The application of these techniques to screening cell substrates would require a great deal of standardization, definition of limits of detection, and validation. Dr. Ganem commented on the difficulty of performing this assay validation in a research laboratory; however, it is fair to say he was intrigued by this application of micro-array technology. Dr. Ganem recognizes risks of PCR cross contamination and has seen contamination with the PCR amplification step. There are some groups who are trying harder to industrialize and standardize these assays. Dr. Ganem mentioned an interest on the part of the Genome Institute of Singapore. Dr. Ganem also mentioned that he is not aware of patent protection in this area and that it would be maybe a plus or minus, but it would certainly leave the door open for this kind of investigation. In summary, he said there would be no way that he could be comfortable in saying definitively either that a virus was present or was not present with the assay in its current form.

Another approach that has been used for microbial detection is B cell-based sensors [5]. A commentary on this approach appeared in the New England Journal [4]. This technique uses a lymphocyte that had been engineered to express antibodies recognizing particular pathogens, in this case vaccinia and anthrax, and has also been engineered to contain aquaporin that emits light when the antibody-signalling

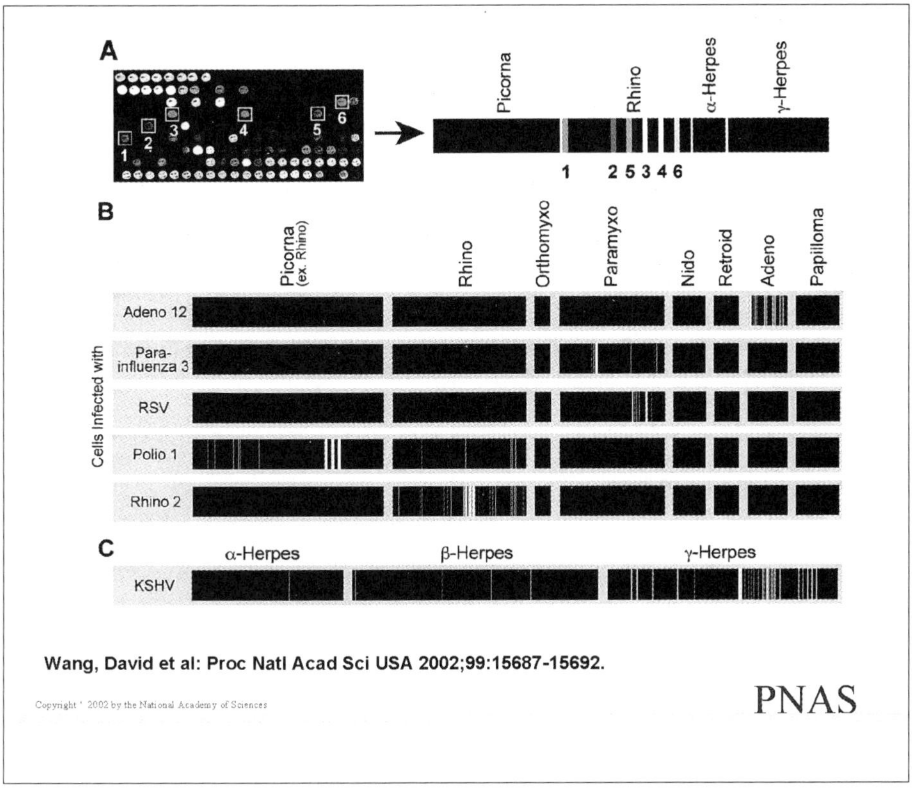

Fig. 1.

pathway leads to an increase in intracellular calcium. This represents another early but very intriguing approach to microbial detection.

Dr. David Relman has also published about the concept that the cytokine profile, and indeed cognate immune response of a cell, may be altered by adventitious agents and that there may be signatures of specific infecting organisms [6]. At present, as with other assays discussed, this is clearly in the developmental stages. It is currently aimed at lymphoid cell responses rather than the cell substrates in which most vaccines are produced. However, it could be adapted to tissue culture and to the detection of innate responses of cytokines and interferon that one could imagine might provide a signature perhaps not of a specific virus, but of the fact that a particular cell or cell line had been perturbed by an infectious agent.

I am not really equipped or prepared to speak about nano-technology, but I am intrigued by the possibilities of tracers in the form of nano-crystals which can readily enter cells and emit a very monochromatic beam when they interact with a specific peptide. Nano-technology is clearly part of a broader scientific and biodefence agenda. Many sensors predictably will be developed for adventitious agents using this technology.

References

1 John V, Williams JV, Harris PA, Tollefson SJ, Halburnt-Rush LL, Pingsterhaus JM, Edwards KM, Wright PF, Crowe JE Jr: Human Meta-pneumovirus and Lower Respiratory Tract Disease in Otherwise Healthy Infants and Children. New England Jour Med 2004; 350: 443-450.

2 Wang D, Coscoy L, Zylberberg M, Avila PC, Boushey HA, Ganem D, DeRisi JL: Microarray-based detection and genotyping of viral pathogens. Proc Natl Acad Sci U S A 2002; 99 (24): 15687-15692.

3. Relman DA, Schmidt TM, MacDermott RP, Falkow S: Identification of the uncultured bacillus of Whipple's disease. N Engl J Med 1992; 327 (5): 293-301.

4 Relman DA: Shedding light on microbial detection. N Engl J Med 2003; 349: 2162-2163.

5 Rider TH, Petrovick MS, Nargi FE, Harper JD, Schwoebel ED, Mathews RH, Blanchard DJ, Bortolin LT, Young AM, Chen J, Hollis MA: A B Cell-Based Sensor for Rapid Identification of Pathogens. Science 2003; 301: 213-215.

6 Relman DA: New Technologies, Human-Microbe Interactions, and the Search for Previously Unrecognized Pathogens. Journal of Infectious Diseases 2002; 186: S254-S258.

Dr. P.F. Wright, Vanderbilt University Medical Center, 1161 21st Avenue South, D-7235 MCN, Nashville, TN 37232, USA.
E-mail: peter.wright@vanderbilt.edu

Petricciani J, Sheets R (eds): Vaccine Cell Substrates 2004.
Dev Biol (Basel). Basel, Karger, 2006, vol 123, p 181.

A Cell-Based Assay for Mouse Prions

Ch. Weissmann

MRC Prion Unit, Department of Neurodegenerative Disease, Institute of Neurology, University College, London, and Scripps Florida Research Institute, FAU Campus, Boca Raton, Florida, USA

Abstract: We have isolated neuroblastoma N2a sub-lines that are highly susceptible to RML scrapie prions as well as those that are relatively resistant. Prion infection of N2a cells results in the accumulation of PrP^{Sc}. We have developed a procedure to identify individual PrP^{SC} – positive cells filtered on to nitrocellulose membranes of an Elispot plate and established conditions under which the proportion of PrP^{SC} – positive cells was a function of the prion titre of the samples used for infection. With this so-called Scrapie Cell Assay (SCA) we were able to quantify RML prion concentrations about as low as those that can be determined by endpoint titration in mice. SCA allows the processing of large numbers of samples using automated evaluation which can be completed in two compared to 20 weeks in the most rapid mouse bioassay and is two orders of magnitude cheaper. A different format, based on the same technology, the Scrapie cell Endpoint Assay (SCEPA), is an order of magnitude more sensitive than the SCA.

Prions are unusually resistant to inactivation. Steel surfaces bind prions tenaciously, even after only five minutes' exposure to infected brain, and transmit them to indicator mice within 30 minutes of contact. This mouse model is being used to assess conditions necessary for the sterilization of surgical instruments and shows that formaldehyde and other products used for sterilization of conventional agents are ineffective against prions. No detectable protein can be eluted from highly infectious wires. Metal-surface-bound prions infect cultured neuroblastoma cells, probably by direct contact between cells and wire, suggesting that immobilized, metal-bound prions can initiate the conversion process at the surface of PrP-bearing cells.

Dr. Ch. Weissmann, National Hospital, MRC Prion Unit, Queen Square, London, WC1N 3BG, U.K.
E-mail: charles.weissmann@prion.ucl.ac.uk.

Petricciani J, Sheets R (eds): Vaccine Cell Substrates 2004.
Dev Biol (Basel). Basel, Karger, 2006, vol 123, pp 183-197.

PANEL III DISCUSSION

Routine Adventitious Agent Tests – Can in Vivo Methods Be Replaced?

Panel Members: *Peter Wright* (Panel Chair), *Rolf Taffs, Rebecca Sheets, John Lewis, Brian Murphy Ray Nims, Mary Whiteman*

PANEL QUESTIONS

Question 1:
Are tests currently available that can replace in vivo tests?

Question 2:
Are we adequately testing for the spectrum of viral agents that might be found in tissue culture?

Question 3:
In investigational lots, exogenous contamination (usually bacterial) is the most common reason for disqualification. Are the assays for this adequate?

Question 4:
Do the tests for vaccine safety (active component of the vaccine) overlap or give additional information on substrate safety?

Question 5:
Are the methods to inactivate active component adequate?

Question 6:
In a vaccine developmental pathway, safety testing can introduce substantial delay. This is as much of a problem as the volume of pool consumed in testing. Is there a way of shortening this process?

Question 7:
Is there a synergy in the work on biosensors and biodefence that can be used to ensure substrate safety?

Question 8:
Do molecular methods, however sensitive, test an adequate volume of vaccine or cell substrate?

Dr. Onions: We have a specific interest in developing chip-based systems for real-time detection of virus infections ultimately in fermenters. The central problem relates to issues of viral concentration. You are dealing with very small samples in these systems. We have done a number of theoretical checks on the way we might do this, and we are going to use our own in-house chip platform. This will involve total genome amplification to obtain the density we need, followed by screening for viruses, but also SOS signals. This will be a message-based system looking for total genome amplification.

I mention this because it is going to be quite a long-term programme, and we would very much like to do this in association with both the regulatory agencies and any interested company. If others are involved in this area, we would certainly like to share experiences and approaches.

Dr. Lewis: I thank David for those comments and, as part of the industrial sector, it is exciting for us to see that alternative technologies are being explored.

In my presentation, although I emphasized PCR, I did not intend that to be an endorsement of a method that is superior to any other. I think we can solve the degeneracy issue for probes, and that would be suitable for any detection system, particularly a high throughput system such as microarrays.

Dr. Carbone: Dr. Whiteman made the point that if the adventitious virus is in low amounts, then you may fail to detect it if you only test a fraction of the production. This is obviously important for polio vaccines. Only 20 or 25 percent of the polio vaccine production is tested for the presence of SV40.

You have seen yesterday from my presentation that some inactivation methods sometimes do not work. The virus that escapes inactivation is going to be in low amounts because even if the inactivation method does not work completely, it will considerably reduce the amount of infectious virus.

Now you are left with the problem that you may fail to detect low amounts, and yet these are most probably the ones that are going to be present. Is there anything that can be done to overcome this problem?

Dr. Whiteman: That is an issue because you cannot test the entire lot. I do not know that there is a solution but it might be to require that a certain amount of material be tested, although that is not really practical.

Dr. Nims: I think that the intention is to ensure that these assays can detect a gross contamination. A very low level contamination would not be detected. Your viral safety is ensured by having demonstrated viral clearance in your process.

Dr. Whiteman: I think the assays are all set up to allow a low-level contaminant to amplify. That is why we do passages, but that still does not solve the problem of not having sampled enough material to ensure that you will pick up that contaminant. I guess viral clearance is the other safety component.

Dr. Sheets: To some extent, that is the issue for this afternoon's panel discussion that will discuss the level of assurance provided by the current testing schema.

Dr. Lubiniecki: A question for Dr. Nims. I believe that you stated in your discussion of the finding of minute virus of mice (MVM) in some of your test systems, that it was believed to be of serum-borne origin. But I think others have implied, especially the group from Genentech, that it really was not clear what the ultimate source was and that therefore other precautions were needed on media preparation and chemical storage to prevent contamination by mice. Do you have additional information that describes this?

Dr. Nims: You are absolutely right. The Cache Valley and the REO were attributed to the serum. The MVM was attributed to environmental causes. I think that the likely scenario was that there were mice in the building that urinated on the packaging which was carried into the process somehow.

The MVM is one of the most stable viruses, and you can imagine it surviving on packaging or other environmental surfaces. It is feasible that it could have come by that route. It was not attributed to the serum.

Dr. Carbone: Once you detect an adventitious agent, what do you do, because the obvious implied answer would be that you have to discuss whatever preparation you have and at the same time, you should clearly identify what adventitious agent it is.

If, for example, a virus is detected, should you sequence it, so that you know exactly what it is, because only when you know exactly, can you possibly understand where it came from? If it happens again, you can prevent it happening.

I understand that many people have been told that, in fact, when something is detected, the preparation is to be discarded. I know for a fact that polio vaccine, for example, is often retested to see whether it happens again. I think that retesting should be absolutely wrong. Something should not be retested if you have the cell preparation in which some adventitious agent is causing vacuoles in cells. You need to identify what they are, and then I think it should be discarded.

Dr. Nims: I agree. My experience has been that that is exactly what is done. The affected lot is definitely discarded although materials are obviously retained for investigation, so that a likely cause can be assigned. Typically, there are bracketing studies done where previous lots are examined, maybe previous sampling points in the manufacturing process, so you can try to find out exactly where in the process the virus might have originated. My experience has been that these are all done.

Dr. Carbone: That was never done for polio vaccines in this report.

Dr. Sheets: I think there is a difference between what was done 40 years ago or what may be done in developing countries versus state-of-the-art manufacturing according to current good manufacturing practices that occur in the U.S. and other developed countries.

Dr. Duchene: Two questions for Dr. Lewis. First, is it irrational to test for bovine viruses when you are cultivating your cells in serum-free conditions?

Secondly, do you not think that it is better to look at the permissivity of your cell line to some virus instead of testing for absence of this virus?

Dr. Lewis: Our rationale for testing bovine and porcine agents was that the cell line carries a legacy of 30 passages in serum, and although we passaged serum-free for nine, we did not feel that we were free of that burden of testing.

Secondly, I would say that we were prepared to recognize that insofar as we had passed them that distance without serum, that PCR test that came up positive was likely to be the consequence of virus that was actively replicating, and not simply a genome that had come in with the serum itself.

We had considered the second question, although theoretically, many times, and we were unable to develop a scheme by which we would reduce it to practise in such a way that it would provide us with a level of confidence that would exclude the need for PCR testing.

We could take a cell line and go to BioReliance and say that we know you have a panel of 35 virus agents, would you systematically test the cell line for susceptibility? It is something we have considered as part of the file for final licensing and continue to contemplate.

When you begin to think about it, you should ask yourself: "Is my assay to detect infection with primary isolates, or to detect sensitivity for cell culture-adapted viruses?"

Most of the viruses that you have in commercial testing laboratories have been expanded in the appropriate tissue culture cell line, and you would expect that it has a tropism. Its preferred tropism is the cell line in which it was propagated, so it would be difficult to extrapolate from those infection data to the sensitivity of, say, the PER. C6 cells for infection with a wild type agent.

Dr. Nims: Sometimes this kind of testing is done on a limited basis, not 35 viruses, but some regulatory person might ask if your cells support the growth of REO type 3 or whatever. Then you would investigate that specific possibility, but not declare that the cell line is a host or not for a variety of viruses. It is really not practical to do that.

Dr. Lewis: When the disclosure of MVM contamination was made, we actually did infectivity studies of MVM on PER. C6 cells, not particularly because we are concerned about the release of the present material, but because if we were to identify the susceptibility to MVM, we would use methods during manufacture to sterilize media, for example, against MVM.

So, in this case, it was a concern about putting 10,000 litre fermenter lots in continued jeopardy without knowing whether MVM was a risk. It was an issue that was too obvious to ignore, and has manufacturing consequences, especially when you are handling large volumes of reagent.

Participant: It seems to me that if you are doing a PCR test, it is quite easy to put in a positive control and say my PCR test has picked up this, that, and the other. If you are doing a sterility test, I think as a matter of routine, you add a bacterium at the end of it to show that your medium was competent to grow it.

I wondered how you deal with that in the virus area. For example, if you are considering a suckling mouse assay, it seems to me that by the time you get to see that they are okay, it is a bit late in the day to inoculate them with a control virus. How do you actually tell that your mice are still susceptible, and how do you know how susceptible they are? Also, that would apply to cell culture assays as well.

Dr. Whiteman: We actually do not run positive controls through our assays. One of the reasons is because of the animal welfare issues. It is based on historical data that certain types of viruses do affect certain animal systems.

But there is the issue that, similar to what Dr. Lewis is talking about, there obviously are tropisms. The same is true in animals as would be true in cells. I guess none of the systems is perfect. We do run positive control viruses through our egg assays, but you are looking at a specific virus that is adapted to eggs in our laboratory.

Dr. Nims: I will add our experience with the in vitro assay which is an interesting philosophical question because it is a screening assay. What does it mean to ask "what is the limit of sensitivity of the in vitro assay"? It is meaningless, so the approach that we have taken is to select some limited number of viruses, do limited detection studies with those viruses and indicator cells for cytopathic effect, haemadsorption and haemagglutination. This is done to gather information typically by inoculating virus into medium. More useful information can be obtained if a client says we are interested in the limit of detection for virus X. That is straightforward, and we can provide a sensitivity for a certain virus by doing the limit of detection.

The data are collected for laboratory-adapted virus, which may not apply to an isolate in the wild. All these matters have to be taken into consideration. When a limit of detection is done, it has to relate to some loading amount. What is that? We do not have absolute numbers of particles but typically titre the virus in an infectivity assay, TCID50 value, or a plaque assay value.

The unsatisfying result is that the assay will detect 10 TCID50s or 10 plaque-forming units. What does that mean? All this comes into question when you talk about validating the in vitro assays [i.e. tissue culture].

Dr. Moore: In the early days of the development of cell characterization guidance, there was a feeling that human cell substrates represented a greater risk for a production substrate, and to some extent that may be related to the wake of surprises with the human retroviruses.

Does that concept still apply today? Is there a heightened concern relating to human production substrates rather than rodent substrates?

Dr. Sheets: The initial thinking was that because viruses that are pathogenic for humans would grow in human cell substrates, they would therefore be risky.

However, the trend moved towards using human diploid cells. Probably in part because they would be susceptible to the human viruses, if they were contaminated, you would probably detect them. The regulatory agencies evaluate whatever cell substrate is proposed for use, and certainly there are many human cells that are used for production.

I think one residual concern about rodent cell substrates is the fact that they spontaneously generate retroviral particles which means that the manufacturing process has to clear them. If you want to make a live attenuated vaccine that is minimally purified, a rodent cell is probably not the best substrate to use.

I think the preferences are driven by the product. If you are trying to grow a vaccine virus, you have to grow it in whatever it will grow in. I think that there are other considerations rather than a priori preferences guiding the decision of what cell substrates are appropriate for your vaccine preparation.

Dr. Taffs: I think Dr. Sheets is absolutely right. Being with FDA, I should say I can only discuss my opinions, and I cannot speak for the agency, but I will say that it is easy to be lulled into a kind of overconfidence that comes from justified trust in our historical evidence of the safety of biological products based on testing for adventitious agents.

But the overconfidence should not prevent us from consistently taking a new look at, for instance, novel tests that are under development, which might actually give us an opportunity to answer exactly the kinds of questions that are being raised at this conference.

If we do not choose to be overconfident, and instead base our clear understanding of the limitations of the tests that we have, we can put these new technologies into context and say whether or not these new technologies can be developed and can be used for regulatory purposes.

I want to move ahead with courage and say that providing the statistical and scientific justifications for the new technologies is where the rubber meets the road.

Participant: In an attempt to address Dr. Lewis' question about whether or not the susceptibility of the PER. C6 cells should be investigated, I would add that about 20 years ago, I was part of a group that was wondering the same things about Chinese hamster ovary cells. We ultimately ended up doing such a susceptibility study, targeting potential human pathogens and also potential bovine pathogens that might arise through raw materials. Basically, we found that not too much grew in CHO cells except a handful of human paramyxoviruses and REO viruses, and we designed all our systems accordingly.

We overlooked, in retrospect, the notion that parvoviruses would be an issue, and so did all the experts. Now, in retrospect, my friends in the veterinary biologicals industry tell me that we should have thought about parvoviruses because they are everywhere.

Susceptibility testing can be useful, but you can only draw the conclusions about the things you deliberately test for. So my advice is that you need to be very careful about what sorts of things you think you want to have concerns about and what kinds of controls you want to put in place based on that information.

Dr. Lewis: I would agree with that entirely. I reviewed taxonomy according to the most recent Fields [Virology], and when you think about the plethora of bunyavirus agents, if one does not replicate, does that mean that another will not? How much probative value is there in that? Directed studies that have manufacturing consequences are clearly of value. We might learn, for example, that HSV-1 would replicate. Could we then conclude that other classes of herpes viruses would also replicate? There is a restriction to the value of a broad survey.

I would agree that an MVM type agent-specific profiling is of advantage. In the development of biologicals, there is an evolution in process development and sophistication.

I described the context of PER. C6 for the manufacture of Ad-5, and I did not tell you that our confidence in product release was related to the fact that we had included several steps that we know to be highly lytic for enveloped viruses. Process validation and process clearance will help to sharpen that panel for which we would look for residual adventitious agents.

If you can get rid of most of the enveloped viruses, not many are left to consider. One can then test for very specific non-enveloped viruses, of which there are far fewer than there are bunyaviruses.

Dr. Wright: Are we at a point where we can begin to replace some of the in vivo tests with new tests? The efforts to try to substitute tests for polio neurovirulence, were somewhat instructive of the difficulty in doing this. There was a long period of development and validation of a neurovirulence test with interspinal injection in mice, and then correlating the results with sequence changes and identifying the percent change.

Dr. Taffs: It took eight years, which is record-breaking time. It is a long process, but it was also fantastic work.

Dr. Wright: There is a certain percentage of failure in the in vivo tests, (probably no more than five percent). I suspect if you go through six tests with any individual vaccine, the likelihood of a vaccine failing one test out of the six is somehow added to the failure rate with the individual tests.

Ms. Whiteman: They are difficult tests, and where we can find ways to get through them, the question is what happens to the sensitivity of the assay? For example, if you have to dilute your material 1 to 1,000 to get through a suckling mouse assay, maybe that is not the assay that you want to use for that particular test material. We try to determine what is appropriate on an individual product basis. Sponsors will often get feedback from the FDA before actually starting an in vivo assay. It is important to recognize that these are screening tests, not necessarily for the viruses that we know exist, but for ones that we do not know exist.

Dr. Sheets: The additional standards for vaccine production were specific for measles, mumps, smallpox, etc., and the test methods are more generally used for cell substrates, viral seeds, and vaccine harvest, etc. The additional standards were revoked because the test methods and the production methods that were promulgated in those regulations were subsumed in the licensing process. The problem is that there is no longer a place where FDA has promulgated the test methods. The EU test methods are slightly different, so sometimes that is a problem.

Phil Krause indicated that the FDA has a draft guidance document that may be published later this year. I believe that document will repromulgate those test methods.

Dr. Sekura: There are slight differences between the European protocols and the U.S. requirements, and I think that might be a good place to start to accumulate some data to harmonize the requirements, because many companies are doing repeat testing to satisfy both sets of requirements. If we could provide data, so that only one set of tests would need to be done, that would be a major improvement.

Dr. Sheets: Because we do not have a good quantitative measure of sensitivity of the tests as performed one way or the other, it is very hard to compare which is the more sensitive method or which is "better". In part, I think it depends on what viruses you are looking for. Is IM inoculation adequate or do you really need to do IC and IP? It depends on whether you are trying to find enteroviruses or something else. I do not think that this has been statistically looked at in the modern era.

Dr. Chumakov: I want to address the question of whether we replace in vivo tests with something more advanced. I think that for the current products, it is not possible. As an example, I can point to the neurovirulence test for polio. Despite more than 10 years of experience working with a number of countries, under a WHO collaborative project, we were unable to convince everyone that this method can be a replacement for the monkey test, even though scientific evidence clearly shows this is more informative and more sensitive.

However, animal tests are done not only for scientific reasons. There are a number of different considerations including historical and legal ones. For current products for which animal tests have been used for many years, it may not be possible to introduce a new method, no matter how informative it is.

Nevertheless, I think that we can use alternatives, but only for new products. Products for which animal tests are being used currently will continue until they go off the market, which hopefully will happen soon with live polio vaccine.

Dr. Murphy: Many vaccines are destined for use in very young children. Do we really **not** want to do an animal study before we give a new product to a human? I certainly would feel it to be safer to have studies showing that a particular preparation has not killed a mouse or a guinea pig before I gave it to a child. I do not think that is an unreasonable thing to do.

We could probably rule out all known agents with sophisticated tests, but you cannot rule out unknown agents unless you do some in vivo screening for such agents as in novel vaccines. Ultimately, Phase I trials are safety tests, not just for known agents, but also anything unknown that is in those preparations. You look at those children for expected reactions, but also for unusual things that might be associated with contaminating agents. You really do want to have some assurances that you have done at least a minimal amount of safety testing before you proceed to that step.

I think the minimal amount of testing that is required currently for the characterization of an investigational agent before it is brought into a susceptible human population is not unreasonable at the present time.

The other reason you do all these tests is that mistakes are made. Much of this testing is done with those unusual possibilities in mind.

Dr. Sheets: Several important issues have been brought up. One is that these tests do have limitations, and we have to acknowledge that any single test is not the end-all and be-all. Another redundancy is that the listing for adventitious agents is applied not only to the characterization of the cell bank, and master cell bank, but to the working cell bank as well.

There is testing of the viral seeds, master viral seeds, working viral seeds, and then of the production lot. Some tests may be repeated as many as five times by the time you have a production lot.

You are counting on redundancy of testing. But once you have established that it is not in the cell bank or the viral seed, and it cannot be introduced during production, then you would not necessarily need to test for it during production, or you might not test the working cell bank or the working viral seed.

I think we really need to be rational about how we are designing our testing systems. Let us not just use a checklist approach, but be rational and think through what is appropriate for any given situation.

While we might not be able to replace the in vivo tests, we might be able to eliminate the need to do them on a lot-by-lot basis. I think that is generally the case for most of the therapeutic biologicals where they only do the in vitro tests during production, so the cell bank and the viral seed might be characterized in these tests, but then lot-to-lot might not have to be done. If we cannot eliminate these tests altogether or replace them altogether, can they be replaced at certain stages of production?

A corollary to that is the issue of where is the most sensitive place to be testing for these agents? Where would they most likely be found? If they cannot be passaged in the cell bank, would they be more likely to be found upstream or at the master cell bank level if they were there at a low level during production?

All those things need to be taken into consideration in putting together a package of what is appropriate testing rather than just a checklist. We are getting more and more specific PCR tests. If we are doing the broad general tests **and** we are doing all these specific PCRs, is that overly redundant? Do we really need to be doing all of them? We are not eliminating any tests, we are just adding to them. It increases the amount that has to be tested, and the likelihood for some kind of failure that has nothing to do with contamination, but simply a failure in the test system. It increases the time that it takes to release lots. We have to be mindful that at some point, we cannot just keep adding on tests. We really need to be rational about which tests are complementary, what is a package that needs to be put together and what things can be left out.

Dr. Lewis: I would encourage an approach that considers the totality of the manufacturing process. For example, if you can manufacture in an entirely closed system from the moment you take your cell vial out of cryopreservation and the vial for seed, and you do it under serum-free conditions, you could ask why should I test the harvested virus bulk using this panel of in vivo tests? Rather than asking, "Can we replace all these in vivo tests throughout the spectrum of testing", you can be much more case-specific and phase-specific and ask at what level does this not contribute to my confidence in the freedom from adventitious agents?

Dr. Wright: That is really a decision the manufacturers make, perhaps in some prediscussion with the FDA. Nothing says that you need to have injected 20 suckling mice. You could conceivably bring a product forward for which you could make a strong argument that it had been tested with other methods. Is that correct?

Dr. Lewis: I agree. Sometimes in this business it is easier to do it than to contest it.

Dr. Wright: We really are dealing with a series of tests that arose from the earliest days of diagnostic virology, and we continue to identify new viruses that either grow very poorly in tissue culture or not at all.

Dr. Lewis: I would just say that in using PER. C6, we made as vigorous and complete an attempt as possible to understand the medical history of the mother and father of the foetus, that was isolated in Europe in 1985. Although the records were available, there were clearly issues of confidentiality that prevented us from knowing, for example, whether the mother was a chronic hep-B carrier at the time. That goes to the issue of the tissue of origin which is a critical aspect in the development of a panel for testing.

Dr. Nims: The beauty of the screening assays is that they are there to catch a new virus when it occurs, assuming that it can grow in the host cells that we use and cause an effect. Take for an example West Nile, which appeared five years ago. West Nile shows up well in an in vitro assay, so you need those kinds of screening assays to pick up a new virus, and then once that is identified, you develop the PCR test. Add that to the spectrum of testing, if you feel it is necessary. I think that the old and new tests are always going to be going hand in hand.

Dr. Onions: Cache Valley virus is a highly lytic virus which has caused crises in big fermenters on three occasions over the past few years. It does worry me in the context of vaccines that might be prepared with unirradiated FBS, because we did a one-step growth curve on this virus, and back-calculated how much virus went into the fermenter. In a 20 litre pool of serum that went into one fermenter, no more than between 10 and 100 virions went into that fermenter. If you then look at what happened, we could detect it by RT-PCR at about day four because there was a defective to infective ratio of about 400 to 1, but we would not detect infectivity until about day five.

Of course, there was infectious virus there on day four, but we would not have picked it up because of sample size. If that had been a live virus vaccine system, you could well have had infectious bunyavirus in your system, and you would not have detected it by the usual routine methodologies. We ought to think of that kind of issue in terms of the way that the virus infections occur in these systems.

We know very little about these circo viruses, but we do know, for instance, porcine circovirus will replicate in HeLa cells. In our experience, it does not appear to replicate in other human cell substrates. We know there is a murine circovirus and, like parvovirus, it is very resistant. They are resistant to irradiation and they are resistant in the environment. So those mice that are busily urinating on the bulk amino acids or on the packaging are probably almost certainly also putting out murine circovirus. Therefore we have another issue that might emerge, maybe not in the vaccine industry, but certainly for the hybridoma industry.

Dr. Wright: I am not absolutely sure that the exogenous contamination is usually bacterial, but I certainly have had significant experience with it. This is more of a potential problem in early development. It certainly can be at a very low level, and might not be picked up in the usual screening.

There really are two things that go on in the testing of a preparation. One is the cell substrate, but there are also tests that are applied, particularly if it is a live viral vaccine, to the vaccine itself, to look at the safety of the active component. Is there some redundancy in the context of toxicity testing related to a vaccine as opposed to adventitious agents related to the substrate ? Neutralizing the active component is important when testing adequately for other agents.

Dr. Nims: That is certainly an issue. Typically, what happens is that a sponsor will suggest a titre of the infectious virus vaccine that can be handled with the antiserum and the procedure. Sometimes that may require a significant dilution. The reality is that you can test, so that has been acceptable. There are limitations to the power of the neutralizing procedures to inactivate the virus.

Dr. Lewis: Although you might think that the antisera are poly-specific, are they really poly-specific ? You never really know how many agents you are looking for that you are neutralizing with the serum that you raised. You can certainly do qualification experiments, but there is a level of practicality that we have never decided how to deal with.

Dr. Nims: You can never state categorically that you are not neutralizing a virus other than your vaccine virus.

Dr. Murphy: Some of the best safety data from the cell substrate concern live attenuated virus vaccines, because these are not very highly purified. When these are given to humans; that is the ultimate safety test. In that circumstance, you have a considerable amount of safety information. I think that is one of the reasons we have considerable data on the safety of Vero cells and other continuous cell lines.

Dr. Wright: A fairly good post-marketing check is in place on safety and sometimes very specific requirements are put on a vaccine to be tested. My impression is that most of them have been aimed towards the safety of the active component of the vaccine, as opposed to the safety of the substrate.

Dr. A. Lewis: The overlap that I think is important to recognize does provide additional information in the sense that any system that allows an infectious virus or contaminant to be amplified will always be more sensitive than a snapshot assay such as the PCR which can only detect something at one particular time. So, amplification is one of the keys to having a successful safety evaluation.

The issue of how you monitor populations of vaccinees over a period necessary to determine whether there is an adverse reaction, especially the development of some neoplastic process, is a difficult one. I think the best example of that is the example of SV40 in polio vaccine. I reviewed those data in 1973. At that time, there was no evidence that there were any problems associated with people who had received polio vaccines. It was not until the early nineties that we began to see the consequences of that process.

So, you are looking at a follow-up period of maybe 20 or 30 years, and there is just nothing in place that I am aware of from an epidemiological or public health point of view, or any other type of monitoring mechanism, to follow up people who have been vaccinated or immunized with these types of products for that length of time.

The mechanisms in our public health system, as far as I am aware, do not exist.

I do not know how to address that, but I think it is an issue that we may have to confront at some time.

Dr. Carbone: About the antiserum inactivation method, some time ago I was reading some papers written in the early sixties that state that antiserum will never inactivate 100 percent. The inactivation at most went to 99 or 98 percent, but there was always a fraction that remained unbound, so it was recommended that, for the inactivation to work, you needed to repeat it. Is this still true 40 years later?

What is normally done is to neutralize the original test article and to include some level of antiserum in the medium during the entire study, so that you can continually neutralize any new virions that are propagating.

That means for example that in a seed which is contaminated, an antiserum has to be added to cells in which the virus could grow, and that you need to continue to add antiserum.

Dr. Nims: What is normally done is to neutralize the load material, the original test article and then to include some level of antiserum in the medium during the entire study, so that you continually neutralize any new vaccine virions that are propagating.

Dr. Wright: From a clinical perspective, a problem is that an HIV vaccine lot may have to wait six to nine months because of the issues involved in safety testing and regulatory clearance before it moves into clinical testing.

Is there any way of shortening this process? I think we all come at it with renewed vigour each time we approach it, and then look back a little later and find that it has taken the same length of time. This is an issue where you are trying not to have many different approaches, and to move them rapidly through an evaluation to discard approaches that are not going to work as quickly as possible while keeping the pathway of vaccine development as linear as it can possibly be.

Dr. Sheets: Part and parcel with considering what tests might supplement or complement each other is to think through the issues of sample size and what is the most sensitive test. Sometimes that may be PCR, albeit you may be detecting non-infectious genomes. Sometimes that may be a biological test. You have to think in terms of what the technical limitations of the test are, what is the volume you can test, and what is the sensitivity of the assay.

Even in considering using PERT versus conventional RT testing for vaccines, one consideration that was taken into account was the fact that both types of assays can be inhibited by components that are in a cell lysate. However, one assay is still a million-fold more sensitive than the other. Even if you have inhibition, you are still better off than if you relied on the less sensitive assay.

Dr. Lewis: In my presentation, I expressed sensitivity as genomes per cell number, and it is clear that the denominator is specific for each vaccine product. If your vaccine dose is 10^{11}, and you have 10^3 particles per cell, you can see you have not even tested one cell dose equivalent of vaccine. By contrast, if your vaccine dose is 10^4, and you have 10^4 replicating infectious viruses per cell, it is a five-fold order of magnitude sensitivity in the declaration of freedom from adventitious agents expressed per vaccine dose rather than cell substrate volume.

Dr. Chumakov: The protocol that we use in biodefense projects for detection and genotyping of smallpox and anthrax is exactly the same as we use for SV40 in vaccines or in cell substrates. There is no difference because you detect something that you know and something that you do not know. It is exactly the same task in both cases.

There are some protocols, for instance, such as molecular colony PCR, that is insensitive to the presence of a huge excess of cellular DNA, so this is something that nobody has explored yet, but I think there are other protocols that can overcome this problem of a needle in a haystack.

Dr. Onions: There are some methodologies now for concentrating virus particles or cells, according to how you use the system. One of these systems that we are looking at is travelling wave di-electrophoresis. This can be implemented on the chip-based system. Eventually we would like to have a chip-based concentration system and then marry that up to an array system.

Dr. Taffs: I would like to reiterate that we are dealing with a number of historical and emerging issues and trying to wrestle with them simultaneously. We have some confidence that comes from our understanding of the safety of the products that we make or that we regulate. However, we are always looking for better tests. These are some problems in trying to achieve acceptance of an emerging technology.

I think we need to be very clear about why any of the new technologies are successful and could be an improvement over the ones we already have in our hands. As these new technologies emerge and move forward, we also need to be clear about the advantages that the new technologies have over our existing methods.

Dr. Murphy: One of the major points that came out today was the fact that the redundancy in the tests provide some reassurance that the products you bring to Phase I trials can be brought there safely.

The combination of bacterial, fungal, mycobacterium, and in vitro tests for viruses, and in vivo tests, provide a relatively large body of information that can be generated relatively safely and reasonably quickly to bring investigation products to Phase 1 trials. There is room for improvement, but it is currently serving us well.

Dr. Whiteman: It is important to remember that not only are these screening assays, but they are amplification assays. That is an important aspect of both the in vitro and the in vivo tests, where we are putting the test material into a system and allowing it to grow to a level where we can detect.

As far as eliminating some of the tests, I think that is an idea that should be looked at in concert by manufacturers and regulatory agencies to determine which tests are appropriate for a particular vaccine or a particular cell line.

Dr. Lewis: It is hard to overestimate the seriousness with which a sponsor approaches a Phase I clinical trial, particularly with a substrate for which he has only a modicum of information, despite his better efforts. In that regard, redundancy of assay testing adds reassurance. Could we accelerate product development by the elimination of some of these tests?

If I were to say that including a test for PCR and a test for mycobacterium with a 50-fold increase in the amount of time it takes to execute the test, it really adds very little to the overall timeline of developing a bank and releasing it, which we estimate at Merck to be at least 18 months. The testing is really a minor percent of the total time required.

So, it is important to keep in mind that when we try to abbreviate tests, we do so, but not because we accelerate the process. The length of that process is determined by the complexity of establishing contractual relationships with contract laboratories, document development, auditing, GMP inspection, and everything else. So, there is an enormous time component which is unrelated to the duration of a specific test.

It is the mechanism of developing the panel, developing a relationship with a sponsor, document exchange, getting the correct wording, and getting lawyers to agree, that really takes time.

Dr. Sheets: One issue is the recognition of the limitations of the testing that we do. For many tests that we have been doing for 50 years, we do not have the sort of validation of sensitivity that we would have for a more modern test.

Our sample sizes are rather limited. We often do not have a susceptibility profile for the substrate to guide us in deciding which agents we really need to focus on. Although that is useful, there are some pragmatic limitations to doing that, and, of course, you are never going to pick up what you did not test for.

I think reaffirming that in large measure we would rely on redundancy of testing, and I think that aspect will be further discussed this afternoon at the level of assurance session.

However, I also want to bear in mind that many of the things we do have been done throughout the history of viral vaccines produced in cell culture, and so we always need to be thinking rationally about why we are doing this, whether we are doing it because it is on a checklist and we have done it for 50 years, or whether this is really contributing and adding something.

My final point is that broad general tests are always going to be a better screening or catch for things than an array of very specific tests.

So, I agree that we cannot get rid of the broad general tests and just go to all PCR, because we are not going to detect the unknown viruses that way, but I think we do have to consider whether each of the various in vivo tests really adds to the tissue culture in vitro tests that are currently done.

Dr. Nims: There is value in redundancy, and there is still some complementarity to the different methods, and that has value as well. There is always going to be a need for both the screening type methodologies, as well as the specific methodologies. A good example of that is when you investigate the occurrence of a positive viral contamination. Typically, what happens is you will detect that agent using a screening assay, and then you will go in with PCR, a quantitative PCR, and look at different parts of the sample to find out where the contamination occurred and how much there is at the next step. Both techniques are very useful.

Summary Statement

The origins of the present in vivo tests go back to the earliest days of diagnostic virology, and many come from attempts to standardize preparations of poliovirus vaccine.

In vivo and tissue culture tests provide an opportunity for biological amplification, which a PCR test will not.

The current systems incorporate redundancy, but do not detect all known viruses. Nevertheless, tests could be looked at critically, and if one test is 50 or 100 times more sensitive than the other, having two tests may not really yield additional benefit.

In addition, it was pointed out that the volume of material that can be tested also limits assurances that all agents are detected. The volume tested is fixed and not related to the size of the bioreactor or the size of the lot being prepared. In general, one tests for agents present in 10^7 cells.

The individual assays have a certain defined failure rate of generally less than 5 %. It is important to recognize as multiple tests are introduced on an individual preparation or cell lot that the chances of disqualifying or having to retest a lot may be greater than five percent because the chances of failing one out of six assays is likely to be more than the chance of failing an individual assay.

The catalogue of potentially contaminating viruses keeps changing and makes certification of a lot more challenging as new agents are discovered and new assays are developed.

In fact, viral contamination of lots has rarely been found. Specific examples of contaminants include reovirus and Cache Valley virus.

It was pointed out that reagent and manufacturing controls decrease the likelihood of contamination, and this is probably at least as important a part of the process as is the actual testing of the end-product.

Testing can and should be targeted at agents that may be present, taking into consideration the history of the cells and the type of preparation being tested. The current menu of tests, either in vivo or in vitro, is not fixed, but rather a smorgasbord that can be altered with validated newer assays. There was a real interest in the introduction of new assays. However, replacing old assays with new ones will require comparison of assay sensitivity and other validation parameters. The lack of validation information on the existing (old) assays limits our ability to do such comparisons.

The history of the current tests is clinically very reassuring, and many new vaccines have been tested successfully based on the safety screening currently in place.

Regulatory requirements and testing vary somewhat between Europe and the United States, and there was strong support for harmonization.

PCR will clearly be a useful adjunct in defining new agents and in providing redundancy or additional assurances on the freedom of recognized agents. It was pointed out that it is possible to screen for almost all the known DNA viruses at present.

It is fairly clear that experience with new cell substrates and perhaps with new vaccines will drive innovation in the field, and it was pointed out how difficult it is to change existing procedures, for example, the testing for polio neurovirulence.

There is not yet a good mechanism in place for long-term clinical safety monitoring of vaccine substrates because it is extraordinarily difficult to devise a way to obtain meaningful long-term safety information on cell substrates or vaccine in the presence of many confounding variables, as well as the length of time such follow-up studies would take.

Petricciani J, Sheets R (eds): Vaccine Cell Substrates 2004.
Dev Biol (Basel). Basel, Karger, 2006, vol 123, p 199.

SESSION IV

Level of Assurance Provided by Guidelines, Especially ICH Q5A and Q5D

Petricciani J, Sheets R (eds): Vaccine Cell Substrates 2004.
Dev Biol (Basel). Basel, Karger, 2006, vol 123, pp 201-205.

Basis for Sample Size in International Conference on Harmonization Guidances

A. Lubiniecki

GlaxoSmithKline Pharmaceuticals, King of Prussia, PA, USA

INTRODUCTION

This article describes the origins of the major recommendations for cell and virus testing, especially from the International Conference on Harmonization (ICH) guidelines. Two ICH documents are relevant: Q5A [1], which is the viral safety document; and Q5D [2], which is the cell substrate document. I was the rapporteur for the cell substrate document Q5D, and Dr. Takao Hayakawa was the rapporteur for Q5A.

An important concept in both documents is the level of assurance. Whenever we perform tests, we are always interested in knowing exactly what the test is telling us both in the way of positive and negative information. The intention of both those documents is to select a battery of tests to minimize the potential for biologically active agents to contaminate products and to prevent the transmission of such agents to product recipients. Notice that this statement of intent is not necessarily that the tests would have the power to preclude it totally, but rather to minimize it.

ICH Q5D, the cell substrate document, became official in 1998. The scope involves the use of animal or microbial cell substrates, which have cell banks, and which are used to prepare biological or biotechnological products.

The cell substrate document was not intended to represent a complete description of in-process testing, lot-release testing, process validation, or cGMP controls that are typically part of the manufacturing process for biological products. Q5D was not intended to address these other means of control.

It was limited to the features of testing and controls that are associated with cell banks and cell banking.

Specifically, Q5D covers a number of concerns such as the potential for bacterial and fungal contaminants, the potential for mycoplasma contaminants, and the potential for other cells that might have been in the research laboratory to contaminate the cultures.

Viral contaminants were specifically not dealt with in this ICH document, but readers were referred to another ICH document, Q5A, as discussed below.

Bacterial, Fungal, and Mycoplasmal Contaminants

The ICH guidance suggests that both the master call bank and the working cell bank should be tested for the presence of bioburden. It suggests that one percent of the containers in the cell bank, but no fewer than two containers should be tested.

This guidance was established because there were interpretations of some of the FDA guidances that were circulating in the middle 1990s stating that as many as 20 containers should be tested. The number 20 came from the number of test articles generally consumed in performing a sterility test on sterile products according to the U.S. Pharmacopeia [3]. However, some cell banks were not very large, and to take 20 away from a low number left very few to use. It was therefore considered that one percent was adequate. However, some modest level of replicate sampling was felt to be useful. So, testing two containers allowed replication without being burdensome or wasteful.

Why was testing two vials or 1 % considered adequate? It is important to note that there is no requirement in this ICH Q5D document to demonstrate the sterility of master and working cell banks, because it was recognized that sterility, viewed in the context of the Food, Drug, and Cosmetic Act, and the United States Pharmacopeia, is not demonstrably achievable with cell banking operations because it is not possible to validate the cell banking operation as aseptic processing. Cell banking is basically a manual operation, and for those types of activities, it is generally considered impossible to achieve aseptic conditions that would normally be a prerequisite for providing a product (in this case, a cell bank) that we would typically term "sterile". Further, cell banks are not intended for parenteral use in humans, but rather as a key starting material for production. That does not mean that the presence of extraneous microbes in cell banks is a good idea, so there certainly was a desire not to have the banks be contaminated. However, bioburden, rather than sterility, was what we were describing in a testing sense, i.e. the presence or absence of small numbers of bacteria or fungi. Therefore, it was agreed to recommend bioburden testing (which was achievable and meaningful), rather than to attempt to demonstrate sterility (which was not really achievable).

It also was stated in the Q5D document that the current methods of the European Pharmacopeia (EP), the Japanese Pharmacopeia (JP), or the U.S. Pharmacopeia (USP), for testing microbial limits or microbial sterility may be considered adequate for that respective regional regulatory authority. Here it is important to distinguish between "sterility" as an intellectual concept or a regulatory requirement versus "sterility" as the name of a compendial test method. The latter meaning of the term is used here. At the time the Q5D document was written, it was necessary to do each different compendial test for each respective region in which one was intending to submit the market authorization.

For mycoplasma, a similar approach was taken. The thought was that we were going to test single containers, that there was not much advantage to testing more than one container and, as with the bioburden testing, it was thought that the current methods of the EP, JP, or the FDA Points to Consider for testing for the presence of mycoplasma would be considered adequate by the respective regional regulatory body. The USP at present has no description of mycoplasma testing.

It was recognized that science would evolve and the methods would evolve. As I mentioned, both the bioburden and mycoplasma test methods were known to be region-specific at the time, but there is also wording in thc Q5D document that states if the pharmacopeia and the regulatory bodies succeed in the future to harmonize methods for testing for the presence of bioburden or mycoplasma, then the scientifically appropriate harmonized assay should be used. So, Q5D does foresee the day when it would be possible to do one bioburden test and one mycoplasma test acceptable in all three regions, and the Q5D recommendation is to use the harmonized protocols when they exist.

Non-Host Cell Contaminants

The Q5D document states that those methods which are used to identify or to assess cell identity are considered adequate to ensure the absence of other oganisms, but care should be taken to prevent cross-contamination.

Those familiar with these cell identity tests know that it is possible to detect some kinds of contaminating cell lines by these methods, but even when that is possible, they are probably not very sensitive.

The real goal is to prevent contamination of the cell banks by other cells, and procedural and engineering controls will be extremely important in achieving this goal, and the value of those controls exceed the value of testing. The test itself should not be relied upon as the only means of minimizing the potential for contamination.

Testing Adequacy

Testing of cell banks or testing of any other test article is really only the beginning, and it is only part of the process of ensuring freedom from contamination and of the product from contamination. It is very important to realize that these tests are not intended to be self-sufficient for protection of the product. For example, there are regulatory requirements for testing pre-harvest cell cultures for mycoplasma, bioburden and viruses. Some processes also look at bioburden levels at many stages of cell culture, e.g., when the cells are being amplified and when the product is in purification. For bioburden, there is also the testing of drug products and drug substance. So, the testing of the cell bank for bioburden or mycoplasma is certainly not the last time that this will occur during processing, and the tests, although limited in their ability to provide assurance of freedom from contamination, are neither the only tests nor the last tests that are being done.

In addition, there are other controls which are possible around the manufacturing process under what we would generally call GMP and, finally, there are multiple levels of testing and protections which exist for preventing final product contamination. Therefore, the rigour of cell bank testing does not need to be exhaustive, and the limited sample sizes recommended are adequate for the purpose of minimizing the likelihood of using a contaminated cell bank in production. In one report of industry testing results over a 15-year interval, Moore [4] found less than 1 % of cell banks submitted for bioburden testing to be contaminated, typically at low levels. So the existing tests seem to work, and occasionally detect what they are designed to find.

Virus Safety

ICH Q5A, the virus safety document, also became official in 1998. The scope is somewhat similar although there is a difference in that Q5D focused on cell banks (for microbes or animal cells), but Q5A focuses on biological and biotechnological products which are derived from animal cells initiated from the cell banks and which have been characterized. The products include purified proteins, natural proteins, recombinant proteins, or monoclonal antibodies (whether these cells are cultivated in vitro or in vivo), and also recombinant-derived vaccines. It does not, however, cover conventional vaccines or gene therapies, but many of the same principles are certainly relevant although they do have to be adjusted because of the nature of the vaccine products or processes.

Viral safety of the products can only be assured by application of three complementary approaches which were well recognized. These include selection and testing of cell lines and raw materials, assessment of the capacity of the production process to clear infectious viruses and, finally, testing the product at an appropriate stage of production. The ICH Q5A approach to virus testing is one based on those tests that were available in the 1990s. Manufacturers were encouraged to be flexible and to base testing plans on the details of the process, the types of viruses likely to be encountered, and the most sensitive point of sampling. Manufacturers were encouraged to be open to newer tests which proved to be sensitive and reliable.

The approach to clearance was that clearance studies would serve as a complementary approach to testing. Together with testing and routine procedural and engineering controls, clearance studies on well-designed processes would ensure freedom from product contamination by infectious viruses. Clearance approaches, like testing approaches, should be tailored to the process and product, and integrated.

For concerns about virus contamination of biological products, testing cell banks is only the beginning. It is part of the regulatory requirements not only to test the cell banks, but also to test pre-harvest cell culture. Raw materials of animal origin should also be tested before use. Some processes also look at the level of contamination of cell culture just before harvest and purification by rapid PCR tests. These are especially used in some organizations for the detection of minute virus of mice [5].

GMP controls are also very important throughout the process and, finally, there are multiple levels of testing and protection which exist to prevent product contamination; therefore, the rigour of cell banking or pre-harvest bulk testing need not be exhaustive.

These guidelines are very effective. The principles are built basically on the last 100 years of history in this business, and the use of these procedures has detected contamination in a few facilities before the products were released for distribution, so the controls work [4,5]. Most importantly, these programmes have allowed tens of millions of product recipients to receive safe and effective biotechnology medicines without evidence of associated infection during the last 20 years.

References

1 Viral Safety Evaluation of Biotechnology Products Derived from Cell Lines of Human or Animal Origin. International Conferences on Harmonization. Federal Register 1997; 63: 51074-51084.

2 Derivation and Characterization of Cell Substrates Used for Production of Biotechnological/Biological Products. International Conferences on Harmonization. Federal Register 1997; 62: 24311-24317.

3 Sterility tests < 71 >. The United States Pharmacopeia XXV, The United States Pharmacopeial Convention, Rockville, Md, 2004, pp 1686-1690.

4 Moore WA: Experience in cell line testing, in Esber EE, Williams M (eds): Continuous Cell Lines – An International Workshop on Current Issues. Dev Biol Stand. Basel, Karger, 1992, vol 76, pp 51-56.

5 Garnick RA: Experience with viral contamination in cell culture, in Brown F, Lubiniecki AS (eds): Dev Biol Stand. Basel, Karger, 1996, vol 88, pp 49-56.

Dr. A. Lubiniecki, Centocor, 145 King of Prussia Road, Radnor, PA 19087, U.S.A.
E-mail: tlubinie@cntus.jnj.com

Petricciani J, Sheets R (eds): Vaccine Cell Substrates 2004.
Dev Biol (Basel). Basel, Karger, 2006, vol 123, p 207.

The Impact of Technology on Assurance

W.A. Moore

Althea Technologies, Inc., San Diego, CA, USA

Abstract: Historical contamination events of biological products, notably vaccines, had a profound impact on the development of substrate characterization batteries and overlapping adventitious agent testing programmes for biopharmaceutical products. Devoid of data on the contamination rate for " new " vaccines and biopharmaceuticals, the best available methods have been applied, providing a very strong record of safety for recombinant biopharmaceuticals and new forms of therapy over the last 20 years. However, it should be noted that the testing approach applied has often reflected the range and practical limitations of the methods and not the over-riding application of quality systems or risk-based evaluation. In this same period, significant progress has been made in the development of rapid PCR-based detection systems capable of sensitive, quantitative and high-throughput sample analysis, micro-array techniques, and the development of sensor-based detection devices. Many of these technologies have been introduced in the testing of high-volume biological products, such as blood, and are currently being explored in the areas of bio-defence. Validation of these methods and regulatory acceptance of these techniques have been accomplished in specific applications.
An overview of the technologies applied, the constraints, practical limitations, and assurances afforded by the present testing regimens is available to explore the potential for the application of new techniques and a systems approach in the mission of advancing new vaccines and biologicals. The potential application of a new tool set as part of the traditional multifaceted approach to biopharmaceutical safety will be reviewed in light of new regulatory bearing.

Dr. W.A. Moore, Althea Technologies, Inc., 11040 Roselle Street, San Diego, CA 92121, USA.
E-mail: amoore@altheatech.com

Petricciani J, Sheets R (eds): Vaccine Cell Substrates 2004.
Dev Biol (Basel). Basel, Karger, 2006, vol 123, pp 209-216.

Statistical Bases for Defining Level of Assurance

T. Schofield

Merck Research Laboratories, West Point, PA, USA

Key words: Quality assurance, quality control, confidence intervals, probability of detection, operating characteristic curve, risk analysis

Abstract: Assurance of a low risk that material produced in cell culture harbours adventitious agents is provided through carefully conducted clearance studies, as well as quality control testing. Clearance studies are a form of quality assurance, and are used to demonstrate the ability of the manufacturing process to minimize or neutralize unwanted contaminants. Quality control testing is confirmation of a low risk that manufactured lots are contaminated with an adventitious agent. Neither of these can guarantee, however, that product released to the customer will be completely free of contaminants. Clearance studies provide information on log reduction of a contaminant, but by their nature cannot demonstrate complete elimination. In quality control testing, the ability to support the claim that there is minimal risk of a lot containing an adventitious agent depends on the quantity of material being tested and on the sensitivity of the detection assay. Statistical methods can be used both to design clearance studies, as well as control tests, to achieve a desirable level of assurance that the contaminant is below a defined threshold in the product. Basic data analysis methods, such as confidence intervals, improve the quality of information, by showing the range of adventitious agent reduction or level.

INTRODUCTION

Quality assurance should be differentiated from quality control, in defining the level of confidence that products produced in cell culture have minimal risk of containing an adventitious agent. Quality assurance is usually associated with process steps and safeguards that are aimed at minimizing the risk of an adventitious agent remaining in the final product, as well as carefully planned experiments that demonstrate the degree of reduction of risk in the manufacturing process. Quality control is associated with periodic testing of the product, either lot-to-lot or over time. As such, assurance is inherent to the manufacturing process, while control confirms that material sent to market is "free" from contamination.

A comprehensive quality system might use both of these levels of management. Neither, however, can completely eliminate the risk of a contaminant existing in the final product. The degree to which we can ensure that process steps and safeguards have reduced contaminant load, as well as that the final container product is "free" of adventitious agents, is inherently associated with the sensitivity of our assays to detect these contaminants. These assays in turn vary in their ability to detect an organism, due to the technology as well as the operating conditions of the procedure. Thus the level of assurance is directly associated with the ability of the analytical method to detect a pathogen of interest, or to reveal a variety of potential pathogens. Also since a contaminant may be intermixed throughout a process intermediate or final product lot, the ability to detect the agent will be linked to the sample volume, and thereby the capacity to capture the agent within the sample.

All the factors that contribute to the degree of risk can be assessed using carefully designed clearance studies on the process, as well as strategic validation experiments of process and product samples. The probabilities that a pathogen remains in an intermediate, or in a production lot, can be calculated using standard statistical tools. These tools are outlined in regulatory guidances such as ICH Q5A. The meaning and implementation of these statistical tools are not always clear, however, to the practitioner. The following sections will describe these tools, providing further clarity on their implementation and interpretation.

Clinical Bases for Assurance

Neither quality assurance nor quality control can provide absolute assurance that a product is free of a potential contaminant. That being the case, it seems prudent to take account of the impact that the contaminant will have on an individual, or the population of individuals receiving products produced in cell culture. Without recognizing this, industry and regulatory agencies must negotiate levels of assurance on behalf of the customer, and have no recourse but to drive down risk, sometimes to arbitrary, even ill-defined levels. The agreed threshold may be defined either qualitatively, such as low, medium, or high risk, or quantitatively, perhaps defining a maximum load or clinical cut-off, that has been demonstrated to result in minimal risk to the individual or a population of recipients of these products.

This customer-defined cut-off might either be obtained through dose exposure studies in animals, or observational studies in humans. Both strategies have their inherent strengths and weakness. Dose-exposure studies in animals can be carefully controlled, providing a precise estimate of the maximal pathogen load. A weakness of such a study is that the animal model may not accurately predict the pathogen load that presents a risk to humans. Observational studies in humans overcome the need to extrapolate from animals to humans, but suffer from their lack of control. These are also useless in finding a maximal tolerated level of pathogen due to lack of control of the level of exposure, and can only result in an assessment of the severity of response upon exposure to the pathogen.

Thus the clinical basis of assurance frequently provides no insight as to the target level of a pathogen that may result in either risk to the individual or population. This factor is nevertheless a critical factor in the assessment of risk for any pharmaceutical or biological product, and should be considered together with the likelihood of concern that a contaminant will occur in cell-based products.

Analytical Bases for Assurance

Two factors that affect the ability to detect a contaminant in a cell-derived product are the limit of detection of the analytical method, and the amount of sample taken of a process intermediate or a production lot. The limit of detection should be the primary parameter of a validation of an analytical method developed to detect pathogens in cell-derived products. The strength of this approach is that the validation can be statistically designed to obtain an estimate of the limit of detection, to a prescribed level of confidence. A weakness comes from the influence of different technologies and techniques upon the limit of detection for various procedures. Thus the exercise of determining the limit of detection of an assay often becomes more a characterization of the procedure, than verification that the procedure has attained a desired level of sensitivity. The amount of sample that is taken of a process intermediate or a production lot influences the probability of detection at low concentrations. Thus while there is a high probability of detecting a pathogen at high concentration in a large sample, there is a low probability of observing the contaminant in a small sample at low concentration.

Limit of Detection

The limit of detection of an assay for an adventitious agent is usually defined to be the level of contaminant that can be reproducibly detected in the assay. The validation study is designed with replicate titrations of one or more representative pathogens, independently diluted across a range of levels that is likely to include the limit of detection of the assay. The limit of detection can be determined as the lowest level that is reproducibly detected (e.g. in greater than or equal to 95 % of runs) in the study. A refinement of this calculation is to model the titration using logistic regression, and then determine the level that has satisfactory reproducibility (e.g. 95%) with suitable confidence (e.g. 95% confidence). While it is beyond the scope of this paper, it should be noted that statistical design criteria can be used to determine the level of detection, with prescribed confidence.

An issue related to defining the limit of detection of an assay for adventitious agents, is that different technologies have different limits of detection, while different laboratories running the same assay may have significantly different sensitivities. This is illustrated by considering the detection limits for two technologies, against a possible distribution of pathogen in a product (Fig. 1). Figure 1a illustrates the determination of the limit of detection for an assay with continuous readout. The limit of detection (LOD_1) is determined as the level of analyte associated with low risks (5% risk) of either observing a false positive or false negative. In Figure 1b, we overlay this with a possible distribution of pathogen in the product. The area under this curve which is greater than LOD_1 represents the proportion of times the assay will detect a pathogen for a product with the given pathogen distribution. This is very small in this illustration. In Figure 1c, we determine the limit of detection (LOD_2) for an assay with greater sensitivity. The area corresponding to the proportion of lots that will be discovered by this assay, and thereby the proportion of lots that will fail, has increased significantly.

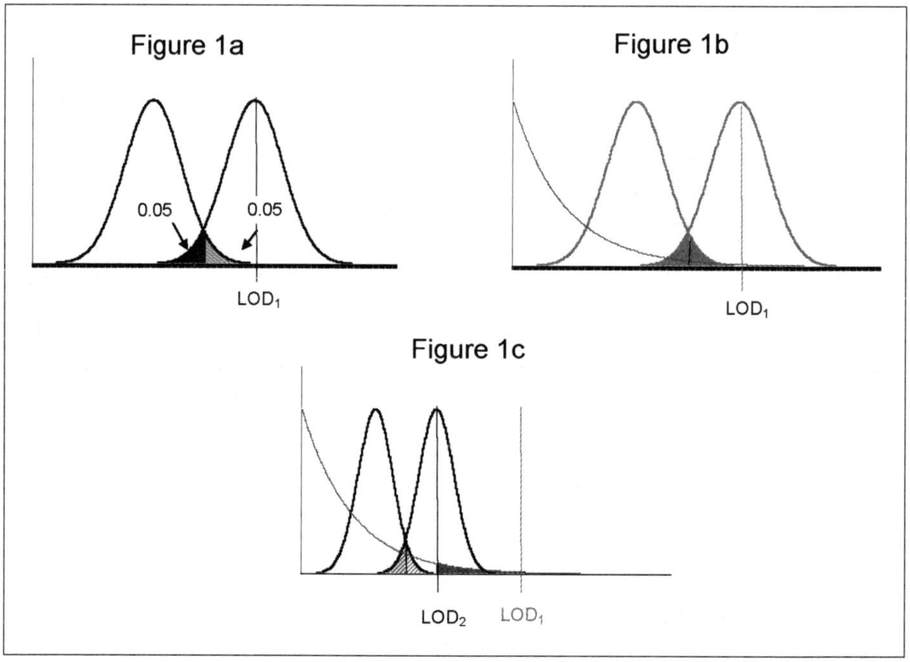

Fig. 1: Impact of the assay limit of detection (LOD) on the probability of detecting an adventitious agent in cell culture products: Figure 1a, determination of the LOD by limiting the false-positive and false-negative rates of an assay; Figure 1b, impact of LOD on the proportion of lots that will be identified as containing a pathogen (area under contaminant load distribution); Figure 1c, increase in the proportion of lots that will be identified using a new technology with lower variability.

A question that arises from this observation is: (i) should the new procedure be used qualitatively, thus resulting in detection of lower levels of pathogen than had been identified by the old procedure, or (ii) should the new procedure by used quantitatively, in conjunction with a limit that fixes the maximum level of pathogen that can be accepted in a lot of product. While assays of this sort are classically operated as detection procedures, the increased sensitivity may be unwarranted, resulting in failure of lots that had previously gone undetected by the old procedure.

Probability of Detection at Low Concentrations

The calculation of probability of detection at low concentrations is outlined in ICH Q5A (pp. 25-6). That guideline introduces the concept that the probability of detection of a pathogen in a cell-derived product is a function of the overall volume of the material being tested (V), the volume of the sample (v), and the number of infectious particles randomly distributed in V (n). Thus the concentration of a pathogen in the product volume is equal to [C] = n/V, and the probability of sampling one or more infectious particles is

$$p = \left(\frac{V - v}{V} \right)^{n}.$$

If the sample volume comprises a small percentage of the overall volume, this can be approximated by the Poisson probability distribution: $p = e^{-[C]v}$. This equation can be used to derive operating characteristic (OC) curves for various concentrations of a pathogen in the material (Fig. 2). Figure 2a illustrates the probabilities of detecting a pathogen as a function of sample volume, for concentrations ranging from 50 infectious particles in a litre, to a single particle in a litre. We notice immediately that there are severe limitations to detecting low concentrations or a contaminant, with low sampling volumes (10% to 20 % of total volume). These OC curves can be used to determine the sample volume necessary to have a high likelihood (perhaps 90%) of detecting specified concentrations in the material. Thus 45 mL (~5% of the total volume) would be necessary to have a 90% chance of sampling an infectious particle, in concentration of 50 particles in 1000 mL (Figure 2b), while 120 mL (12% of the total volume) is necessary when the concentration is 20 particles in 1000 mL.

To use these OC curves effectively, however, the practitioner should have some idea of a "meaningful" concentration to detect. Lacking this, these graphs can be used to describe the properties of the sampling plan. In general, given the low likelihood of detecting low concentrations of a pathogen, these tests might be better viewed as precautionary, rather than definitive.

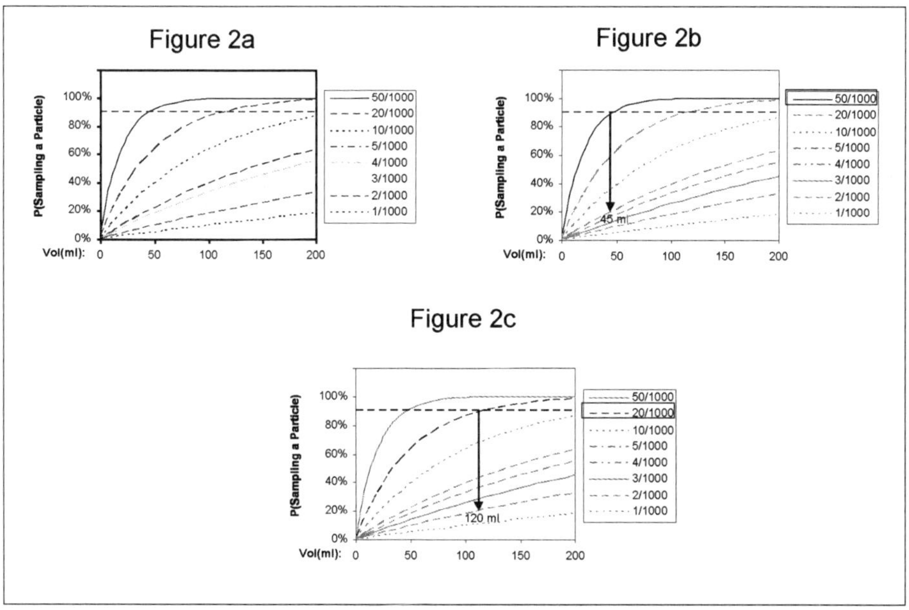

Fig. 2: Operating characteristics curves showing probability of detection: Figure 1a, probability of detection for different contaminant loads (one particle per 1000 mL, two particles per 1000 mL, etc.) and different sampling volumes (5 to 200 mL); Figure 2b, volume necessary to have a high probability of detection (90%), with 50 particles in 1000 mL; Figure 2c, volume necessary to have a high probability of detection (90%), with 20 particles in 1000 mL.

Clearance Studies

Clearance studies are used to provide a high level of assurance that the risk of a contaminant has been adequately reduced through the manufacturing process. The strengths of such studies are that they can be designed to measure clearance of an expected contaminant reproducibly. They are unable, however, to show complete clearance since the clearance curves are typically expressed in the logarithmic scale.

Clearance studies are typically designed to determine the clearance rate (log loss/clearance increment) for a particular contaminant. That rate can be addressed over time, over a number of treatment steps, or over the concentration of a particular reagent that is known to inactivate the specific pathogen. Using time as an example, the impact of the combination of the amount of time in the process that is devoted to clearance, together with the number of replicates of the clearance experiment, can be depicted through basic operating characteristic curves (Fig. 3). Any experiment has associated uncertainty, which is a function of the experimental design, and is inversely proportional to the amount of work that is performed. Given the desire to have a reliable estimate of the clearance rate, the scientists can design their study to achieve either a specific level of variability, or to obtain maximum discrimination at minimum cost. The proportional decrease in variability that is achieved for various designs is depicted in Figure 3a. Designs range from holding material from 5- to 50-hours and measured in 5-hour increments, run once (1X) versus multiple times (2X, 3X, etc.). Thus if one were to want to measure the

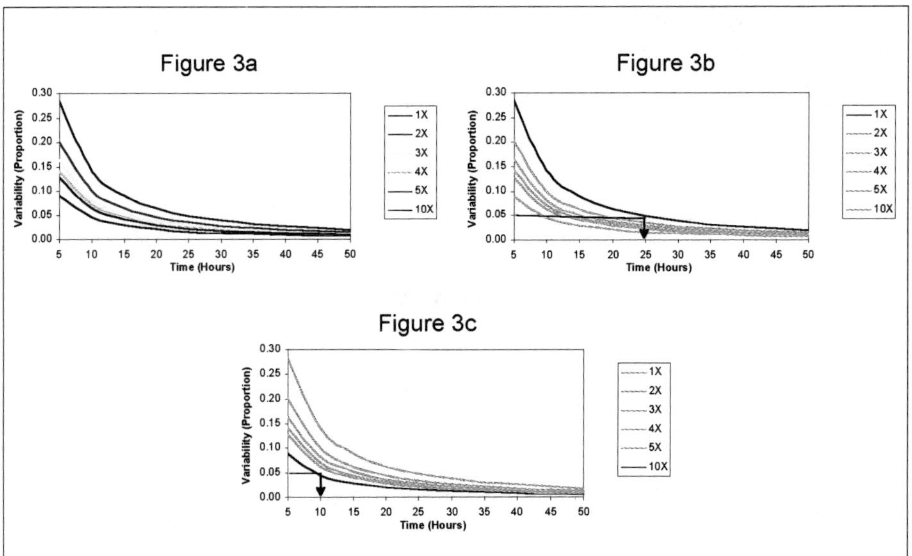

Fig. 3: Operating characteristics curves for clearance studies: Figure 3a, variability of loss rate estimate for various designs (5 through 50 hours treatment in increments of five hours), in single (1X) and replicate experiments (2X, etc.); Figure 3b, time required to obtain a variability equal to 0.05 in the estimate of clearance rate, with a single replicate at each time point; Figure 3c, time required to obtain a variability equal to 0.05 in the estimate of clearance rate, with 10-replicates per time point.

clearance rate with a variability that is 5% of the method variability, the experiment may be carried out through 25-hours in a single replicate (Figure 3b), or through 10-hours in 10-replicates. In this manner, the experimental design can be adjusted to the experimental conditions. Based on the amount of input contaminant, and the resolution of the assay, 10 replicates must be performed if the assay is capable of detecting a level of contaminant that remains through 10-hours, while a single replicate can be performed if the assay resolution supports an extended study duration.

The choice of design may be made on the basis of the reduction in the margin of error of the clearance rate to some desirable level. The proportional decrease in variability illustrated above can be expressed as the margin of error (MOE) of the clearance rate:

$$\text{Margin of Error} = 2 \cdot \text{Sigma}$$
$$= 2 \cdot \sqrt{\frac{s^2}{n \cdot (t-1) \cdot \text{var}(x)}}$$

where "s" is the variability of the assay, "n" is the number of replicate experiments, "t" is the number of time point measurements, var(x) is the variance of the time points, and "2" places a bound on the MOE, that associates the estimate with 95% confidence (note that 2 is an approximation to this factor, which is actually a function of the study design). Alternatively a design can be selected on the basis of the "law of diminishing returns", to a level beyond which there is little improvement in the margin of error.

RISK ANALYSIS

In a risk analysis, the laboratory typically enumerates the factors which contribute to the reduction of an adventitious agent throughout the process. Those factors include terms for the acquisition of the agent during procurement and addition of a particular process component, as well as terms that represent clearance and dilution steps in the process. These can be accumulated mathematically, to establish the overall reduction of the contaminant. If each individual reduction factor is represented as F_i, then the overall reduction can be calculated as:

$$\text{Reduction} = F_1 \cdot F_2 \cdots F_K,$$
$$\text{or log Reduction} = LR = \log(F_1) + \log(F_2) + \cdots + \log(F_K).$$

If the variabilities for each factor can be ascertained, either theoretically or experimentally, a confidence interval (CI) can be placed upon the overall reduction, which represents the lower and upper bounds for the contaminant after processing.

$$\text{Var}(LR) = \sum \text{Var}(\log(F_i))$$

$$\text{and CI} = LR \pm 2 \cdot \sqrt{\text{Var}(LR)}$$

Here as above, "2" places a bound on the MOE, that associates the estimate with 95% confidence.

Discussion

Several factors affect our ability to control and detect adventitious agents in a product produced in cell culture. Statistics provide both an awareness of the capability of our procedures to detect a contaminant, as well as designs that improve our estimates of key process and analytical parameters. Clearance studies are performed to evaluate the rate of elimination or inactivation of a pathogen through a clearance step in the process, and can be designed to achieve an acceptable level of certainty in the final loss rate estimate. Strategic testing involves the combination of knowledge concerning the detection limit of the assay, as well as the impact of the amount of sample taken. These taken together describe the overall assurance that the final product will contain no more than a particular level of the adventitious agent. These calculations teach us that no procedure or experiment can guarantee that the product is completely free of the contaminant. In addition to the assurance obtained for individual batches and clearance studies, the overall manufacturing record contributes to the assurance against the systematic presence of contaminant in the process. A risk analysis can be improved by incorporating the combined variabilities of the risk factors, to derive a confidence interval on the total reduction of a contaminant in a process.

An appropriate use of statistical design and analysis aids the manufacturer in better understanding their processes, and communicating the characteristics and results from their experiments and detection procedures. The combination of good science, and better understanding from well designed experiments and comprehensive analysis, provide better protection to the customer of contaminant in products produced in cell culture. Depending on the specific concern, its likelihood, the impact of the concern actually occurring, and the technology, one may choose to do QC or QA or a combination of the two. However, one size does not fit all occasions, and a variety of approaches is typically needed across a family of products.

References

1. ICH Q5A, Viral Safety Evaluation of Biotechnology Products Derived From Cell Lines Of Human Or Animal Origin, Recommended for Adoption at Step 4 of the ICH Process on 5 March 1997 by the ICH Steering Committee.

Mr. T. Schofield, Merck Research Laboratories, P.O. Box 4, WP37C-305, West Point, PA 19486, USA.
E-mail: tim_schofield@merck.com

Petricciani J, Sheets R (eds): Vaccine Cell Substrates 2004.
Dev Biol (Basel). Basel, Karger, 2006, vol 123, pp 217-222.

PANEL IV DISCUSSION

Level of Assurance Provided by Guidelines Especially ICH Q5A and Q5D

Panel Members: *Anthony Lubiniecki (Panel Chair), Alan Moore, Tim Schofield, Peter A. Lachenbruch*

Panel Questions

Question 1:
What amount of a cell bank should be tested in each characterization test?

Question 2:
What amount of a serum lot should be tested?

Dr. Arnold: In light of all the PCR technology we have heard about, why are we still using the Points to Consider mycoplasma method, and should we be using PCR if we could get more primer pairs?

Dr. Egan: This was something that came up about a year ago at a workshop at the Well Characterized Biologics meeting in San Francisco. Some preliminary results were presented on PCR-based methods by a number of people. This was followed up by a workshop in Philadelphia in September 2003. It looks like a viable method. Within our own laboratories, we have been working on a micro-array method for the detection of mycoplasmas in Konstantin Chumakov's group and Vladamire Chizakoff in our laboratory. Currently, there are about 60 mycoplasmas on the chip. They are sequencing about 40 more, and presumably will have all 100 on the test.

Dr. Chandler: One of the questions is the validation of any methodology compared to the tried-and-true methods, the culture, broth and agar media.

It is important to have a side-by-side, head-to-head comparison of PCR studies with the traditional assays.

At FDA, we have known for a long time that, in theory, other new methods should be usable. We still have the old 610.30 21 CFR requirement for mycoplasma testing that has been obsolete for a number of years because it does not have the indicator cell method for non-cultivatable organisms.

We need data to show that any new method is going to be comparable to the old methods. The question is not whether you can completely substitute the PCR for the culture methods, but rather under what conditions one method is appropriate for use rather than another.

The other issue, of course, is the sample size for PCR. When you have a 10,000 litre fermenter and you are going to take the equivalent of a few microlitres to do the PCR testing, it may not be as sensitive on a volume basis, even though the technology is quite good and sensitive.

Dr. Onions: We are trying to look at a hybrid method at the moment where we are doing shorter term cultures then a pan-PCR screen. This enables you to pick up some of the agents that we might miss just by PCR alone because of sensitivity issues.

Dr. Egan: There are about 100 different mycoplasmas known, of which about 10 to 20 have been isolated commonly as contaminants during manufacturing.

The PCR method, the micro-array method, is based on the constancy of the DNA for the 16 sRNA and the 23 sRNA. The PCR primers are from these two genes that are very highly conserved among the mycoplasma. Then there is an intergenic space of about 400 base pairs, and these vary by groups and by species.

The technique is based on having these PCR primers, and then the sequences for group and species differences that are seen for different mycoplasmas.

There are about 30 mycoplasmas where the sequence is known, at least for this intergenic region, and we have sequenced another 30.

So, at this point, we have a chip that has about 220 oligo probes for 60 mycoplasma species. The sensitivity is about 10 organisms, or 10 copies of DNA, which is virtually the same for all organisms. When we finish the other 40, we will have all 100 on one chip.

Dr. Lubiniecki: Our assigned questions for this section are: (i) what amount of the cell bank should be tested in each characterization test, and (ii) what amount of the serum lot should be tested?

I would argue that the answer to Question 1 is already specified in the ICH Q5A and Q5D. Depending on the test, the guidelines tell you what should be tested.

The second question is far more complex. I think part of the answer lies in what are you testing for. It also relates to the talk that Dr. Schofield gave. It not only relates to what you are testing for, but how much of it you expect to be there, because if you only expect a low-level contamination, and if this was the only means you had of ensuring quality, you might be wanting to test more rather than less.

The classic example is BVD testing in serum. It is pretty well documented in a number of laboratories that if you double the test volume, you typically double the incidence of positivity, which tells you that the level of the agent is probably hovering somewhere around the limit of detection. The result from any test is not an end in itself, but it is part of a control procedure that integrates with other procedures in the process and in the plant to protect the product. The same result in a given test may have a different meaning, depending on how it fits into the overall control scheme.

Dr. Onions: We are developing a 384-well PCR plate that will cover all the viruses we have been discussing, including quite a few of the bunyavirus variants. If we can get the assay costs down to a few U.S. dollars, which we think we can with PCR, we could actually produce a certified FBS that would have every single donation tested. That would get rid of the BVD issue, because we would pick it up at the single donor level. That, of course, would add a considerable cost to the end product. It is really a question of whether there is a need in the industry to do this.

Dr. Tente: I would speak for the cell and gene therapy industry where foetal bovine serum use actually appears to be continuing. Many stem cell CNS cell lines or isolates depend on FBS. Investigators have been unable to eliminate the use of foetal bovine serum.

There are some very promising new therapies that are being developed that unfortunately use foetal bovine serum very late into the processes, so speaking for those sponsors, I would say there is an interest in that type of approach.

Many of the decisions and guidances that come out of meetings like this on vaccines translate directly to some of the newer therapies that are under development.

Dr. Sheets: Given that the sample volumes may be as little as 3 ml out of a bioreactor that might be 2,000 litres or 10,000 litres of cell supernatant or cell lysate, is that amount sufficient? Also, for cell bank characterization we test 10^7 cells, but the number of human doses that comes out of that number of cells is going to vary widely. So, is it adequate to be taking this generic approach, or should we calculate this for each and every process and product? How do we approach this?

Dr. Lubiniecki: My view is that each market application should contain some sort of analysis that determines and justifies what control procedures are appropriate and why those are chosen. It is up to the regulatory agencies to agree or disagree with the adequacy of the approach.

It is very difficult to have a 'one size fits all', which is one of the reasons the regulatory community has always dealt with issues on a case-by-case basis. It is difficult to make a single formula that will work for every product and every use of every product.

Dr. Moore: I think one of the things that we have tended to dance around is this concept of assessing relative risk and the tool set that we have available to us.

Given that we do not have generic universal virus detection methodologies, we have to accept limiting both the samples that we test and the expense and practical implementation of this testing in the face of the benefit that these products provide.

That is a very difficult concept to work with and to establish hard and fast rules, and I think the FDA has done a good job in terms of looking at risk assessment and trying to develop an approach.

We should recognize that we are testing a woefully inadequate sample volume, probably across the board, and that we are doing qualitative testing often when we believe we may be doing quantitative testing.

It may not be comfortable for us to accept that, but it is, in many cases, a reality. Now, to address that, I think one has to take a look at increasing the numbers of samples that you take through a process, or perhaps designing programmes that are statistically or systematically driven. But until we have the right tool sets, we are going to be straddled with the practical limitations of the tools that we have.

Dr. Lachenbruch: One of the things that we statisticians always look for is a random sample. But the question is, a random sample of what?

Dr. Lubiniecki: In the days before we had ICH to give us harmonized data requirements, many of us did three separate tests of essentially the same thing in slightly different ways to meet the regulatory requirements of the three regions.

I do not remember hearing of a case or seeing in our own data that there was any discrepancy among the data packages. This could be interpreted as saying that if you do three times the testing, results will still be the same, which leads to the conclusion that if you believe in a world of finite resources, doubling or tripling the resource by testing additional amounts is necessarily going to give you a higher level of assurance.

However, it could also be argued that that is a relatively small sample, and you probably could not rely on that approach in every case, but at least at some level, it appears to be adequate.

The other thing to remember is that most of these assays are inherently screening assays, and they also involve biological amplification, which allow an enhanced ability to find things. While they may not be perfect, they are almost always not the only way of ensuring quality for a given issue. One could argue that they are probably good enough most of the time.

Dr. Schofield: If we did the calculation, what would you do with the number? The answer to that question is that there is no expectation. If you want to ask the question, you should start with having an expectation, and then you can work from that expectation.

So the calculation is fine, but I do not know what I would do with the number once I got it. I do not say that there should be one, but the key problem that we are dancing around is that there is no expectation other than zero, and you cannot prove zero.

Dr. Lubiniecki: The one area where there is a formal analysis typically is in recombinant products that come from rodent substrates where there is some estimation of retrovirus burden and safety factors, so there is a fairly formal estimation for that one purpose, but that is usually the only one.

Participant: The other extreme is a small investigational lot which may have only 500 or 1,000 doses in which you end up testing a quarter or a third of the volume in qualifying the vaccine and the cell substrate.

Is there any reason either in terms of risk and distribution of that lot, numbers of people involved, or in terms of any other considerations for thinking that there may be instances in which you test smaller volumes although still a very high percentage of the lot that you are administering?

Dr. Lubiniecki: Well, one thing that is typically different about investigational products versus commercial materials is that the commercial process has been validated, and the investigational process, especially for early investigational studies, is typically not coming from a validated process, so a certain level of relying on the assurances given by testing is perhaps normal for investigational materials, because you just do not have process validation to rely on. But beyond that, the considerations are probably the same.

Participant: In terms of the serum lot percentage testing issue, there were some studies that assessed foetal bovine serum contamination using some of the more established in vitro tests at the time, and now, with the new test methods that we have been discussing here from different providers are assessed, so we can get to the bottom of whether or not we are looking at a very high percentage of foetal bovine serum that is contaminated at a very low level, or if it is sporadic.

Dr. Lachenbruch: If you want to show that you are below a certain limit, your null hypothesis would be that the level within the sample or within the product is too high, and then you conclude the right thing that way. You can extend that to confidence intervals so that the upper bound of the confidence interval is below that level. It is important to distinguish between the point estimate and the upper limit of the confidence interval, and that is particularly important in the previous context.

The more statistical point that I was concerned about in reading ICH Q5A is the fact that we are doing a considerable amount of testing when you do multiple testing. If you do 100 tests, you would expect to see five tests significant at the 0.05 level.

So, we should not become excited if we see five tests although one or more of them may, in fact, be a real difference. There had been much thought given to these in terms of micro-arrays when we are doing perhaps 10,000 tests on a relatively small number of expression estimates. Therefore it is appropriate to think about false discovery rates.

Dr. Moore: The BacT Alert system that I mentioned earlier was approved by FDA for the early release of product to the clinic. It did not displace the full 14-day test. Also, I would like to suggest that FDA recognizes that there may be companies interested in participating in developing PCR Mycoplasma technology.

Summary Statement

There was a general sense that testing for adventitious agents is part of an integrated programme to ensure product safety and quality and that there are other elements which are also important, including procedural and engineering controls generally classified under GMP and viral clearance.

While there was concern that the powers of detection of any tests performed are limited by sample volume, by statistical considerations, and by the nature of the sample, testing is not the only means of assurance of freedom from contamination. The inherent limitations of testing do not appear to have resulted in reports of the administration of contaminated products to recipients and, in fact, this very testing has identified a few failures and prevented those products from being released.

New technologies are certainly becoming available and may help to improve the limits of detection, improve the ease with which this can be done, or even expand the scope of testing. New technologies should be carefully evaluated before replacing existing technology in terms of the validation of those assays, including a careful analysis of their statistical power and how they fit into an integrated control strategy before they are actually used as routine tests.

Nearer term, data around PCR technology for testing for mycoplasma suggest that these may be a practical alternative to cultivational methods for mycoplasma for some uses. Longer term, there are certainly other genetic, genomic and microarray technologies which may become available.

Harmonized approaches to regulatory acceptance are quite useful and they should be considered wherever this is appropriate.

Present test recommendations seem to be adequate to ensure product quality and safety, but we should remain vigilant for issues and watch the evolution of new technologies. Interest was expressed in compiling existing information regarding testing for viral contamination of bovine serum. For example, the U.S. Department of Agriculture has tested bovine serum for several decades.

Petricciani J, Sheets R (eds): Vaccine Cell Substrates 2004.
Dev Biol (Basel). Basel, Karger, 2006, vol 123, p 223.

BONUS SESSION

Experience with Novel Cell Lines

Petricciani J, Sheets R (eds): Vaccine Cell Substrates 2004.
Dev Biol (Basel). Basel, Karger, 2006, vol 123, pp 225-233.

Comprehensive Characterization of the 293-ORF6 Cell Line

B.T. Butman[1], A. Lizonova[1], D.E. Brough[1], J.M. Sowers[1], R. Sheets[2], J. Gall[1], P. Newton[1], P. Gomez[2]

[1] GenVec, Inc., Gaithersburg, MD, USA.
[2] Vaccine Research Center, NIAID/NIH, Bethesda, MD, USA.

INTRODUCTION

Genetically engineered viruses (vectors) provide an efficient system for gene delivery and expression of medically-relevant proteins. GenVec's platform technology uses replication-deficient Adenoviral vectors (Adenovectors) for both therapeutic and vaccine applications. The "GV11" Adenovector, which has complete deletions of both the E1 and E4 domains and partial deletion of the E3 domain is, by design, incapable of replicating in normal human cells. Replacement of the Adenovirus E1 domain with a protein expression cassette enables safe delivery and expression of proteins of interest. This concept is presented in Figure 1.

We have developed a proprietary cell line for the efficient production of these replication-incompetent Adenovectors. The 293-ORF6 cell line was established to complement the missing E1 and E4 gene functions, thereby enabling Adenovector production in vitro [1]. The parental 293 HEK cell line [2], which already possessed Adenovirus E1a and E1b domains, was further engineered through the addition of an Adenovirus E4 open reading frame 6 (ORF6) domain to provide adequate complementation for E1 and E4-deleted GV11 vectors. The E4 ORF6 domain, which is under the control of a zinc-inducible metalothionine promoter, enables optimization of expression of an E4 protein necessary for complementation. This new cell line efficiently generates large quantities of GV11 Adenovectors in serum-free suspension culture (up to 10^5 viral particles per cell). Figure 2 demonstrates the deletion of the E4 domain from the GV11 vector and the insertion of an E4 open reading frame 6 (ORF6) expression cassette into the 293 cellular genome, resulting in the 293-ORF6 cell line.

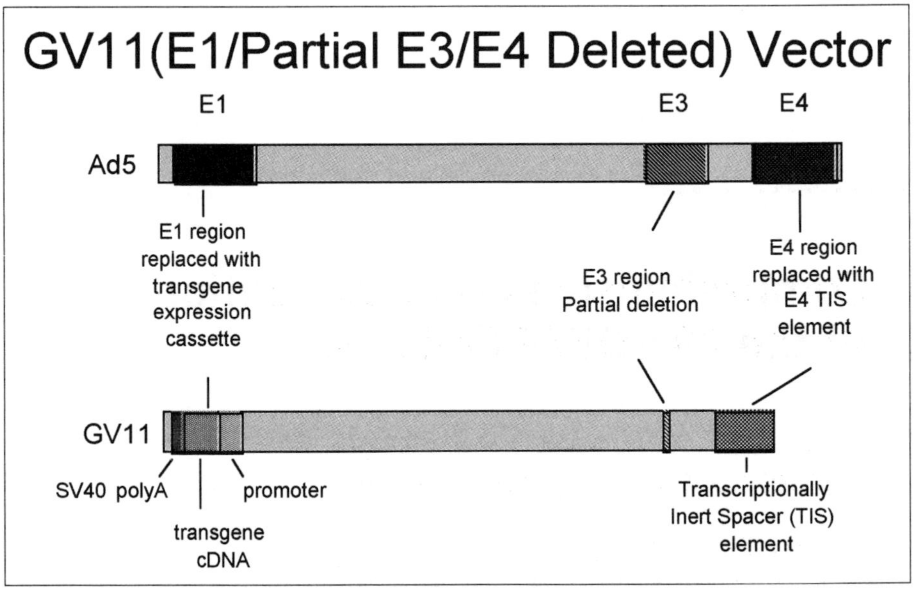

Fig. 1: GenVec's proprietary "GV11" vector has deletions in the E1, E3 and E4 domains of the Adenovirus genome. The expression cassette is inserted in place of the E1 domain and a transcriptionally inert spacer (TIS) is inserted in place of the E4 domain.

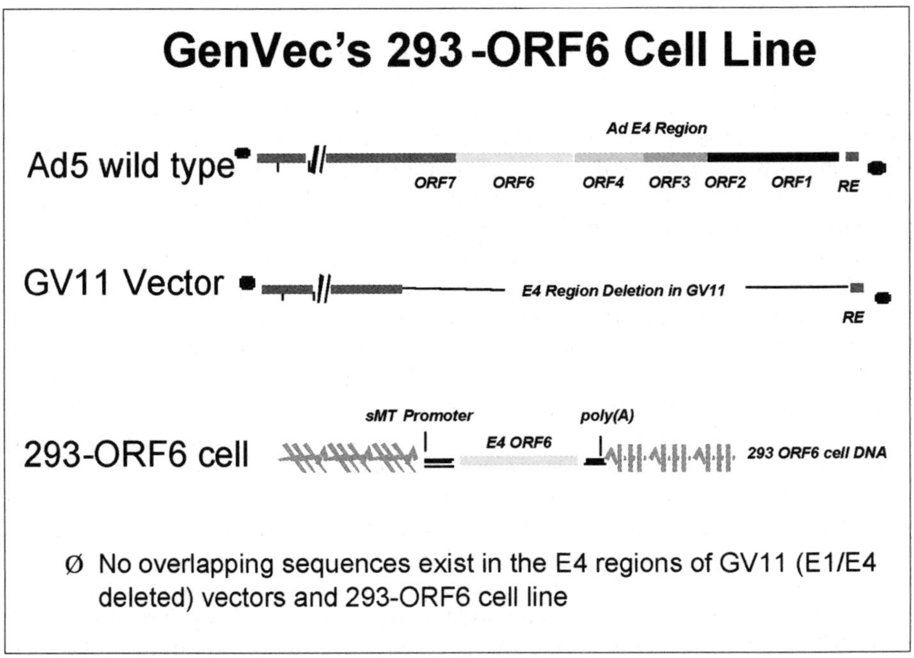

Fig. 2: Complementation for E4 deletion by E4 ORF6 Expression Cassette in the 293-ORF6 cell line.

For vaccine applications, full characterization of this vaccine cell substrate is required to ensure safe and consistent production of GV11 Adenovectors. In addition to standard viral and microbial safety assays, we have conducted extensive molecular characterization of the 293-ORF6 Master Cell Bank (MCB).

For this purpose a high passage "aged cell bank" (ACB) was created, representing 16 passages beyond the MCB. This ACB represents the equivalent of expansion of a vial from a 293-ORF6 Working Cell Bank (WCB) to the level of a 10,000 L suspension bioreactor. Characterization of the MCB and ACB in parallel provides assurance that expansion of the MCB during large-scale production will not result in either expression of latent pathogenic agents or in the loss of efficient vector production.

A comparative characterization study of the MCB and ACB included: (i) cytogenetic analysis (karyotyping); (ii) molecular characterization of E1 and E4 copy number (Q-PCR) and chromosome localization (FISH); (iii) sequencing of the protease resistant protein (PRP) gene to assess deletions associated with Prion diseases; (iv) evaluation of protease sensitivity of the PRP gene product (Western blot) and (v) evaluation of cell production characteristics such as cell doubling time and vector productivity. Furthermore, because the parental 293 HEK cell line is known to be tumorigenic in nude mice, the high passage ACB was also evaluated in a tumorigenicity assay to ascertain whether either the addition of the E4 ORF6 domain or high passage resulted in altered tumorigenicity.

Cell Line Characterization Studies

Establishment of 293-ORF6 master cell bank and aged cell bank

Creation of the 293-ORF6 cell line is described in Brough et al [1]. Clone A232 was adapted to serum-free (GIBCO SFM-II medium) suspension culture over 10 passages in shaker flasks. The MCB was established at passage 27 and the WCB at passage 32. Starting at passage 17 the ACB was derived in parallel by passaging 293-ORF6 cells in serum-free suspension culture to passage 43. At passage 27 (MCB), passage 32 (WCB) and passage 43 (ACB), cells were cryopreserved in SFM-II medium with 10% glycerol. All vials of cryopreserved cells were subsequently stored in the vapour phase of a liquid nitrogen freezer.

Microbial and viral safety testing panel

Safety testing of the MCB included standard microbial assays for sterility and Mycoplasma, in vitro assays for human, bovine and porcine adventitious viruses and in vivo assays for inapparent viruses [3-7]. Furthermore, we evaluated the MCB for a comprehensive panel of known human pathogens by PCR. This included PCR assays for AAV, CMV, EBV, HBV, HCV (by RT-PCR), HHV-6/7/8, HIV-1/2, HPV, HTLV-I/II, Parvo B19 and SV40. The results of testing on the MCB are presented in Figure 3. In summary, no bacterial, fungal or viral pathogens were detected by any of these methods.

- **Microbial**: Mycoplasma, Sterility (B&F)
- **In Vitro**: Adventitious viruses (3 cell lines), Bovine Viruses (9 CFR), Porcine Parvovirus
- **In Vivo**: Inapparent viruses in animals
- **PCR**: AAV, CMV, EBV, HBV, HCV (RT-PCR), HHV-6/7/8, HIV-1/2, HPV, HTLV-I/II, Parvo B19, SV40
- **Biochemical**: Isoenzyme analysis (human origin), FPERT (Retroviruses)
- **Other**: TEM

Fig. 3: Comprehensive safety testing of 293-ORF6 MCB.

Biochemical and electron microscopic characterization

Iso-enzyme analysis of the MCB confirmed human origin. To ensure the absence of retroviral contamination further, the MCB was assayed by fluorescent product-enhanced reversed transcriptase assay (F-PERT). No reverse transcriptase activity was detected. Finally, electron microscopic analysis of the MCB did not result in detection of retroviral morphology.

Demonstration of cell line genetic stability

To demonstrate the molecular stability of the MCB, additional characterization was performed on both the MCB and the ACB. For cytogenetic analysis, both lines were expanded in culture, treated with colcemid (metaphase block), hypotonically lysed, fixed and stained with Wright's stain. For karyotyping, 100 metaphase spreads from each cell line were microscopically examined for chromosome number and chromosomal abnormalities. The results are shown in Table 1. Both cell lines showed a median of 72 chromosomes per metaphase spread. The MCB ranged from 66 to 80 and the ACB ranged from 54 to 78. These values were not statistically significantly different.

Table 1: Cytogenetic analysis of 293-ORF6 cells.

Results		
	P27	**P43**
Median#Metaphase	72	72
Range	66-80	54-78
Aberrations/cell	0.02	0.00

For determination of the copy number of E1 and E4 domains per cell, Q-PCR analysis was performed on both cell lines. The E1 PCR assay targeted Adenovirus Type 5 bases 3343 to 3411, whereas the E4 PCR assay targeted Adenovirus Type 5 bases 34007 to 34073. PCR amplicons were quantified using fluorescent-labelled plasmids encoding the full length E1 and E4-ORF6 domains as probes. Human Glyceraldehyde 3-Phosphate Dehydrogenase (GAPDH), which has a known copy number in human cells, was used as an index for cell number. The results are presented in Table 2. The E1 and E4 copy numbers were the same for both cell lines. The Dixon Q test demonstrated that there were no outliers for either of the samples.

Table 2: E1 and E4 copy number in 293-ORF6 cells.

	Results	
	P27	**P43**
E1 Copy #	6.5 ± 0.8	5.1 ± 0.7
E4 Copy #	2.6 ± 0.2	2.4 ± 0.4

E1 and E4 chromosome localizations were determined for both cell lines by fluorescent in situ hybridization (FISH) assay. Plasmids encoding full length E1 and E4-ORF6 domains were labelled with Texas Red 5'UTP by nick translation. P27 and P43 cells were expanded, treated with Colcemid, lysed, fixed and then stained with the labelled probes and counterstained with DAPI. Both cell lines revealed the presence of E1 on chromosome #19 at 19q13.3 locus and E4-ORF6 on chromosome #6 at 6q16-21 locus. These results are presented in Table 3. Based upon the Q-PCR results, it can be concluded that the E1 domain is present in multiple copies at the chromosomal insertion site. These studies reveal that the chromosomal insertions of both E1 and E4 are stable.

Table 3: FISH analysis for E1 and E4 chromosome localization in 293-ORF6 cells.

	Results	
	P27	**P43**
E1 localization	Chromosome #19 at 19q13.3	Chromosome #19 at 19q13.3
E4 localization	Chromosome #6 at 6q16-21	Chromosome #6 at 6q16-21

Characterization of Protease-Resistant Protein (PRP)

To characterize the PrP open reading frame found on chromosome 20, both cell lines were subjected to PCR sequencing using cellular DNA as template. The PCR primers used were AP-1 [8] and AP-2 [9]. The resultant amplicons were subsequently purified and sequenced. Based on the GenBank accession #U29185, the expected amplicon was 894 base pairs, coding for 245 amino acids. Both the MCB and ACB generated PCR amplicons of 918 bp, representing 253 amino acids. Sequence analysis revealed that this discrepancy from the expected sequence was due to the insertion of 24 bp between bp 339 and 340, coding for a known repeated octapeptide that is not associated with abnormal PrP. Furthermore, full sequence analysis did not reveal the presence of any mutations that are known to be associated with infectious PrP^{SC}.

Absence of PrP mutations associated with Transmissible Spongiform Encephalopathies (TSE) was further demonstrated by analysis of the protease sensitivity of the PrP gene product in a Western blot assay [10]. Both P27 and P43 cells were expanded in culture and the PrP protein was extracted, digested with Proteinase K and subjected to Western blotting. A monoclonal antibody (Mab 3F4) that recognizes both human and hamster scrapies PrP-res was used to probe the blots. Bound antibodies were detected using alkaline-phosphatase-labelled anti-mouse antibodies followed by chromogenic substrate. The positive control, hamster scrapies agent, was tested in parallel, both alone and spiked into the test articles. The positive control and spiked samples demonstrated the presence of protease resistant PrP. However, there was no evidence of protease-resistant PrP in either the P27 or P43 samples.

Tumorigenicity studies

The parental 293 HEK cell line was previously established as being a moderately tumorigenic cell line in the nude mouse model. We conducted a study to determine whether either addition of the E4 ORF6 domain or continued passaging would result in a significant change in the tumorigenic profile of the 293-ORF6 cell line. For this purpose, both the parental 293 cells and the 293-ORF6 ACB were assayed in parallel. HeLa cells (which are known to be strongly tumorigenic) were chosen as the positive control for the assay. Syrian Hamster Embryo (SHE) cells were selected as the negative control. The parental 293, 293-ORF6 ACB and HeLa cells were injected at 10^7, 10^5 and 10^3 cells per animal by intraperitoneal route. SHE cells were only tested at the highest cell number (10^7 cells/animal). For each study group, 20 Balb/c nu/nu mice were injected. The study duration was 140 days and a positive endpoint was defined as a histopathically-confirmed neoplasm. In this study, the HeLa positive control cells were positive at 10^5 and 10^7 cells, whereas the negative control SHE cells were negative at 10^7 cells/animal. Both the parental 293 and 293-ORF6 cells were positive only at the 10^7 cells/animal level. This demonstrated that the addition of the E4 ORF6 domain to the 293 cells did not significantly change the tumorigenic potential of the resultant 293-ORF6 cell line.

Given that the 293-ORF6 cells are tumorigenic in nude mice, further studies are warranted to assess the oncogenic potential of the cell line. The purpose of this study is to demonstrate the absence of previously undetected oncogenic viruses or viral genes within the cell line. In the proposed study, both cell lysate and purified DNA from the 293-ORF6 cells will be injected into newborn mice, rats and hamsters. The negative control cell line will be MRC-5. It is acknowledged that a relevant positive control cell line for this study does not exist. The study duration will be five months and the endpoint will be histopathically-confirmed neoplasms.

Cell identity PCR assay

To distinguish the 293-ORF6 cells from the parental 293 cells, a PCR assay was developed that targets both the E1 (1207 bp amplicon) and E4 ORF6 (307 bp amplicon) domains. Controls in the assay included A549 cells (which are negative for both E1 and E4), 293 cells (which are positive for E1 and negative for E4) and J1-20 cells (which are negative for E1 and positive for E4 ORF6). Table 4 demonstrates the design of this assay.

The results of this assay are presented in Figure 4 and demonstrate the specificity of the assay for 293-ORF6 cells (both P27 and P43).

Table 4: Design of PCR identity assay.

Cell line	**E1**	**E4 ORF6**
293 cells	+	–
293/ORF6 (P27)	+	+
293-ORF6 (P43)	+	+
A549 cells	–	–
J1-20 cells	–	+

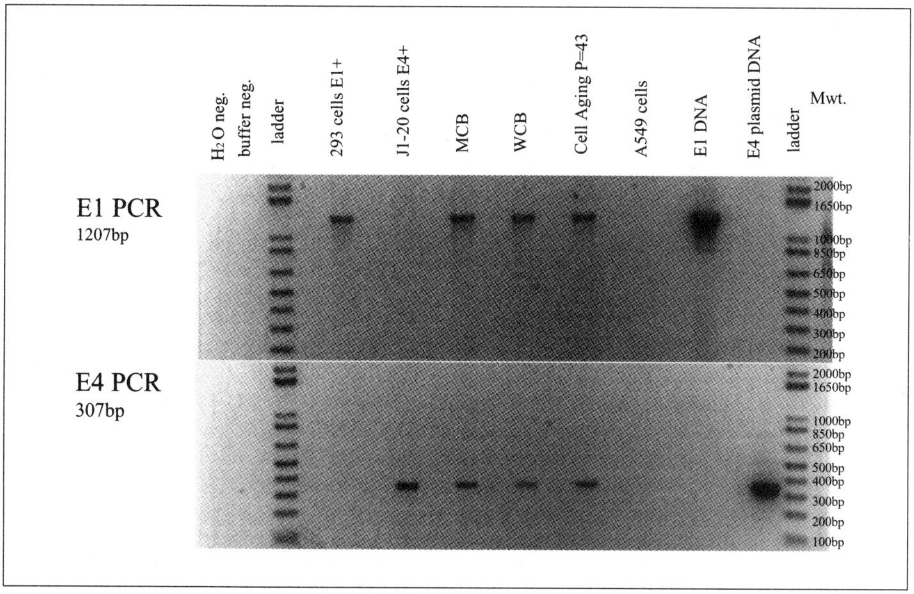

Fig. 4: Agarose electrophoresis gel of PCR products from E1 and E4 PCR identity assay.

Cell line stability study

Two cell aging studies have been conducted by passaging the 293-ORF6 cells to a level of expansion that is equivalent to a 10,000 L bioreactor production run. The purpose of these studies was to monitor cell doubling time and vector productivity as a function of cell passage. The 293-ORF6 cells were stable for both parameters over the course of the study.

We are routinely monitoring the viability of the cryopreserved MCB. To date, over 20 months the cryopreserved cell bank appears to be stable (data on file).

Conclusion

GenVec's 293-ORF6 cell line (MCB) has been thoroughly characterized with respect to safety, genetic and molecular characteristics and stability. The results of these studies have been submitted to the U.S. Food and Drug Administration (FDA) in a Biological Master File supporting the initiation of a Phase I human clinical trial using a multi-clade Adenoviral-based vaccine against HIV. This production cell line is also being used for the development of a broad array of Adenovector-based vaccine and therapeutic products and is being evaluated for use in the production of other viruses.

Acknowledgements

The characterization studies described in this paper are based on design and execution by numerous collaborating scientists. The author would like to acknowledge the following key contributors:

- Viral PCR Analyses – Audrey Chang, BioReliance, Inc.
- Cytogenetic analysis – Carlos Sanchez, Applied Genetics Laboratories, Inc.
- E1 and E4 copy number and localization – Elena Jouravleva, Lark Technologies, Inc.
- PrP sequencing – Heath Knight, Lark Technologies, Inc.
- Protease resistance WB – Byron Caughey, Rocky Mountain Laboratories, NIAID
- Tumorigenicity – Mary Whiteman and Dan Ragland, BioReliance, Inc.
- Cell identity – Gerald Lohnas, GenVec, Inc.
- Project Management – Victoria Haque, GenVec, Inc.

References

1 Brough DE, Lizonova A, Hsu C, Kulesa V, Kovesdi I: A gene transfer vector-cell line system for complete functional complementation of adenovirus early regions E1 and E4. Journal of Virology 1996;70(9):6497.

2 Graham FL, Smiley J; Russell WC, Nairn R: Characteristics of a human cell line transformed by DNA from human adenovirus type 5. Journal of General Virology 1977;36(1):59-74.

3 CBER's Points to consider in the characterization of cell lines used to produce. Biologicals, 1993, http://www.fda.gov/cber/gdlns/ptccell.pdf.

4 CBER's: Points to consider in the production and testing of new drugs and biologicals produced by recombinant DNA technology, 1985, http://www.fda.gov/cber/gdlns/ptcdna.pdf.

5 CBER's: Guidance for human somatic cell therapy and gene therapy, 1998, http://www.fda.gov/cber/gdlns/somgene.pdf.

6 International Conference on Harmonisation (ICH) Q5A Document: Viral Safety Evaluation of Biotechnology Products Derived from Cell Lines of Human or Animal Origin. Federal Register 1998;63(185):51074-51084.

7 International Conference on Harmonisation (ICH) Q5D Document: Quality of Biotechnological/Biological Products: Derivation and Characterization of Cell Substrates Used for Production of Biotechnological/Biological Products. Federal Register 1998;63(182):50244-50249.

8 Palmer MS, van Leeven RH, Mahal SP, Campbell TA, Humphreys CB, Collinge J: Sequence Variation in Intron of Prion Protein Gene, Crucial for Complete Diagnostic Strategies. Human Mutation 1996;7(3):280-281.

9 Windl O, Dempster M, Estibeiro JP, Lathe R, de Silva R, Esmonde T, Will R, Springbett A, Campbell TA, Sidle KC, Palmer MS, Collinge J: Genetic basis of Creutzfeldt-Jakob disease in the United Kingdom: A systematic analysis of predisposing mutations and allelic variation in the PRNP gene. Human Genetics 1996;98(3):259-264.

10 Caughey B, Horiuchi M, Demaimay R, Raymond G: Assays of protease-resistant prion protein and its formation. Methods in Enzymology 1999;309:122-133.

Dr. B.T. Butman, GenVec, Inc., 65 W. Watkins Mill Road, Gaithersburg, MD 20878, USA.
E-mail: bbutman@genvec.com

Petricciani J, Sheets R (eds): Vaccine Cell Substrates 2004.
Dev Biol (Basel). Basel, Karger, 2006, vol 123, pp 235-241.

Insect Cells as a New Substrate for Vaccine Production

W. Dalemans

Regulatory Strategy and Development, GlaxoSmithKline Biologicals, Rixensart, Belgium

Abstract: Vaccine antigens can be expressed in several expression systems, of which expression in eukaryotic cell lines is a well-known example. Whereas for most current applications cell substrates such as Vero, MRC-5 or chicken embryo fibroblasts are used, cell lines derived from insects present an alternative for the future production of vaccines. Since such substrates are not yet extensively used, knowledge about their feasibility and safety for vaccine antigen production is limited. In the current example, an insect-derived cell line Hi-5, used for the expression of the human papilloma virus L1 antigen, was extensively characterized, in particular for aspects of viral safety. A comprehensive QC testing programme, complemented with additional characterization tests relevant to the insect nature of the cell line, was developed. Current results indicate the satisfactory quality and safety of the Hi-5 cell line. The proposed testing programme could stand as an example for the release testing and further characterization which would be required for insect cell lines.

INTRODUCTION

Vaccines, the most efficacious preventive medicines against infectious diseases, are mainly biological products and exist under different forms, such as live attenuated micro-organisms, inactivated or killed pathogens, or subunit vaccines. The last are either extracts from the pathogen itself or can be derived by recombinant expression of the antigen(s) of interest. For the expression of recombinant antigens, several expression systems exist. Viral antigens are often expressed in and subsequently purified from eukaryotic cells. Well-known examples of such producer cells or cell lines are primary chicken embryo fibroblasts, MRC-5, Vero, or CHO cells.

Recent advances in cell culture technology allow the use of new cell substrates and the implementation of suspension and animal-free culture systems. One such new cell culture system is the Baculovirus expression system (BVES), where infection of a susceptible insect cell line with a recombinant baculovirus carrying an expression cassette for the vaccine antigen leads to high levels of antigen expression. Insect cells are an interesting alternative to producing vaccines on eukaryotic cell substrates as the cells are expressing proteins at high level, perform post-translational modifications and are easily amenable to upscaling and growth in suspension [1,2]. Several insect cell lines such as Sf9, Sf21, or Hi-5 can be used for baculo-driven expression. In the current example, the *Trichoplusia ni* derived Hi-5 cell line was used to produce the L1 protein of the human papilloma virus [3]. In the absence of specific guidelines for the evaluation of the quality of biotechnology products derived from insect cell lines, a comprehensive testing programme was designed. This QC testing and characterization programme was based on existing guidance for testing cellular substrates for the manufacturing of vaccines or biologicals, and complemented with specific tests was considered relevant for insect cell substrates. The proposed testing could constitute a framework for defining quality and safety testing requirements for insect cell lines. The results of the tests performed indicate a satisfactory quality and viral safety profile of the BVES in the Hi-5 cell line in view of future vaccine production.

Testing Programme: General and Viral Safety Tests

Several guidance documents give details of the testing requirements for cell substrates. These documents describe classical tests for investigating the general quality and (viral) safety of cell lines, and such tests will also generically apply to novel cell substrates, regardless of their origin. Examples of such tests are the purity and sterility tests, the screening for the presence of adventitious agents by culture on susceptible cell lines, and the tests for adventitious agents in animal models. The following guidance documents were considered to be applicable and served to design the basic QC testing programme of the Hi-5 cell line.

- ICH Q05A Guideline – Quality of biotechnological products: viral safety evaluation of biotechnology products derived from cell lines of human or animal origin - CPMP/ICH/295/95 (1997).
- ICH Q05D Guideline – Quality of biotechnological products: Derivation and characterisation of cell substrates used for the production of biotechnological/ biological products - CPMP/ICH/294/95 (1998).
- WHO Technical report Series No. 878, 1998 – Annex 1: requirements for the use of animal cells as in vitro substrates for the production of biological products.
- CBER's Points to Consider in the Characterisation of Cell Lines used to Produce Biologicals (1993)
- CBER's letter to manufacturers dated 14 December 1998: request for PCR-based reverse transcriptase (PBRT) assay to characterise cell substrates.
- Ph. Eur. 5.2.3. Cell substrates for the production of vaccines for human use.

With respect to testing for the presence of adventitious viruses and based on these guidelines, the in vitro testing programme included a human cell line (MRC-5, human diploid cells), a non-human primate cell line (Vero, Cercopithecus kidney cells), an animal cell line (BHK21, baby hamster kidney cells) and a cell line of the same species as that used for production (High Five insect cells). The three first cell lines are mammalian cell lines meant to detect a wide range of human and animal adventitious viruses. In the context of the use of insect cells for vaccine antigen production, the use of BHK21 and Vero cells as detector cells present an additional benefit given their known permissivity to insect-borne arboviruses like Flaviviridae (e.g. Dengue or West Nile viruses). The Hi-5 cell line is aimed at detecting viruses of insect origin. Given the novelty of the use of insect cells as vaccine production cells, three additional cell lines of insect origin were also used as detector cells for adventitious viruses. These cell lines were C6/36, Sf-9 and DL-2, known to be broadly permissive to insect viruses.

The detector cell lines were incubated with the test article and then cultured and observed for at least 14 days. The culture incubations were performed either at 37°C (mammalian cell lines) or at 27°C, being the production temperature as well as optimal for growth of insect cells (applied for the insect cell lines). Detection of any potential adventitious agents was conducted by monitoring the appearance of cytopathic effect in the detector cell cultures or by testing the detector cells ability to haemadsorb with guinea pig red blood cells at the end of the incubation period.

As specified in the general guidance documents, the cell line was also tested in in vivo models. According to the standard testing protocol, cells and culture supernatants were inoculated in embryonated eggs, suckling and adult mice as well as in guinea pigs; eggs and animals were observed for adverse effects.

Given the insect origin of the Hi-5 cell line, testing for Spiroplasma has also been included in the QC testing programme. This test is part of the general sterility testing, and is already required in some guidelines (CBER's Points to Consider in the Characterisation of Cell Lines Used to Produce Biologicals (1993)).

Table 1 summarizes the testing programme for the Hi-5 cell line. All the tests performed did "Pass", illustrating the quality and viral safety of the cell line. The complementarity of the different test substrates (seven in vitro cell substrates as well as four in vivo models) in their capacity to detect adventitious agents, indicates the absence of infectious viral agents.

As regulatory guidance also requires extensive testing on viral seeds derived on such cell lines, a similar testing programme, based on that of the cell substrate, has been designed for the testing of the seeds (summarized in Table 2). Also in this testing, all results were "Pass", indicative of the quality of the seeds and the absence of adventitious agents.

Insect Virus-Specific Characterization Tests

Given the insect origin of the cell line, additional characterization tests were performed to screen specifically for the presence of insect viruses. In defining the viral families to be screened for, two types of viruses were considered as being relevant: human pathogenic viruses which are transmitted by insects, and insect viruses which could potentially infect the insect host from which the cell line was derived, i.e. *Trichoplusia ni*. With respect to the first category, viruses being members of the *Flaviviridae, Togaviridae, Rhabdoviridae* or other families are known to be replicating in several of the indicator cell lines used in the testing plan (Vero, BHK21, C6/36). These viruses are also known to be infectious in the animal models, and should therefore be detected by these complementary testing methods. For the insect viruses which could potentially infect the insect host, detection by specific PCR before and after culture on insect cells was performed. A literature analysis described seven viral families to infect *Trichoplusia ni,* or more broadly the Lepidoptera insect family to which it belongs. These seven viral families are

Table 1: Testing programme of the Hi-5 cell line.

Tests	Hi5		
	MCB	WCB	EPC
• **Identity**			
– morphology (OM + EM)	X	–	X
– iso-enzyme analysis	X	X	X
• **Purity** (DNA fingerprinting)	X	X	X
• **Sterility**			
– bacteria	X	X	–
– fungi	X	X	–
– mycoplasma	X	X	–
- spiroplasma	X	X	–
• **Adventitious agents in cell cultures**			
ON DISRUPTED CELLS			
– MRC5 37°C	X	X	X
– VERO 37°C	X	X	X
– BHK21 37°C	X	X	X
– Hi5 27°C	X	X	X
– Sf-9 27°C	X		X
– C6/36 27°C	X		X
– DL2 27°C	X		X
ON CO-CULTURE			
– MRC5 37°C	–	–	X
– VERO 37°C	–	–	X
– BHK21 37°C	–	–	X
– Hi5 27°C	–	–	X
• **Adventitious agents in animals and eggs**			
(cell fluid and living cells)			
– adult mice	X	–	X
– suckling mice	X	–	X
– guinea-pigs	X	–	X
– embryonated eggs	X	–	X
• **Retroviruses**			
– detection by electron microscopy	X	–	X
– CoPERT	X	–	X

Abbreviations : MCB, master cell bank; WCB : working cell bank; EPC : end of production cells.

Nodaviridae, Entomopoxviridae, Tetraviridae, Picornaviridae, Baculoviridae, Ascoviridae and *Reoviridae*. Specific and consensus PCR primers for these virus families were designed based on conservative regions in the respective genomes and were dependent on the available sequence data in the databases. PCR analysis was done before and after amplification of Hi-5 cell extracts on Sf-9, Hi-5 or DL-2 insect cell lines. In none of the analyses could a PCR signal indicative for the presence of the targeted viral genomes be detected. These PCR results complement the in vitro and in vivo analyses and provide additional data to support the safety of the Hi-5 cell line further.

Table 2: Testing programme of the recombinant baculovirus viral seeds.

Tests	Baculo virus MS	Baculo virus WS
CELLS		
• **Identity**		
– iso-enzyme analysis	X	X
• **Sterility:**		
– mycoplasma	X	X
• **Adventitious agents on cell lysate**		
– MRC5 37°C	X	X
– VERO 37°C	X	X
– BHK21 37°C	X	X
– Hi5 27°C	X	X
– Sf-9 27°C	–	X
– C6/36 27°C	–	X
– DL2 27°C	–	X
HARVESTS		
• **Identity**	X	X
• **Potency**	X	X
• **Sterility**		
– bacteria	X	X
– fungi	X	X
– BK	X	X
– mycoplasma	X	X
– spiroplasma	X	X
• **Adventitious agents**		
– MRC5 37°C	X	X
– VERO 37°C	X	X
– BHK21 37°C	X	X
– Hi5 27°C	X	X
– Sf-9 27°C	–	X
– C6/36 27°C	–	X
– DL2 27°C	–	X
• **Adventitious agents in animals**		
– adult mice	–	X
– suckling mice	–	X
– guinea-pigs	–	X
Retroviruses:		
– CoPERT	–	X

Abbreviations : MS : master seed; WS : working seed.

Retroviral Signature

Part of the classical testing requirements for cell substrates is testing for the presence of retroviruses (ICH Q05A Guideline – Quality of biotechnological products: viral safety evaluation of biotechnology products derived from cell lines of human or animal origin - CPMP/ICH/295/95 (1997)); CBER's Points to Consider in the Characterization of Cell Lines Used to Produce Biologicals (1993); CBER's letter to manufacturers dated 14 December 1998: request for PCR-based reverse transcriptase (PBRT) assay to characterise cell substrates). Insect cells are described to be positive for reverse transcriptase (RT) activity, so further testing to discriminate for the presence of infectious retrovirus is necessary. In the case of the Hi-5 cell, a retrotransposon named TED has been described to be present in the insect [4], and the cell line was thus as expected positive for RT activity. Electron microscopy analysis did not reveal the presence of retroviral particles. This analysis was complemented by a Co-PERT test. Co-cultivation of Hi-5 cells with human 293 cells for several passages did not allow any RT signal by PERT to be detected, indicating the absence of infectious retrovirus in the Hi-5 cell line. This represents a similar situation as for the CHO and chicken cell substrates which are used for production of biologicals and vaccines. Moreover, viral clearance studies of the vaccine antigen purification protocol show a high clearance factor for retroviruses.

Safety Aspects of the Baculovirus

A particularity of each recombinant virus-based expression system with respect to viral safety, is the inherent presence of the "producer" viruses at high titre in the production process. Indeed, high level of antigen expression is achieved by efficient infection, replication and propagation of the recombinant virus. Therefore, the purification process needs to be designed to include highly efficient viral removal and/or inactivation steps to clear this initial viral load. With respect to the process applied for the purification of the L1 protein, different steps in the purification protocol have been validated for their potential viral clearance. The cumulative viral clearance factor for Baculovirus was shown to be >12 logs. Importantly, and further documenting the efficacy of the viral clearance, absence of infectious baculovirus has been demonstrated for the purified bulk of the vaccine.

It is also worthwhile noting that Baculovirus is considered to be safe for humans as it is non-infectious for mammals and therefore is used as a biopesticide [5].

Conclusions

Insect cells are an interesting alternative for producing vaccines on eukaryotic cell substrates as these cells are expressing proteins at high level, perform post-translational modifications and are easily amenable to upscaling and growth in suspension. In contrast to well-established cell substrates for vaccine production, insect cell lines have until now limited documentation on their viral safety aspects. Despite the lack of specific guidance on the control testing of insect cell lines for the manufacturing of biologicals or vaccines, a specifically adapted testing package was defined. Most of the tests applied were based on in vitro and in vivo tests already described for cell substrates since they are of general applicability for quality and characterization testing as well as for viral safety. With respect to the in vitro testing, cell lines of insect origin and with a broad permissivity to insect viruses were added to the mammalian cell lines classically used for this purpose. Tests for Spiroplasma do complement the general sterility tests. In addition to the in vitro and in vivo infectivity tests performed, specific PCR analyses for detection of insect viruses potentially infecting the insect host species were added to the cell line characterization. The combination of all the generic and more specific tests should eventually allow to comprehensively document the quality and safety of new insect cell lines, as was done in this case for the Hi-5 cell line. In addition, safety of vaccine antigens and their final product is further ensured by the demonstration of robust viral clearance of the purification process, raw material testing and strict adherence to GMP compliance.

The approach described illustrates that a comprehensive definition of the quality and safety testing programme for insect cell lines is feasible and practicable. It remains to be defined which of these tests would be required for routine QC release testing and which tests would be part of cell line characterization. A specific guidance document for insect cell substrates might not be needed, provided that current guidance documents are updated with specific requirements for insect cell lines.

Acknowledgements

I would like to thank Isabelle Knott, Jean-Paul Cassart, Michel Deschuytteneer, Donatienne Lefebvre, Isabelle Pierard, Anne-Françoise Macq, and Jean-Marc Wilkin, as well as many other colleagues from the R&D, QC and RA department for their invaluable help in establishing this programme and generating its results.

References

1 Vlak JM, de Gooijer CD, Tramper J, Miltenburger HG (eds) : Insect cell cultures, fundamental and applied aspects. Kluver Academic Press, 1996, pp 1-309.

2 Sandig V, Lieber A, Strauss M: Vectors for gene transfer and expression in animal cells, in Hause H, Wagner R (eds): Mammalian cell biotechnology in protein production. Berlin, Walter de Gruyter, 1997, pp 65-85.

3 Li GX, Hashimoto Y, Granados RR : Growth characteristics and expression of recombinant proteins by new cell clones derived from Trichoplusia ni (BTI Tn5B1-4) High Five ™ cells. Bioprocessing J 2003;2:35-38.

4 Miller DW, Miller LK: A virus mutant with an insertion of a copia-like transposable element. Nature 1982;299:562-564.

5 Consensus document on information used in the assessment of environmental applications involving baculoviruses. Series on harmonization of regulatory oversight in biotechnology, No.20. OECD, Paris, 2002.

Dr. W. Dalemans, GlaxoSmithKline Biologicals, Rue de l'Institut 89, BE-1330 Rixensart, Belgium.
E-mail: wilfried.dalemans@gskbio.com

Petricciani J, Sheets R (eds): Vaccine Cell Substrates 2004.
Dev Biol (Basel). Basel, Karger, 2006, vol 123, pp 243-250.

Safety Assessment of Madin Darby Canine Kidney Cells as Vaccine Substrate

J.K. Medema[1], J. Meijer[1], A.J. Kersten[1], R. Horton[2]

[1] Solvay Pharmaceuticals BV, Weesp, The Netherlands
[2] Solvay Pharmaceuticals Inc, Marietta GA, USA

Abstract: Conventional influenza vaccines are manufactured using embryonated chicken eggs, a substrate with little flexibility and vulnerable to extraneous agents. Solvay Pharmaceuticals developed a production technology based on the continuous cell line Madin Darby Canine Kidney (MDCK) as vaccine cell substrate. A risk-based safety assessment of MDCK, with respect to tumorigenicity of intact cells and oncogenicity of cellular components, cellular DNA and adventitious agents, shows that this substrate is as safe as other substrates and therefore without increased risk to the vaccine recipient.

INTRODUCTION

For decades, both research activities in the field of influenza and activities for surveillance of circulating influenza viruses have used cell culture systems. These range from primary cells to continuous cell lines, such as Madin Darby Canine Kidney (MDCK) and VERO, which are used most often. The influenza vaccine industry on the contrary has always used embryonated chicken eggs as substrate for influenza virus propagation, because of the lack of technologies to use such cell culture systems on a larger scale and relative lack of commercial incentive to develop new influenza vaccines. Therefore, industry addresses the rise in vaccine demand by increasing the number of embryonated eggs to be processed.

Excess 'all-cause' mortality attributable to influenza is estimated for example in the United States at approximately 0.01% or one death per 10^4 inhabitants, whereas vaccination is shown to reduce all-cause mortality by 39-59% [1]. In the mid 1990s, growing awareness of the medical and social impact of influenza has led to expanded guidelines for the use of inactivated vaccines by the World Health Organisation (WHO) and different public health authorities, thereby acknowledging the benefit of vaccination. As a consequence, steep increases in the use of influenza vaccines over the last few years have been observed, a change in market expectations to which industry cannot adequately respond; due to the reliance on embryonated eggs, industry needs to plan years ahead in manufacturing capacity and the breeding of flocks. This inflexibility has driven the search for more flexible vaccine production technologies. Most new technologies use mammalian cell culture systems, mainly for two reasons: the economy of scale of cell culture and the option of stocking all materials at the outset, which largely improves manufacturing logistics and makes the long- term planning for the supply of eggs obsolete. This led WHO to identify an urgent need for cell culture to control the public health burden of influenza in 1995 [2]. Likewise, Solvay Pharmaceuticals has developed Influvac® TC, an influenza surface antigen vaccine and the first biological product for human use that applies MDCK as cell substrate. After the initial marketing approval in an EU Member State in 2001, the first for a cell culture-based influenza vaccine, Solvay is currently pursuing market approval beyond The Netherlands after completion and validation of the fully dedicated production facility in 2005.

Historically, some theoretical quality concerns for cell-based biological products have originated from the presence of adventitious contaminants or from specific properties of the cells used for production. This led the U.S. Armed Forces Epidemiology Board in 1954 to recommend using "normal" cells rather than human tumour-derived cell lines for adenovirus vaccine development, mainly because of concerns about the transmission of oncogenic agents. The recommendation created a precedent that still affects the regulatory assessment of the use of continuous cell lines for vaccine development [3]. However, knowledge on the use of and risks associated with continuous cell lines as vaccine cell substrate has increased significantly over the last 50 years and justifies the continuous revisiting of the regulatory approach toward vaccine cell substrates. Regulatory guidance on the use of cell substrates in general and on the risks associated with several specific substrates does exist, and therefore the applicability of MDCK for the production of influenza vaccines is evaluated based on an assessment of the absence of risks to the vaccine recipient originating from the vaccine cell substrate.

Materials and Methods

The original MDCK cell line was established from the kidney of a normal, adult, female cocker spaniel in 1958, by S.H. Madin and N.B. Darby, who submitted this line to the American Type Culture Collection (ATCC) for deposition in 1964 at passage 49 [4,5]. It was used primarily for virological studies in the early 1960s, followed by characterisation studies of the cell line itself as early as 1966 [6]. It is one of the most studied polarised cell lines and already by 1979 considered one of the best characterised epithelial cell lines, still closely resembling kidney epithelia after almost 20 years of culture in vitro [7]. In 1991, the ATCC prepared a Working Cell Bank, designated MDCK CCL-34, at passage 54 by direct sub-culturing from the passage 49 depositor ampoules. All animal-derived materials used were compliant with TSE guidelines. Solvay Pharmaceuticals has acquired ATCC's MDCK CCL-34 and adapted this to serum-free medium (MDCK-SF). The MDCK-SF cell line was banked by BioReliance Biotech (Rockville, Maryland, U.S.A. and Stirling, U.K.) at passages 56-58, again only using TSE-compliant animal-derived materials. This Master Cell Bank/Working Cell Bank system was subcultured to passage 96 and designated Extended Cell Bank, using methods representative for the intended scale of commercial production. All subculturing for cell banking and commercial production is free of animal components that do not comply with current TSE guidelines. Both passages 56-58 and passages at or over 96, i.e. above those intended for production, were characterised and validated in accordance with European [8] and United States [9] regulatory guidelines. From the Extended Cell Bank, viable intact cells were used for tumorigenicity and oncogenicity studies.

The safety of the MDCK-SF cell line for vaccine production for human use is evaluated by a risk-based assessment, which is divided into the perceived risks of (i) intact, viable cells, (ii) cellular components and adventitious agents and (iii) cellular DNA. The theoretical risk associated with each of these is evaluated both for the substrate itself and for the inactivated, surface antigen influenza vaccine produced on this cell substrate.

Risk assessment of intact, viable cells

The tumorigenic potential of the MDCK-SF cell line is quantified by inoculating four-week-old athymic nude mice subcutaneously with 10^7, 10^5, 10^3 and 10^1 viable cells at a passage level higher than the maximum intended for production. Cells were grown from the ECB in serum-free medium and harvested so that viability was maintained. Control groups were subcutaneously inoculated with 10^7 viable HeLa cells or 10^7 viable cells of the parent MDCK cell line (ATCC CCL-34). The observation period reported here spans five months, during which animals were checked regularly for nodule development and clinical symptoms.

The risk associated with tumorigenic potential of the vaccine cell substrate to recipients of the MDCK-based, purified surface antigen influenza vaccine is further assessed by evaluation of the clearance of viable, intact cells and of the tumorigenic potential by the vaccine purification process.

Risk assessment of adventitious agents

The MDCK Master Cell Bank (MCB) and Extended Cell Bank (ECB) were assessed for the presence of extraneous and endogenous adventitious agents. This was done by:

(i) tracking the passage history of the MDCK cell line, including assessing applied materials for the possible introduction of transmissible spongiform encephalitis (TSE) agents;

(ii) non-specific adventitious agent testing of MCB and ECB in accordance with regulatory guidance;

(iii) specific adventitious virus testing with a Canine Antibody Production test based on the assessment and prioritisation of MDCK susceptibility to these viruses.

The oncogenicity of cellular components of MDCK and of potentially present tumourviruses or tumourvirus sequences in the cell substrate is evaluated by inoculating four-week-old athymic nude mice, newborn rats and newborn hamsters subcutaneously with a lysate of 10^7 viable cells at a passage level higher than the maximum intended for production. The observation period reported here spans five months, during which animals were checked regularly for nodule development and clinical symptoms.

The risk associated with adventitious agents in general to recipients of the MDCK-based, inactivated, purified surface antigen influenza vaccine is further assessed by evaluation of the clearance of model viruses. This is complemented with evaluation of tumorigenic potential clearance during vaccine purification by including crude, clarified harvest of 10^7 cells and inactivated virus concentrate in the oncogenicity studies described above.

Risk assessment of cellular DNA

The oncogenic potential of MDCK-DNA is evaluated by inoculating four-week-old athymic nude mice, newborn rats and newborn hamsters subcutaneously with at least 100 micrograms of DNA extracted from viable cells at a passage level higher than the maximum intended for production. The observation period reported here spans five months, during which animals were checked regularly for nodule development and clinical symptoms.

The risk associated with the possible oncogenic potential of the vaccine cell substrate DNA for recipients of the inactivated, purified surface antigen influenza vaccine produced on this substrate is further assessed by evaluation of the clearance of DNA by the vaccine purification process.

Results

Risk assessment of intact, viable cells

We previously reported that in the artificial tumorigenicity model using immunodeficient mice, 10^7 viable cells of the MDCK-SF cell line led to nodule formation at the injection site only [10]. This was part of a comprehensive preclinical safety evaluation, based on which a full clinical development programme was accepted and performed. The safety of the MDCK-SF-based vaccine was confirmed in these clinical studies, shown to match the excellent safety profile of egg-derived influenza inactivated surface antigen vaccines and reported already from 1997 onwards [11,12]. Here we report the further quantification of the tumorigenic potential of the cell line by subcutaneous inoculation of immunodeficient mice (four-week-old athymic nude) with a dilution series of 10^7, 10^5, 10^3 and 10^1 viable MDCK-serum free cells, derived from the ECB. Results are depicted in Table 1.

Table 1: Nodule development in tumorigenicity studies of intact cells.

	% positive nodule development at any time during observation period	**day of first observation of nodule development**
MDCK-SF, 10^1	0	none
MDCK-SF, 10^3	0	none
MDCK-SF, 10^5	73.1	22
MDCK-SF, 10^7	100	5
MDCK-ATCC, 10^7	96.2	5
HeLa, 10^7	100	5

The theoretical risk of using such neoplastic cells for vaccine production has been extensively discussed [13 and current proceedings of Vaccine Cell Substrate 2004], and can be further assessed by evaluation of the chance of exposure of a vaccine recipient to a tumorigenic dose of the vaccine cell substrate. This risk is assessed by a validation of the capacity of the influenza subunit vaccine purification process to clear intact cells and thus the tumorigenic potential. The clearance of intact cells has been validated on a pilot scale, showing a high redundancy in clearance capacity of intact cells by clarification, chemical inactivation, solubilisation, ultracentrifugation and final sterile filtration procedures. This redundant cell substrate clearance will be confirmed on the commercial scale intended. The capacity of the vaccine purification process to clear the tumorigenic potential observed with intact cells is furthermore shown by including early in-process materials generated during vaccine purification as a control group in the oncogenicity lysate studies in four-week-old athymic nude mice, newborn rats, and newborn hamsters. None of the early in-process materials induces nodule development in any of the species in five months, which shows that the earliest purification procedure of clarification centrifugation of a crude, homogenised harvest of 10^7 cells is sufficient to remove the tumorigenic potential of the vaccine cell substrate. Subsequent stringent purification procedures ensure sufficient redundancy and fully eliminate the risk of exposure of a vaccine recipient to the cell substrate tumorigenic potential.

Risk assessment of adventitious agents

The MDCK-SF cell bank system was assessed for absence of adventitious agents in accordance with European [8] and U.S. [9] guidelines, which is summarised in Table 2: The assessment includes

(i) non-specific tests for adventitious viruses in vitro by co-cultivation in six cell lines of human, simian, bovine, canine and rodent origin, in vivo in suckling

Table 2: Tests performed on MDCK cell banks, other than tumorigenicity/oncogenicity.

Test	**passage 56-58**	**passage 96+**
Identity		
– Iso-enzyme analysis	canine	canine
– Genetic analysis		
Chromosome count	canine	canine
Chromosome aberration	canine	canine
Karyology	canine	canine
Extraneous and endogenous agents		
– Mycoplasma		
Agar cultivable and non-cultivable	negative	negative
Aerobic and non-aerobic	negative	negative
– Bacterial and fungal sterility		
Growth in different broths	negative	negative
– Endogenous and adventitious viruses		
In vitro in different cell lines	negative	negative
In vivo in suckling mice, adult mice, guinea pigs, rabbits and embryonated eggs (allantoic, chorioallantoic and yolk-sac)	negative	negative
Canine Antibody Production test	negative	negative
– Retrovirus testing		
Induction reverse transcriptase	negative	negative
Electron microscopy	negative	negative

mice, adult mice, guinea pigs and rabbits and in ovo in embryonated chicken eggs via three different inoculation routes;

(ii) specific retrovirus testing;

(iii) microbiological testing for the presence of bacteria, fungi and mycoplasmas.

No evidence was found for the presence of any adventitious agent in the MDCK-SF cell banks.

Furthermore, the specific in vitro and in vivo assays for adventitious viruses were assessed for their suitability to detect canine, bovine and porcine viruses, which might be present in the cell bank system originating from the original cell line seed or from contaminated animal sera, culture media or trypsin used during subculturing of the original cell seed. This resulted in several viruses that were not likely to be detected by the applied non-specific assays; therefore these were included in a Canine Antibody Production test, in which lysate of MDCK-SF cells were inoculated in six different breeds. No evidence for the presence of any of these viruses was found. This confirms that the MDCK-SF cell banks can be considered free from endogenous and extraneous adventitious viruses.

The oncogenicity of cellular components of MDCK-SF and of possibly present tumourviruses or tumourvirus sequences in the cell substrate was evaluated by inoculating four-week-old athymic nude mice, newborn rats and newborn hamsters subcutaneously with cell lysate from 10^7 cells at a passage level higher than the maximum intended for production. Nodule development was not observed in any of these groups up to five months; thus cellular components of MDCK-SF do not lead to transforming events even in immunodeficient animals and evidence for the presence of oncogenic viruses is not found.

Overall these studies demonstrate that the MDCK-SF cell bank is free of adventitious agents. Moreover, the risk associated with adventitious agents to recipients of the MDCK-based inactivated, purified surface-antigen influenza vaccine is fully diminished by validation of the capacity of the purification process to remove or inactivate any type of virus. This is performed by a viral clearance validation using model viruses that represent a range of size, morphology and genome types, including enveloped and non-enveloped viruses [10]. It can be concluded that the risk of exposure to an adventitious virus by the MDCK-derived influenza subunit vaccine is negligible, because of the double safeguard that the vaccine cell substrate was shown to be devoid of virus and the purification process has redundant capacity to clear any potentially present virus.

Risk assessment of cellular DNA

To show that nodule development observed in immunodeficient mice inoculated with viable MDCK-SF cells is not caused by a transforming event originating from oncogenic genomic sequences, the oncogenicity of MDCK-SF DNA has been assessed as well. This was done by inoculation of at least 100 micrograms of high molecular weight DNA, extracted from cells derived from the Extended Cell Bank, in four-week-old athymic nude mice, newborn rats and newborn hamsters. Nodule development was not observed in any of the inoculated animals up to five months.

The Influvac® TC purification process has been validated on clearance of DNA, typically reducing levels of residual MDCK-DNA to levels below 100 picograms per dose. This gives a safety margin for absence of an oncogenic event shown in the in vivo oncogenicity models of 10^6; risk events associated with MDCK-DNA are thus less than one in a million per vaccine dose.

Discussion

The detailed historical track record of Solvay's MDCK-SF cell line currently dates back to 1964. Gaps in the track record before that date are not considered as critical for risks associated with the use of MDCK for human vaccine production, because it is unlikely that TSE agents were introduced in that era and the absence of other adventitious agents is shown during the extensive testing programme.

Subcutaneous inoculation of the immunodeficient mice with 10^1 and 10^3 viable MDCK-SF cells at a passage level above those intended for production did not lead to nodule formation at the injection site during the observation period. Inoculation with 10^5 and 10^7 viable MDCK-SF cells does induce nodules at the injection site in 70-100% of inoculated animals. This results in a TPD_{50} (dose at which 50% of immunodeficient mice develop tumours, calculated by Spearman-Karber) for the MDCK-SF cell line of approximately 4.5, and therefore this cell line should not be considered nor classified as highly tumorigenic [3]. The onset of nodule development was clearly later in the 10^5 viable cells group compared to all (MDCK-SF, MDCK-ATCC and HeLa) 10^7 groups. Nodules induced in the MDCK-SF inoculated animals are somewhat more persistent than in the MDCK-ATCC group, in line with the adaptation of MDCK-SF to more stringent, serum-free conditions. As determined visually, microscopically and partly histopathologically, all nodules observed are of benign phenotype, moderate in size and non-progressive, in contrast to the HeLa inoculated mice in which tumour development was significantly more aggressive. Progressive tumour growth exceeding 100 mm^2 was only observed in HeLa inoculated mice. No evidence of the formation of secondary tumours has been found in any of the MDCK inoculated mice specifically sacrificed for this purpose; nodules at the injection site were fluid filled with monolayer epithelial sheets around the periphery of the nodule.

The benign character of nodules observed at the injection site when immunodeficient mice are inoculated with $\geq 10^5$ viable MDCK cells, combined with the absence of secondary tumour induction, clearly indicates that the MDCK cell line is able to grow under artificial conditions, both in vitro and in the tumorigenicity model of immunodeficient mice, but not under natural conditions, as no evidence for nodule development was found in any immuno-competent species (mice, guinea pigs and rabbits).

The observation that a lysate of 10^7 viable cells does not lead to nodule development indicates that even in immunodeficient animals no evidence for the presence of oncogenic viruses or a transforming event by cellular components of MDCK is found. This further supports the observation that a positive result in a viable cell tumorigenicity study is a mere characterisation of the phenotype of that particular cell growing under artificial conditions, not indicative for an oncogenic event

Risks attributable to residual vaccine cell substrate DNA are a topic of continuing discussions, as actual transforming events by DNA in vivo appear to be difficult to achieve in practice. Earlier estimations of risks for the recipients of a vaccine containing 100 pg of residual DNA range from 1 in 5×10^8 to as low as 1 in 10^{12} [14]. The generation of experimental proof to support these estimations is difficult, because of practical limitations to the amount of DNA that can be inoculated in oncogenicity/infectivity assays. Although considered acceptable in the CBER-defined risk approach [3], the risk of oncogenic events attributable to residual DNA in the MDCK-derived vaccine of less than 1 in 10^6 presented here probably still overestimates the actual risk by several orders of magnitude, because of the same practical limitations of the oncogenicity assay.

Even though the studies of intact cells, adventitious agents and cellular DNA reported here are limited by practicalities of artificial models, they do demonstrate that risks associated with the MDCK-SF cell substrate are remote and negligible. Compared to other substrates, such as embryonated chicken eggs and non-tumorigenic cell substrates, the use of MDCK-SF as influenza vaccine substrate therefore is not associated with any increased risk to the recipient. Moreover, the use of embryonated eggs as vaccine substrate is accompanied with the practical impossibility of testing each egg for the absence of extraneous agents, in contrast to the MDCK-SF cell bank. This makes MDCK-SF a very suitable, safe cell substrate for influenza vaccine manufacture, which effectively addresses the urgent need identified by WHO for cell culture to control the burden of influenza.

Conclusion

The studies presented that use MDCK-SF as cell substrate for the production of a purified, surface antigen influenza vaccine comprehensively assess the risks associated with intact cells, cellular components, cellular DNA and potentially present adventitious agents. Although the absence of remote theoretical risks associated with continuous cell lines used as a vaccine cell substrate can never be shown, these studies clearly demonstrate that these theoretical risks indeed are acceptable and well and redundantly controlled. Compared to other substrates, such as embryonated chicken eggs and other continuous cell lines, the use of MDCK-SF as influenza vaccine substrate is as safe and therefore without increased risk to the recipient.

References

1 Poland GA, Rottinghaus ST, Jacobson RM: Influenza vaccines: a review and rationale for use in developed and underdeveloped countries. Vaccine 2001;19:2216-2220.

2 Editorial: Cell culture as substrate for the production of influenza vaccines: Memorandum from a WHO meeting. Bull WHO 1995;73:431-435.

3 Lewis AM Jr, Krause P, Peden K: A defined-risk approach to the regulatory assessment of the use of neoplastic cells as substrates for viral vaccine manufacture, in Brown F, Lewis AM, Peden K, Krause P (eds): Evolving Scientific and Regulatory Perspectives on Cell Substrates for Vaccine Development. Dev Biol. Basel, Karger, 2001, vol 106, pp 513-535.

4 Madin SH, Darby NB: Proc Soc Exp Biol Med 1958;98:574-576.

5 U.S. Dept. HEW, P.H.S. Registry of Animal Cell Lines, 1965, 1st Ed., 1st Suppl.

6 Gaush CR, Hard, WL, Smith TF: Characterization of an established line of canine kidney cells (MDCK). Proc Soc Exp Biol Med 1966;122:931-935.

7 Taub M, Saier, MH Jr: An established but differentiated kidney epithelial cell line (MDCK). Methods Enzymol 1979;58:552-560.

8 European Pharmacopoeia 5.2.3: Cell substrates for the production of vaccines for human use. PharmEuropa 1998;10(4): 629-632.

9 Food and Drug Administration: International Conference on Harmonisation: Guidance on Quality of Biotechnological/Biological Products: Derivation and Characterisation of Cell Substrates Used for Production of Biotechnological/Biological Products; Availability. Federal Register 1999;63(182):50244-50249.

10 Brands R, Visser J, Medema J, Palache AM, van Scharrenburg GJM: InfluvacTC: A Safe Madin Darby Canine Kidney (MDCK) cell culture-vased influenza vaccine, in Brown F, Robertson JS, Schild GC, Wood JM (eds): Inactivated Influenza Vaccines Prepared in Cell Culture. Dev Biol Stand. Basel, Karger, 1999, vol 98, pp 93-100.

11 Palache AM, Brands R, van Scharrenburg GJ: Immunogenicity and reactogenicity of influenza subunit vaccines produced in MDCK cells or fertilized chicken eggs. J Infect Dis. 1997;176 Suppl 1:20-23.

12 Palache AM, Scheepers HSJ, de Regt V, van Ewijk P, Baljet M, Brands R, van Scharrenburg GJM: Safety, Reactogenicity and Immunogenicity of Madin Darby Canine Kidney Cell-Derived Inactivated Influenza Subunit Vaccine. A Meta-Analysis of Clinical Studies, in Brown F, Robertson JS, Schild GC, Wood JM (eds): Inactivated Influenza Vaccines Prepared in Cell Culture. Dev Biol Stand. Basel, Karger, 1999, vol 98, pp 115-125.

13 Brown F, Lewis AM, Peden K, Krause P (eds): Evolving Scientific and Regulatory Perspectives on Cell Substrates for Vaccine Development. Dev Biol. Basel, Karger, 2001, vol 106, pp 1-535.

14 Petricciani J, Loewer J: An overview of cell DNA issues, in Brown F, Lewis AM, Peden K, Krause P (eds): Evolving Scientific and Regulatory Perspectives on Cell Substrates for Vaccine Development. Dev Biol. Basel, Karger, 2001, vol 106, pp 283-290.

Dr. J.K. Medema, Solvay Pharmaceuticals BV, PO Box 900, NL-1380 DA Weesp, The Netherlands;
E-mail: jeroen.medema@solvay.com

Petricciani J, Sheets R (eds): Vaccine Cell Substrates 2004.
Dev Biol (Basel). Basel, Karger, 2006, vol 123, pp 251-263.

Tumorigenicity Assessments of PER.C6® Cells and of an Ad5-Vectored HIV-1 Vaccine Produced on this Continuous Cell Line

B.J. Ledwith[1], C.L. Lanning[1], L.A. Gumprecht[1], C.A. Anderson[1], J.B. Coleman[1], N.T. Gatto[1], G. Balasubramanian[1], G.M. Farris[1], R.K. Kemp[1], L.B. Harper[1], A.B. Barnum[1], S.J. Pacchione[1], K.L. Mauer[1], P.J. Troilo[1], E.R. Brown[2], J.J. Wolf[1], J.A. Lebron[1], J.A. Lewis[2], W.W. Nichols[1]

[1] Department of Safety Assessment and

[2] Department of BioProcess Research and Development, Merck Research Laboratories, Merck & Co., Inc., West Point, Pennsylvania , USA

Abstract: PER.C6®, a cell line derived from human embryonic retinal cells transformed with the Adenovirus Type 5 (Ad5) E1A and E1B genes, is used to produce E1-deleted Ad5 vectors such as the MRKAd5 HIV-1 gag vaccine. While whole, live PER.C6® cells are capable of growing as tumours when transplanted subcutaneously into immunodeficient nude mice at a high dosage, the process for vaccine production includes filtration steps and other methods which effectively preclude contamination by intact viable substrate cells. However, because of the neoplastic nature of this cell line, we carried out a series of investigations to assess the tumorigenic risk posed by residuals from the cell substrate in a vaccine. To address concerns about transmission of oncogenic DNA, we demonstrated that purified PER.C6® cellular DNA does not induce tumours in newborn hamsters or nude mice. To address concerns about other potential residuals, including hypothetical adventitious tumour viruses, we demonstrated that a PER.C6® cell lysate and a MRKAd5 HIV-1 gag vaccine produced on PER.C6® cells do not induce tumours in newborn hamsters or newborn rats. These results, in conjunction with the wide panel of viral safety tests performed on these cells, support the safety of the PER.C6® as a cell substrate for vaccine production.

INTRODUCTION

PER.C6® cells are used to produce Adenovirus Type 5 (Ad5) vectors, including an experimental MRKAd5 HIV-1 gag vaccine currently in clinical development [1]. Ad5 vectors are replication-defective due to deletion of the Ad5 E1A and E1B genes, and consequently must be propagated using a helper (complementing) cell line. The PER.C6® cell line was derived from human embryonic retinal cells that were transfected with an expression plasmid containing the E1A and E1B genes of Ad5 [2]. Unlike other helper cell lines such as HEK293, the Ad5 sequences in PER.C6® cells include only the coding sequences for the E1 proteins and have no overlapping sequences with E1-deleted Ad5 vectors – thus, there is no potential for homologous recombination that could produce replication-competent adenovirus (RCA) [2,3].

Transfection with the Ad5 E1 genes (E1A and E1B) is known to cause cellular transformation, and PER.C6® cells are immortalized (capable of continuous growth in vitro), heteroploid (a typical characteristic of continuous cell lines), and tumorigenic (capable of forming tumours when transplanted subsequently into immunodeficient mice) [3]. We carried out a series of investigations to characterize the tumorigenic phenotype of the whole live cells and to assess the tumorigenic risk posed by cellular residuals in a vaccine. The tumorigenic phenotype of the cells was characterized by subcutaneous transplantation in immunodeficient nude mice, titrating the dosage of both the PER.C6® cells and the positive control HeLa cells. The oncogenic potential of PER.C6® DNA was tested in a nine-month tumorigenicity study in nude mice, to supplement previously reported data from five-month studies in newborn hamsters and nude mice [3]. To address theoretical concerns about tumorigenic risk from other PER.C6® cell residuals, including hypothetical adventitious tumour viruses, we carried out six-month tumorigenicity studies of both a concentrated PER.C6® cell lysate and a vaccine (MRKAd5 HIV-1 gag) produced on PER.C6® cells, each tested in both newborn rats and newborn hamsters.

Materials and Methods

Test articles

PER.C6® cells were derived from the Crucell research Working Cell Bank (rWCB) [3], A068-043W, which is at Passage 33 and arbitrarily designated to be population doubling level (PDL) = 0. The cells were grown from the rWCB to the indicated passage or PDL in suspension culture using serum-free medium.

PER.C6® DNA was extracted and purified from PER.C6® cells (harvested at a PDL of 63.9 relative to Crucell's rWCB) by lysis with sodium dodecyl sulphate, treatment with RNase A and proteinase K, extractions with phenol:chloroform:isoamyl alcohol and chloroform:isoamyl alcohol, and precipitation with isopropanol. The DNA was resuspended in Tris-EDTA-Saline (10 mM Tris-HCl [pH 8.0], 1 mM EDTA, and 150 mM NaCl) to a target concentration of 1.0 mg/mL (actual final concentration = 1.17 mg/mL) for a target dose of 250 mg (actual dose = 293 mg). The PER.C6® DNA preparation was shown to be free of RNA and protein contamination by UV spectrophotometry (A260 nm/A280 nm ratio >1.8). Analytical agarose gel electrophoresis confirmed the absence of RNA contamination. The average size of the DNA was approximately 50-100 Kb based on analytical pulse-field gel electrophoresis.

The PER.C6® cell lysate was obtained by treating a cell suspension (harvested at PDL 63.3 from the Crucell rWCB), at a concentration of 10^8 cells/mL, to three cycles of freeze-thaw lysis. The lysate was stored at approximately –70°C. Since one hypothetical concern to be addressed by this study was the potential presence of unknown tumour viruses, it was important to demonstrate that the lysate preparation conditions were compatible with maintaining viral activity. To do this, control viruses (MMR vaccine strains) were spiked into PER.C6® cells, subjected to the freeze-thaw lysis procedure and storage conditions, assayed for infectivity in vitro, and were found to exhibit only a slight decrease (0 to 0.4 logs) in potency (which ranged from 4.2 to 6.1 logs for control samples).

PER.C6® cells: tumorigenicity study in nude mice

Female nude mice (Harlan Sprague Dawley [Hsd]: athymic nude-nu) were injected subcutaneously with a dose volume of 0.2 mL of either cell culture media (Complete 293SFM II; negative control arm), HeLa cells (Lot #HELA ABF 29; positive control arm), or PER.C6® cells (Lot # F01-040, Passage 58, PDL 64 from the Crucell rWCB). Palpations were performed on all surviving mice on post-injection days 3, 7, 10, 14, 18, 21, 28, 35, 39, 42, 49, 56, 63, 70, 77, 81 and 84. Necropsies were performed at 21, 42 and 84 days post-injection (10 mice/group/time point), except for the HeLa positive control mice which were only necropsied at 42 days post-injection. At necropsy, the injection site, thoracic and abdominal viscera, and regional lymph nodes (when detectable) were examined. Histological examinations were performed on the injection sites, regional lymph nodes (when possible), and lungs, as well as on any masses detected in thoracic and abdominal viscera.

PER.C6® DNA: Thirty-nine-week subcutaneous tumorigenicity study in female nude mice

Female Hsd: Athymic Nude-nu mice, 49-57 days old, were injected subcutaneously with a dose volume of 0.25 mL. Two negative control groups of 100 mice each received the vehicle control, Tris-EDTA-Saline. The treated group of 100 mice received PER.C6® DNA (250 µg/mouse). Mice were observed daily for physical signs and palpated for masses. Necropsy with complete gross examination was performed at study termination (39 weeks post-injection). Histological examination was performed on the injection site, gross nodules, and tissues with grossly noted changes from all animals. Tumours from the PER.C6® DNA-treated mice, and appropriate control samples, were assayed by real-time quantitative PCR for PER.C6®-derived DNA sequences (Ad5 E1A and E1B and human *Alu* sequences).

PER.C6® cell lysate: Twenty-seven week subcutaneous tumorigenicity study in newborn hamsters

Newborn hamsters (LAK: LVG (SYR) BR), approximately 12- to 30-hrs old, were injected subcutaneously into the intrascapular region with a dose volume of 0.1 mL. Two groups of 150 control hamsters (gender undetermined) received the vehicle control, A105 placebo (5 mM Tris base, 75 mM NaCl, 1 mM $MgCl_2$, 0.005% Tween-80 and 5% sucrose). A group of 150 treated hamsters received the PER.C6® cell lysate (10^7 cells/animal). While with the dam, the pups were observed daily for mortality until study day 21. After study day 21, the pups were housed individually and observed weekly for mortality, physical signs, palpable masses and body weights. After weaning, each group was culled to 100 animals (approximately 50 animals/sex). Complete necropsies, including examination and collection of tissues from an extensive list, were performed on all hamsters approximately 27 weeks after dosing. Histological examination was performed on the injection site, gross nodules, and tissues with grossly noted changes from all animals. Tumours from the PER.C6® cell lysate group, and appropriate control samples, were assayed by real-time quantitative PCR for PER.C6®-derived DNA sequences (Ad5 E1A and E1B and human Alu sequences) to rule out the involvement of PER.C6® DNA in the cell lysate in the process of tumorigenicity.

PER.C6® cell lysate: Twenty-eight week subcutaneous tumorigenicity study in newborn rats

Newborn rats (Crl:CD®(SD)IGS BR), approximately 12- to 30-hrs old, were injected subcutaneously into the intrascapular region with a dose volume of 0.1 mL. Two groups of 118 control rats (approximately 59 per sex/group) received the vehicle control, A105 placebo. A group of 118 treated rats (approximately 59/sex) received the PER.C6® cell lysate (10^7 cells/animal). While with the dam, the pups were observed daily for mortality until study day 21. After study day 21, the pups were housed individually and observed weekly for mortality, physical signs, palpable masses and body weights. Each group was culled to 102 animals (approximately 51 animals/sex) after weaning. Complete necropsies, including examination and collection of tissues from an extensive list, were performed on all rats approximately 28 weeks after dosing. Histological examination was performed on the injection site, gross nodules, and tissues with grossly noted changes from all animals. Tumours from the PER.C6® cell lysate group, and appropriate control samples, were assayed by real-time quantitative PCR for PER.C6®-derived DNA sequences (Ad5 E1A and E1B and human Alu sequences) to rule out the involvement of PER.C6® DNA in the cell lysate in the process of tumorigenicity.

MRKAd5 HIV-1 gag vaccine: Twenty-eight week subcutaneous tumorigenicity study in newborn hamsters

Newborn hamsters (LAK: LVG (SYR) BR), approximately 12- to 30-hrs old, were injected subcutaneously into the intrascapular region with a dose volume of 0.1 mL. Two groups of 150 control hamsters (gender undetermined) received the vehicle control, A105 placebo. A group of 150 treated hamsters received the

MRK Ad5 HIV-1 gag vaccine (1×10^{10} viral particles/animal). While with the dam, the pups were observed daily for mortality until study day 21. After study day 21, the pups were housed individually and observed weekly for mortality, physical signs, palpable masses and body weights. Each group was culled to 100 animals (approximately 50 animals/sex) after weaning. Complete necropsies, including examination and collection of tissues from an extensive list, were performed on all hamsters approximately 28 weeks after dosing. Histological examination was performed on the injection site, gross nodules, and tissues with grossly noted changes from all animals. Tumours from the vaccine-treated group, and appropriate control samples, were assayed by real-time quantitative PCR for vector-derived DNA sequences (Ad5 or gag sequences) to rule out the involvement of MRKAd5gag vector in the process of tumorigenicity.

MRKAd5 HIV-1 gag vaccine: Twenty-seven week subcutaneous tumorigenicity study in newborn rats

Newborn rats (Crl:CD®(SD)IGS BR), approximately 18- to 23-hrs old, were injected subcutaneously into the intrascapular region with a dose volume of 0.1 mL. Two groups of 150 control rats (approximately 75 per sex/group) received the vehicle control, A105 placebo. A group of 150 treated rats (approximately 75/sex) received the MRK Ad5 HIV-1 gag vaccine (1×10^{11} viral particles/animal). While with the dam, the pups were observed daily for mortality until study day 21. After study day 21, the pups were housed individually and observed weekly for mortality, physical signs, palpable masses and body weights. Each group was culled to 100 animals (approximately 50 animals/sex) after weaning. Complete necropsies, including examination and collection of tissues from an extensive list, were performed on all rats approximately 27 weeks after dosing. Histological examination was performed on the injection site, gross nodules, and tissues with grossly noted changes from all animals. Tumours from the vaccine-treated group, and appropriate control samples, were assayed by real-time quantitative PCR for vector-derived DNA sequences (Ad5 or gag sequences) to rule out the involvement of the MRKAd5gag vector in the process of tumorigenicity.

PCR analysis

DNA samples were analyzed by "TaqMan®" real-time quantitative fluorescence PCR in several of the tumorigenicity studies. The E1A and E1B segments were targeted to Ad5 sequence inserts within PER.C6®, and the ALU segment was targeted to the human *alu* repetitive elements. The AFLG3-1 and GAG segments were targeted to the MRK Ad5 HIV-1 gag vector (bases 30,726 to 30,785 and bases 2073 to 2158, respectively, of the 34,445-bp vector). The oligonucleotide primers, the fluorescent TaqMan® probe, and the expected product size are listed below (Table 1).

Table 1.

Segment	Oligonucleotide primers/probes	Name	Product size
E1A	5'-TGC CCC ATT AAA CCA GTT GC 5'-TCC TCG ATA CAT TCC ACA GCC fam-TGA GAG TGG GTG GGC GTC GCC-BHQ1	Ad5-F1 Ad5-R1 Ad5-BHQPr1	65 bp
E1B	5'- TTG GGT ATC CTG TCT GAG GGT AA 5'- ATG AAG CAA CCA CAG TCG GAG fam-TCC AGG GTG CGC CAC AAT GTG-BHQ1	E1B-F3 E1B-R3 E1B-Pr1	66 bp
ALU	5'- TGG TGA ACC CCC GTC TCT ACT AA 5'- CCT CAG CCT CCC GAG TAG C fam- AAA ATT AGC CGG GCG TGG TGG C-BHQ1	ALU-F1 ALU-R1 ALU-BHQPr1	88 bp
AdFL3-1	5'- GGC ATG CTT GCG CTC AA 5'- GGT AAG GTT GCC GGC CTC fam-TGG GCA ACG GCC TCT CTC TGG A-tamra	AFLG3-F1 AFLG3-R1 AFLG3-Pr1	60 bp
GAG	5'- CAA GAT TGT GAG GAT GTA CTC C 5'- ACC TGT CCA CAT AGT CCC TGA fam-ATC CTG GAC ATC AGG CAG GGC-tamra	GAG-F GAG-R GAG-pr	86 bp

Results

Characterization of the tumorigenic phenotype of PER.C6® cells

The tumorigenic phenotype of PER.C6® cells was previously demonstrated in two different 84-day subcutaneous tumorigenicity studies in female nude mice [3]. In the first study, cells at passage 64 (PDL 67.7 from the Crucell rWCB) were positive for tumour formation when 10^7 cells were injected. In the second study, cells at passage 59 (PDL 56.3 from the Crucell rWCB) were positive for tumour formation when 10^7 cells were injected, but negative for tumour formation when $\leq 10^5$ cells were injected.

Subsequently a third 84-day study was performed in which both PER.C6® cells (passage 58; PDL 64 from the Crucell rWCB) and HeLa cells were titrated. Mice were injected subcutaneously with either PER.C6® cells (at 10^7, 10^5, or 10^3 cells/animal), HeLa cells (at 10^6, 10^5, or 10^3 cells/animal), or cell culture medium (negative control). The tumour incidences in the third study are summarized in Table 2.

Table 2: PER.C6® cells: 84-day tumorigenicity study in female nude mice.

		Tumour incidence (Fraction of mice with tumours)		
Cell type	**No. of cells**	**Day 21**	**Day 42**	**Day 84**
HeLa Cells	$1{\times}10^6$	NA	10/10	NA
	$1{\times}10^5$	NA	9/10	NA
	$1{\times}10^3$	NA	0/10	NA
PER.C6® Cells	$1{\times}10^7$	5/10	10/10	7/10[a]
	$1{\times}10^5$	0/10	0/10	0/10
	$1{\times}10^3$	0/10	0/10	0/10
Medium control	–	0/10	0/10	0/10

[a] One of the seven animals with masses was sacrificed and necropsied post-injection day 64 and three of the seven remaining animals were sacrificed and necropsied on post-injection day 72.

None of the mice receiving 10^3 HeLa cells had palpable masses at the injection sites from the first palpation day (day 3 post-injection) until they were necropsied, and none of these mice had gross or microscopic evidence of tumour formation. Nine out of 10 animals receiving 10^5 HeLa cells and 10 out of 10 animals receiving 10^6 HeLa cells developed large, persistent neoplasms at the injection sites; none had neoplasms in other tissues examined.

None of the mice receiving 10^3 or 10^5 PER.C6® cells had palpable masses at the injection sites from post-injection day 3 until they were necropsied. In addition, none of these mice had gross or microscopic evidence of tumours. Palpable nodules were present at the subcutaneous injection sites in 30 of 30 animals receiving 10^7 PER.C6® cells on post-injection day 3. These nodules regressed in size, and in some cases disappeared. In most mice the regression was transient, and the nodules subsequently enlarged. In several mice in which the nodules had disappeared, they subsequently reappeared and became enlarged. Histologically, these nodules were compressive and were composed of sheets of large, pleomorphic cells with numerous, sometimes abnormal mitotic figures.

Thus, PER.C6® cells form tumours when injected subcutaneously in immunodeficient mice at a dose of 10^7 cells, but are negative for tumour growth at doses of $\leq 10^5$ cells. In contrast, HeLa cells are positive for tumour growth at 10^5 cells, though negative at 10^3 cells.

Investigation of the oncogenic potential of PER.C6® cellular DNA

The oncogenic potential of residual DNA from PER.C6® cells was previously investigated in newborn hamsters and nude mice [3]. PER.C6® DNA was negative for tumorigenicity when approximately 100 µg of DNA (isolated from passage 61 cells; PDL 61.4 from the Crucell rWCB) was injected as a single subcutaneous dose in newborn hamsters followed by a five-month observation period [3]. Similarly, no tumours were detected at or near the injection sites when approximately 225 µg of DNA (same preparation) was injected as a single subcutaneous dose in female nude mice followed by a five-month observation period [3]. However, one of the 20 treated female nude mice exhibited a lymphoma, whereas none of the 40 control female nude mice had tumours. The lymphoma was negative for PER.C6®-derived E1A sequences when analysed using real-time quantitative PCR, and its incidence was consistent with the occurrence of spontaneous lymphoma in various strains of nude mice.

Subsequently, to confirm the spontaneous nature of the lymphoma, the nude mouse study was repeated using a longer (nine-month) study duration and a larger number of nude mice. Two groups of 100 female nude mice received a single subcutaneous dose of the control article, and one group of 100 female nude mice received a single subcutaneous dose of 250 µg PER.C6® DNA (isolated from cells at PDL 63.9 from the Crucell rWCB). A summary of the neoplasms observed in this nine-month study is provided in Table 3. As expected, spontaneous lymphomas were observed in control mice and there was no evidence for a treatment-related effect on tumorigenesis.

Tumours observed in the PER.C6® DNA-treated mice were assayed by real-time quantitative PCR for the potential presence of PER.C6®-derived DNA sequences, including Ad5 E1A and E1B sequences and human *alu* sequences. The latter is a

Table 3: PER.C6® DNA: 39-week tumorigenicity study in female nude mice.

Summary of primary neoplasms	Control group 1	Control group 2	PER.C6® DNA group
Number of mice	100	100	100
Skin, malignant fibrosarcoma	0	1	0
Skin, benign papilloma	1	1	1^
Lung, benign adenoma	1	1	0
Malignant lymphoma*	1	4	2^

* Primary site undetermined.
^ Not considered to be treatment-related since incidence not significantly different from control groups.

ubiquitous contaminant in the laboratory and, therefore, the tumours were compared to similarly handled control samples. The results of these PCR analyses were negative for E1A and E1B, and within control levels for *alu*, confirming that the tumours observed in PER.C6® DNA-treated mice were not treatment related (data not shown).

Thus, PER.C6® DNA is not tumorigenic in either newborn hamsters or nude mice.

Tumorigenicity studies of a PER.C6® cell lysate

To address theoretical concerns about tumorigenic risk from other PER.C6® cell residuals, including hypothetical adventitious tumour viruses, we carried out tumorigenicity studies of a concentrated PER.C6® cell lysate. The PER.C6® cell lysate, at a subcutaneous dose of approximately 10^7 lysed cells, was tested in six-month tumorigenicity studies in both newborn rats and newborn hamsters and had no treatment-related effects on tumorigenesis in either study.

In hamsters, there were no treatment-related neoplasms at the injection site of any animal, nor were there treatment-related neoplasms in any tissue examined. One of 100 PER.C6® cell lysate-treated hamsters had an enlarged mesenteric lymph node that was a histiocytic sarcoma based on light microscopy (Table 4). Because of the low, single incidence, and the fact that this neoplasm can occur spontaneously in hamsters, this tumour was considered to be unrelated to treatment. To rule out a further treatment-related effect, specifically due to DNA in the cell lysate, the tumour and representative controls were assayed for the presence of PER.C6® DNA sequences. The tumour was negative for the Ad5 E1A and E1B segments, and had levels of *alu* contamination within the range of that observed in the control samples when analysed using real-time quantitative PCR.

Table 4: PER.C6® cell lysate: 28-week tumorigenicity study in newborn hamsters.

Summary of primary neoplasms	**Control group 1**		**Control group 2**		**PER.C6® cell lysate group**	
	Female	Male	Female	Male	Female	Male
Number of hamsters [a]	50	50	50	50	54	46
Tumours at injection site	0	0	0	0	0	0
Mesenteric lymph node, Histiocytic sarcoma	0	0	0	0	1^	0

[a] Number necropsied post-weaning

^ Not considered to be treatment-related due to low (single) incidence, and since this tumour type is observed in hamsters.

In rats, several tumours were observed in both control and treated animals (Table 5). There were no treatment-related neoplasms at the injection site of any rat, nor was there treatment-related neoplasms in any tissue examined. Again, when alanysed using real-time quantitative PCR tumours from the PER.C6® cell lysate-treated animals were negative for the Ad5 E1A and E1B segments, and had levels of *alu* contamination within the range of that observed in the control samples, confirming that the tumours did not contain PER.C6®-derived DNA.

Table 5: PER.C6® cell lysate: 28-week tumorigenicity study in newborn rats.

Summary of primary neoplasms	Control group 1		Control group 2		PER.C6® cell lysate group	
	Female	Male	Female	Male	Female	Male
Number of rats [a]	50	52	50	52	51	51
Tumours at injection site	0	0	0	0	0	0
Mammary adenocarcinoma	2	0	0	0	1^	1^
Malignant lymphoma	0	0	1	0	0	0

[a] Number necropsied post-weaning.
^ Not considered to be treatment-related since incidence not significantly different from control groups.

Thus, the PER.C6® cell lysate is not tumorigenic in newborn rats and newborn hamsters.

Tumorigenicity studies of a MRK Ad5 HIV-1 gag vaccine produced on PER.C6® cells

To address further theoretical concerns about tumorigenic risk from other PER.C6® cell residuals, including hypothetical tumour viruses, we carried out tumorigenicity studies of a vaccine, MRKAd5 HIV-1 gag, produced using PER.C6® cells. The MRKAd5 HIV-1 gag vaccine, tested at a subcutaneous dose of 10^{11} and 10^{10} viral particles in newborn rats and newborn hamsters, respectively, had no treatment-related effects on tumorigenesis.

Both studies involved a single subcutaneous injection in newborn animals, using the maximum-tolerated dose of vaccine (10^{11} and 10^{10} viral particles in newborn rats and newborn hamsters, respectively) as determined in exploratory studies for both species. In hamsters, there were no treatment-related neoplasms at the injection site of any animal, nor were there treatment-related neoplasms in any tissue examined. One female vaccine-treated animal died in drug week 25 due to myeloid leukaemia (Table 6). This neoplasm was not considered to be treatment-related due to its low, single incidence (1/100 treated animals). As additional evidence supporting the spontaneous nature of this myeloid leukemia, neoplastic lymph node tissue from the affected animal was analysed by real-time quantitative PCR and found to be negative for the presence of the MRK Ad5 gag vector DNA.

In rats, there were no treatment-related neoplasms at the injection site of any animal, nor were there treatment-related neoplasms in any tissue examined. One control male rat had an enlarged thymus, lymph nodes and spleen attributed to a poorly differentiated lymphoma (Table 7).

Thus, the MRK Ad5 HIV-1 gag Vaccine produced on PER.C6® cells is not tumorigenic in newborn rats and newborn hamsters.

Discussion

Whole, live PER.C6® cells are capable of forming tumours when transplanted subcutaneously at a high dosage in immunodeficient nude mice. This does not pose

Table 6: MRKAd5 HIV-1 gag vaccine: 28-week tumorigenicity study in newborn hamsters.

Summary of primary neoplasms	**Control group 1**		**Control group 2**		**MRKAd5 HIV-1 gag vaccine (1×10^{10} vp)**	
	Female	Male	Female	Male	Female	Male
Number of hamsters [a]	49	51	51	49	50	50
Tumours at injection site	0	0	0	0	0	0
Myeloid leukemia	0	0	0	0	1^	0

[a] Number necropsied post-weaning.
^ Not considered to be treatment-related due to low (single) incidence. Also, tumour was negative for the MRKAd5 gag vector by PCR analysis.

Table 7: MRKAd5 HIV-1 gag vaccine: 28-week tumorigenicity study in newborn rats.

Summary of primary neoplasms	**Control group 1**		**Control group 2**		**MRKAd5 HIV-1 gag vaccine (1×10^{11} vp)**	
	Female	Male	Female	Male	Female	Male
Number of rats [a]	50	50	50	50	50	50
Tumours at injection site	0	0	0	0	0	0
Malignant lymphoma	0	1	0	0	0	0

[a] Number necropsied post-weaning.

a risk to vaccine recipients because the production of the adenovirus-vectored vaccine includes filtration steps and other process methods which effectively preclude contamination by intact viable substrate cells. However, the neoplastic nature of the cell substrate raises two theoretical concerns about tumorigenic risk. First, there is the potential for residual cell substrate DNA in the vaccine to contain an activated oncogene that could be transmitted to the vaccine recipient. Second, there is the hypothetical concern that the cell substrate may contain an adventitious tumour virus that contributes to the neoplastic phenotype of the cells.

The oncogenic risk of residual DNA from neoplastic cell substrates has been the subject of theoretical risk assessments and experimental research for approximately two decades [4,5]. This risk is remote, since it would require that the complete oncogene sequence from the cellular DNA be taken up by host cells, target the nucleus, integrate into the host cell genome, and be expressed in a cell type sensitive to transformation. To date there is no evidence that DNA from neoplastic cell lines can induce tumours when injected in rodent models that are known to be highly sensitive to tumour viruses, tumour virus DNA, and chemical carcinogens. On the contrary, numerous studies summarized in Table 8 have uniformly demonstrated that large amounts of cellular DNA, even when isolated from highly tumorigenic cell lines, do not produce tumours when injected into animals [4,6-10]. Our studies demonstrating that PER.C6® cellular DNA does not induce tumours in newborn hamsters or nude mice are consistent with these findings. These negative results are also consistent with the WHO risk assessment which suggests that the theoretical Tumour-Producing Dose (TPD) for cellular DNA from neoplastic cells is approximately 2 grams (derived by using an experimentally determined TPD for tumour virus DNA in rodents by correcting for relative genome size, and assuming one oncogene per cellular genome) [4]. Based on this risk assessment, the WHO recommended a limit of

Table 8: Lack of tumorigenicity of cellular DNA isolated from neoplastic cell lines.

Source (Cell line) of DNA	Amount of DNA	Route	Animal model	Study duration	Tumour incidence[a]	Ref.
Balb/c B-cell hybridoma line 3-17	50 µg	s.c.	Newborn rats (+/- immuno-suppression)	1 year	0/9	6
Balb/c B-cell hybridoma line 3-17	250 µg	s.c.	Female Balb/c mice	2 years	0/200[b]	6
HeLa tumorigenic cell line	100 µg	s.c.	Newborn rats (+ immuno-suppression)	28 days	0/35	7, 8
CHO-K1 tumorigenic cell line	100 µg	s.c.	Newborn rats (+ immuno-suppression)	28 days	0/20	7, 8
Rat EJ-6 tumorigenic cell line	500 µg	s.c.	Newborn hamsters	5 months	0/8	9
PER.C6® tumorigenic cell line (E1-transformed human retinoblasts)	100 µg	s.c.	Newborn hamsters	5 months	0/20	3
PER.C6® tumorigenic cell line (E1-transformed human retinoblasts)	250 µg	s.c.	Nude mice	9 months	3/100 [c]	Note [d]
Human T24 tumour cell line	10 µg	s.c.	Newborn rats (+ immuno-suppression)	28 days	0/20	7, 8
Human T24 tumour cell line	1000 µg as chromatin	i.m., i.v., i.c.	Rhesus monkeys (+/- immuno-suppression)	8 years	0/10	10

[a] Fraction of animals which developed tumours.
[b] One animal had a haemangioma at the injection site which was not attributable to DNA injection. A similar lesion was observed in one of 50 control animals, and the neoplastic nature of the haemangioma was "questionable".
[c] No treatment-related tumours. The tumour incidence in treated mice (1/100 mice with skin benign papilloma and 2/100 mice with malignant lymphoma) was the same or lower as the tumour incidence in control animals (2/200 with skin benign papilloma, 1/200 with skin malignant fibrosarcoma, 2/200 with lung benign adenoma, 5/200 with malignant lymphoma).
[d] Study reported here.

s.c. = subcutaneous; i.m. = intramuscular; i.v. = intravenous; i.c. = intracerebral

10 ng per dose for the level of residual DNA from a continuous cell line in a parenteral vaccine, providing a safety margin over the theoretical TPD of approximately 2 x 10^8 [5]. The dose of PER.C6® DNA (250 μg) shown to be negative for tumorigenicity in the nude mouse study reported here is more than 25,000-fold greater than the WHO recommended limit of 10 ng, and more than 2.5 x 10^6 -fold greater than the level of residual PER.C6® DNA in the MRKAd5 HIV-1 gag Vaccine (< 100 pg).

The risk of adventitious viral contamination of cell substrates and biological products is addressed by rigorous sterility and viral safety testing in accordance with the ICH Guidance (11). A comprehensive panel of in vitro and in vivo tests that are known to be sensitive to a variety of viral agents are used, including primary and continuous cell lines of mammalian and human origin, embryonated eggs, and small animals. These classical viral propagation methods are supplemented with agent-specific testing, using Polymerase Enhanced Reverse Transcriptase (PERT) and PCR assays. The extensive panel of viral safety tests performed on Merck's PER.C6® Master Cell Bank is listed in Table 9. The PCR tests included

Table 9: Safety testing of the PER.C6® Master Cell Bank.

Test	Method
Sterility	Broth and agar for cultivation of Bacteria, Fungi, Mycoplasma In vitro cell culture testing for detection of Mycoplasma using Hoechst stain
Viral safety **In vivo eggs**	Eggs (allantoic, amniotic, yolk sac)
Viral safety **In vivo animals**	Guinea pig Adult and suckling mouse
Viral safety **In vitro cell culture**	Vero, MRC-5, 293, RK (Rabbit kidney) Bovine turbinate, Porcine kidney
Viral safety **agent-specific testing**	Transmission electron microscopy Tests for retroviruses: PERT for RT applied to PER.C6® supernatants and supernatants of PER.C6® cells cocultivated with Raji, RD, and H9 cells PCR tests for HBV, HCV, CMV, EBV, HHV-6,7,8, HPV, SV40, HIV-1,2, human and bovine polyoma, HTLV-1,2, SFV, measles, generic enteroviruses, generic adenoviruses, AAV, B19, BPV, BHV, BIV, BLV, bovine and porcine circoviruses, bovine rotavirus

Abbreviations:

PERT = Polymerase Enhanced Reverse Transcriptase
RT = Reverse Transcriptase
RD = Human rhabdomyosarcoma cell line
H9 = Human T cell line
HBV = Hepatitis B virus
HCV = Hepatitis C virus
CMV = Cytomegalovirus
EBV = Epstein-Barr virus
HHV = Human Herpes virus
HPV = Human papillomavirus
SV40 = Simian virus 40
HIV = Human immunodeficiency virus
HTLV= Human T-lymphotropic virus
SFV = Simian foamy virus
AAV = Adeno-Associated virus
B19 = Human parvovirus B19
BPV = Bovine parvovirus
BAV = Bovine adenovirus
BHV = Bovine herpesvirus
BIV = Bovine immunodeficiency virus
BLV = Bovine leukemia virus

viral agents with oncogenic potential in humans. However, because the PER.C6® cell line is a new, human, neoplastic cell substrate to be used for manufacturing of a vaccine, we performed additional studies to rule out further the presence of adventitious tumour viruses that hypothetically could have contributed to the neoplastic phenotype of these cells (in addition to the known effect of transformation by the Ad5 E1A and E1B genes). These studies included tumorigenicity studies in both newborn rats and newborn hamsters for both a PER.C6® cell lysate and a vaccine produced on PER.C6® cells, as described in this report. The negative results in these studies, together with the negative results in the comprehensive panel of viral safety tests described above, support the safety of PER.C6® cells as a substrate for vaccine production.

Conclusions

Whole (live) PER.C6® cells are capable of growing as tumours when transplanted subcutaneously into immunodeficient nude mice at a high dosage (10^7 cells). However, residuals from PER.C6® cells (tested as purified DNA, a concentrated cell lysate, and a vaccine produced on the cells) do not exhibit tumorigenic activity. These results, in conjunction with the wide panel of viral safety tests performed on these cells, support the safety of PER.C6® as a cell substrate for vaccine production.

Acknowledgements

We thank Weichang Zhou, Wayne Herber, and John Aunins for producing the PER.C6® cells for these experiments; Lori Mixson and Keith Soper for biostatisical support for tumorigenicity study designs; and Lyudmila Denisova, Chanda Johnson, Thomas Griffiths, Nathalie Innocent, Krista Maurer, Sue Morgan-Hoffman, Zhutian Niu, Xin Wang and Zhibin Wang for technical assistance with the DNA isolation and PCR analyses of tumours.

References

1 Shiver JW, Emini EA: Recent advances in the development of HIV-1 vaccines using replication-incompetent adenovirus vectors. Annual Review Medicine 2004;55:355-372.

2 Fallaux FJ, Bout A, Van Der Held I, Van Den Wollenberg DJM, Hehir KM, Keegan J et al: Human Gene Therapy 1998;9:1909-1917.

3 Nichols WW, Lardenoije R, Ledwith BJ, Brouwer K, Manam S, Vogels R et al: Propagation of adenoviral vectors: use of PER.C6® cells, in Curiel DT, Douglas JT (eds): Adenoviral Vectors for Gene Therapy. San Diego, Academic Press, 2002, pp 129-166.

4 WHO Technical Report Series: Acceptability of cell substrates for production of biologicals. Report of a WHO Study Group. World Health Organization, 1987, No. 747.

5 WHO Technical Report Series: WHO requirements for the use of animals cells as in vitro substrates for the production of biologicals (Requirements for Biological Substances No. 50), 1998, No. 878. Reproduced in Biologicals 1998;26:175-193.

6 Dortant PM, Claassen IJ, van Kreyl CF, van Steenis G, Wester PW: Risk assessment on the carcinogenic potential of hybridoma cell DNA: implications for residual contaminating cellular DNA in biological products. Biologicals 1997;25(4):381-390.

7 Palladino MA, Levinson AD, Svedersky LP, Obijeski JF: Safety issues related to the use of recombinant DNA-derived cell culture products: I. Cellular Components, in Spier R, Hennessen W (eds): Advances in Animal Cell Technology: Cell Engineering, Evaluation and Exploitation. Dev Biol Stand 1987, vol 66, pp 13-22.

8 Levinson AD, Svedersky LP, Palladino MA Jr: Tumorigenic Potential of DNA derived from mammalian cell lines. In vitro monograph 1985;6:161-165.

9 Mufson RA, Gesner T: Lack of tumorigenicity of cellular DNA and oncogene DNA in newborn Chinese hamsters. In vitro monograph, 1985;6:168-169.

10 Wierenga DE, Cogan J, Petricciani JC: Administration of tumour cell chromatin to immunosuppressed and non-immunosuppressed non-human primates. Biologicals 1995;23:221-224.

11 International Conference on Harmonization: Q5A. Guidance on Viral Safety Evaluation of Biotechnology Products Derived From Cell Lines of Human or Animal Origin, 1998.

Dr. B.J. Ledwith, Merck Research Laboratories, WP 45-215, 770 Sumneytown Pike, P.O. Box 4, West Point, PA 19486-0004, USA.
E-mail: brian_ledwith@merck.com

Petricciani J, Sheets R (eds): Vaccine Cell Substrates 2004.
Dev Biol (Basel). Basel, Karger, 2006, vol 123, pp 265-266.

BONUS SESSION DISCUSSION

Experience with Novel Cell Lines

Panel Members: *Robert Johnson (Session Moderator), Brian Butman, Wilfried Dalemans, Jereon Medema, Brian Ledwith*

Participant: How easy is it to establish insect cell lines? Do they contain viruses? What is the origin of the Hi-5 line?

Dr. Dalemans: The Hi-5 cell line was derived many years ago and I do not recall whether it was easy to establish it.

It is known that insect cell lines can carry viruses, and this was the challenge at the start of the use of these cell lines, because they are positive for retroviral signals.

I think it is the cumulative data from different testing systems which give you a reliable feeling on safety. It is also important to recall that these systems are used to produce purified antigens. We heard earlier that you can use continuous cell lines despite the fact that they are tumorigenic. If viral clearance is documented and they pass various tests, the level of assurance regarding safety increases.

Dr. Cichutek: What kind of experiments should be done to make sure that the theoretical assumptions that have been made are right? For example, could polyoma DNA be used to produce tumours, titrate them and look at which level a tumour no longer develops?

Dr. Ledwith: This raises two points. First, all the studies that I described were done concurrently with clinical development, and in no way precluded the development of our vaccine. I do not see this as a disadvantage. We are all trying to do the best studies to assure safety and do risk assessments. We want to develop models that enhance risk assessment, and I am very encouraged by the studies of Keith Peden because the results provide us with a more useful positive control.

There is great reservation about using tumour viruses as a positive control because they are infectious. Cloned oncogenes would be an alternative, and probably are more relevant since we are talking about cellular DNA.

My personal feeling about positive controls is their real value is in characterizing and validating the model. I do not know that they necessarily have to be run by every manufacturer every time they run a study, because once you have the model validated and characterized, you should be able to proceed using GLP procedures as we do for drug carcinogenicity studies. We certainly do not use chemical carcinogens every time we test a drug.

I think the new studies with the ras plus myc support the original conclusions of the WHO with the 10-nanogram limit, because if the same type of calculation based on those results is done, the same safety margin results. That type of work is valuable.

Dr. Ackland: Looking at those last results, have you any advice or learned any lessons from those studies that you would do differently if you had to do the studies again?

Dr. Ledwith: Animals are prone to spontaneous tumours, and experimenters are prone to bad luck in having them fall in the treated group. The answer is that the study may need to be repeated. Alternately, the studies could be scaled to capture spontaneous lymphoma. But then more rare events will start to be captured, and the same problem could arise. The study may have to be repeated to confirm that a certain tumour was spontaneous.

Genetic Studies of the DNA and of our vector can be used to show: (i) that a tumour was not caused by the vector because it is not integrated into the vector DNA and (ii) that there is no PER. C6 cell DNA present.

Petricciani J, Sheets R (eds): Vaccine Cell Substrates 2004.
Dev Biol (Basel). Basel, Karger, 2006, vol 123, p 267.

SESSION V

Bovine (and Porcine) Viruses

Petricciani J, Sheets R (eds): Vaccine Cell Substrates 2004.
Dev Biol (Basel). Basel, Karger, 2006, vol 123, pp 269-272.

Bovine Serum – Regulatory Issues

J.S. Robertson

Division of Virology, National Institute for Biological Standards and Control, Potters Bar, U.K.

Key words: Bovine serum, 9CFR, EU guidelines, bovine polyoma virus

Abstract : For many years users of bovine serum in the manufacture of human and veterinary biological medicinal products have relied upon USDA 9CFR to ensure the viral safety of their serum. Recently, EU regulators have formalised their position by issuing guidelines on the use of bovine serum during manufacture. Additionally, the European Pharmacopoeia has drafted a monograph on bovine serum. There is good harmonisation among the recommendations and requirements although the EU CHMP guideline calls for greater attention to be paid to the potential presence of infectious bovine polyoma virus. The EU guidelines also call for various tests to assess the effect of BVDV antibodies in the detection of BVDV. However, in response to criticisms from serum suppliers and users, the stringency of these recommendations is being relaxed. The overall viral safety of bovine serum should be subject to a risk analysis as infectious virus will invariably be present in some batches of serum but remain undetected. Other factors such as the geographical source of the serum, the efficacy of viral inactivation/removal steps and the ability of specific viruses to grow in the production cells being used, should be taken into consideration.

Historical Background

Animal serum and other complex biological preparations have been a requirement for tissue and cell culture from their origins approximately 100 years ago. At that time, small pieces of tissue such as lymphosarcomas or sections of embryonic medullary tube were suspended in biological extracts such as lymph or blood clots, ascitic fluid, blood or embryo extracts. It was often not certain if the samples of tissue were actually growing or whether they were simply being kept alive. From these beginnings, researchers gradually turned their attention to the growth of individual cells, rather than tissue clumps, and also to their understanding of the nutritional needs of the cells. By the middle of the 20th century, both cell culture and defined media were well established. However, cell culture media still required a natural biological extract e.g. plasma clots, serum or embryo extracts, to provide ill-defined nutrients. These biological preparations are now recognised as providing many discrete functions such as protease inhibitors, adhesion factors, growth factors, and essential nutrients such as trace elements and hormones.

Use of Bovine Serum

Nowadays, many cell lines used for the production of biological medicinal products have been adapted to growth in media in which every component is defined and can be sourced animal-free. Such media may still contain complex (but defined) macromolecules, but these are now being supplied from recombinant sources, e.g. bacterially expressed protein, or may contain plant-derived material. For other cell lines, however, serum remains a necessary source of complex nutritional requirements, especially those cell lines used in the production of human vaccines, with foetal bovine serum being the preferred choice of extract. The drive to eliminate the use of complex animal-derived materials is to avoid potential viral contamination and, in the case of bovine or ruminant sourced materials, to remove concerns regarding potential BSE contamination. With regard viral contamination, this drive is not unwarranted. Several instances of viral contamination of bioreactors have occurred and in most cases the source of contamination has been proven, or was highly suspected, to be the bovine serum supplement. Besides foetal bovine serum, newborn calf serum, calf serum, adult bovine serum and donor herd serum, as well as serum from other animal sources, are available.

Assuring the Viral Safety of Bovine Serum (U.S.A. Regulations)

For many years, the accepted standard for ensuring the viral safety of bovine serum used in the manufacture of medicinal products has been the USDA code of federal regulation 9CFR 113.53 [1]. This code stipulates the requirements for ingredients of animal origin, and not just serum, that are used in the production of biological products. It requires that each lot of material is tested and shown to be free of mycoplasma, bacteria and fungi and that tests are performed for viral contamination. For this latter analysis, the test material should be applied to at least two cell lines, viz., Vero cells and a primary cell or cell line of bovine origin (recognising the widespread use of bovine serum, to which the code was clearly written in the first place). The cells should then be subcultured at least twice and for at least 21 days, after which time they should be tested for cytopathic agents, haemadsorbing agents and for specified viruses by fluorescent antibody tests. Requirements for the detection of cytopathogenic and/or haemadsorbing agents are further defined in the sub-section of the code 113.46, while the requirements for the fluorescent antibody techniques are defined in sub-section 113.47. This latter section of the code also specifies the particular bovine viruses that have to be tested for, and these include BVDV, Reo, Rabies (FA testing to be applied to all cells), bluetongue, bovine adenovirus, bovine parvovirus and bovine RSV (in bovine cells only). Trypsin, another animal sourced material used in cell culture during cell passaging, is also regulated by 9CFR 113.53. Trypsin is typically sourced from porcine and specific tests for porcine parvovirus are stipulated (if the preparation has not been specifically treated to inactivate PPV).

ENSURING THE VIRAL SAFETY OF BOVINE SERUM (EUROPEAN REGULATIONS)

In Europe and elsewhere, such detailed viral safety requirements for bovine serum per se have not, until recently, been available, and both serum suppliers and vaccine manufacturers have generally followed the USDA code in ensuring the quality of bovine serum. The European Pharmacopoeia provides some, but very limited, information on the use of bovine serum in the cultivation of cell substrates for human vaccine production [2], while guidance for the manufacture of veterinary vaccines within the EU can be found in Eudralex Vol. 7B – Immunologicals, Quality [3]. This also provides a long list of viruses that should be considered as potential contaminants of animal-derived starting materials (section 7Blm10a).

Recently, however, the EU Committee for Human Medicinal Products (CHMP – formerly the CPMP) and the EU Committee for Veterinary Medicinal Products (CVMP) both issued specific guidelines on the use of bovine serum in the production of human [4] or veterinary [5] biological medicinal products, respectively. Both guidelines provide information relating to the collection, use and testing of all types of bovine serum with special emphasis within the guidelines on ensuring viral safety. The viral guidance provided generally mimics that of 9CFR, including the list of specified viruses for which tests are to be applied.

The CHMP guideline also highlights the need to gather information on the presence of bovine polyoma virus (BPyV) in batches of serum. This virus has not previously been specified as a target for testing and the CHMP guideline discusses it separately from its list of specified viruses. It has been known for some time that bovine serum is often PCR-positive for BPyV DNA and one recent study found 90% of batches PCR-positive. The absence of a suitable infectivity assay for BPyV has resulted in a paucity of data regarding the infectivity of the DNA being detected by PCR. Polyoma viruses are often oncogenic when inoculated into a foreign host species or can transform cells whose origins are distinct from the usual host of the virus, although BPyV itself has not been capable of transforming cells experimentally entirely on its own. BPyV antibodies are prevalent in humans with close contact with bovine species such as abattoir workers or veterinarians although it is not totally clear if the assays for BPyV antibodies can cross-react with antibodies to the human polyoma viruses JC and BK. Consequently, although no human disease has been associated with BPyV, the prevalence of the virus in bovine serum and its apparent zoonotic activity raises cause for concern and, at minimum, the need to determine to what extent batches of serum contain infectious virus.

Both the human and the veterinary EU guidelines recommend that no viruses should ultimately be detected in a batch of serum. In addition, serum should generally not be used for the manufacture of human biological products if an infectious virus contaminant is detected before inactivation procedures have been applied. The one exception is if BVDV is detected, in which case rigorous and validated virus inactivation/removal procedures must be in place.

The European Pharmacopoeia has also drafted a specific monograph for bovine serum [6]. Initial versions of this draft called for the testing of a greater list of specified bovine viruses than appears in USDA 9CFR; however, in the most recent draft version, this list had been reduced to be in harmony with 9CFR and the EU guidelines. A feature of the European guidelines/monograph is the emphasis given to the detection of antibodies to BVDV and the role they may play in masking the detection of infectious BVDV. However, criticism of the tests and the possible resulting loss of large quantities of bovine serum through the presence of BVDV antibodies have resulted in a re-think of the various guidelines. At the point of writing, a revision of the EU veterinary guideline has been released for consultation. In this revision, there is no specified list of viruses of concern and less emphasis is placed on the testing for and the effects of antibodies to BVDV. The drafting groups for the CHMP guideline and the Ph. Eur. monograph are also reconsidering the guidance being given or the requirements being made on tests for BVDV antibodies, and their implications.

Other regulatory documents that concern bovine material are the Japanese Pharmacopoeia (14th edition) [7] in which general viral safety requirements for biological medicinal products of animal origin are provided and an EU document issued by DG Sanco [8] from the Scientific Committee on Animal Health and Animal Welfare provides recommendations on virus inactivation steps that should be applied to bovine blood and blood products.

References

1 U.S. Code of Federal Regulations, Title 9, Part 113.53, Requirements for ingredients of animal origin used for production of biologics.

2 European Pharmacopoeia, 4th Edition, Chapter 5.2.3, Cell Substrates for the Production of Vaccines for Human Use.

3 The Rules Governing Medicinal Products in the European Union, Volume 7B, Guidelines for Veterinary Medicinal Products, Immmunologicals, Quality (http://pharmacos.eudra.org/F2/eudralex/download/volpdf/vol7/vol7ben.pdf).

4 CPMP/BWP/1793/02. CPMP Note for Guidance on the Use of Bovine Serum in the Manufacture of Human Biological Medicinal Products (http://www.emea.eu.int/pdfs/human/bwp/179302en.pdf).

5 EMEA/CVMP/743/00. CVMP Guideline on Requirements and Controls Applied to Bovine Serum Used in the Production of Immunological Veterinary Medicinal Products. (http://www.emea.eu.int/pdfs/vet/iwp/074300.pdf).

6 European Pharmacopoeia PA/PH/Exp. BSR/T (03) 1 COM, Bovine Serum (draft).

7 Japanese Pharmacopoeia, 14th edition. Basic requirements for viral safety of biotechnological/biological products listed in Japanese Pharmacopoeia. JP Forum 2003;12(1):3-17.

8 Sanco/C3/AH/R21/2000. European Commission, DG Sanco, Scientific Committee on Animal Health and Animal Welfare, "Virus inactivation in bovine blood and blood products".

Dr. J.S. Robertson, Division of Virology, National Institute for Biological Standards and Control, Blanche Lane, South Mimms, Potters Bar, Herts EN6 3QG, U.K.
E-mail: jrobertson@nibsc.ac.uk

Petricciani J, Sheets R (eds): Vaccine Cell Substrates 2004.
Dev Biol (Basel). Basel, Karger, 2006, vol 123, pp 273-280.

Vaccine Cell Substrates: Bovine and Porcine Virus Considerations

S.J. Wessman

USDA, APHIS, VS, Center for Veterinary Biologics, Ames, Iowa, U.S.A.

Abstract: The use of materials of animal origin to supplement cell cultures used in vaccine production, viral diagnostic testing, or materials testing may lead to contamination of the vaccines, with seroconversion or disease in the vaccinated animals, and possible misdiagnosis of diagnostic samples or incorrect test results. The methods used by the Center for Veterinary Biologics to monitor serum and cell cultures are described. Considerations for the use of animal origin materials, especially bovine and porcine, as substrates or additives, plus the possibility of crossovers to humans are discussed.

INTRODUCTION

The substrates used in the production of vaccines can become contaminated in multiple ways. The most frequent are viruses which may be introduced in serum, media additives, or materials used in the processing steps, such as trypsin for cell dissociation. The original material may also have been contaminated with the contaminant remaining undetected for a long period of time. Cross contamination with other cell lines, which may contain viruses that can cross species lines, is also possible. Other considerations are mycoplasmal and bacterial contamination, although advances in filtration and other processing steps, plus insistence on good laboratory procedures, have reduced these as sources.

The use of virally contaminated serum, especially foetal bovine serum (FBS), is still a major source of contamination of cell cultures, and always has the potential to contaminate vaccines or cause misdiagnosis of various animal diseases. The Center for Veterinary Biologics (CVB) uses a combination of procedures to test lots of FBS before they are used in the production of cell cultures used to validate master seed and cell testing, vaccine serial testing, and support of diagnostic testing for the National Veterinary Services Laboratories [1-3].

Testing of FBS has resulted in CVB rejection rates of serum lots of from 38 to 75 percent over the last 19 years. Others have reported similar results [4,5]. While most of the contamination is due to infection with bovine viral diarrhoea virus

(BVDV), the presence of other viruses such as infectious bovine rhinotracheitis, parainfluenza 3, bovine adenovirus and bovine rotavirus has been detected. In addition, not all lots of serum are satisfactory for all uses or all cells, as evidenced by interference with serum neutralization tests for canine virus and rabies, even though no titre or virus can be detected.

Many cell lines, including non-bovine origin cells, can be infected with BVDV, as previously reported [6-9] (Table 1). Vaccine producers use many of these as substrates for vaccine production, so it is necessary to require testing to detect any extraneous agents which might be infected with BVDV, as well as any other agents. This testing is described in Title 9 of the Code of Federal Regulations (9CFR) Section 113, especially 113.50-113.55 covering ingredients [10]. Infection with any agents may cause alterations in cell growth patterns, may change viral yield and may induce disease in the inoculated animals.

Table 1: Cell lines reported as susceptible to bovine viral diarrhoea virus.

Bovine turbinates	Porcine kidney (PK-15 and others)
Madin-Darby bovine kidney(MDBK)	Minipig kidney
Bovine kidney-primary and continuous	Swine testis
Bovine lung	Goat kidney
Bovine trachea	Goat esophagus
Bovine aortic endothelium	Rabbit kidney (RK-13)
Bovine testis	Cat lung
Sheep choroid plexus	Crandell feline kidney (CRFK)
Lamb kidney	Cat tongue
Ovine turbinates	Feline embryo
Monkey kidney (Vero and others)	Mosquito cells
Bighorn sheep kidney	Chinese hamster ovary (CHO)
Chick embryo fibroblasts	Others

[1,11]

Materials and Methods

The techniques used to test FBS, master seeds and cells are stated in Title 9 CFR and have been previously described [2]. Serum testing is conducted, with any positive samples being eliminated as material under test. The testing includes antibody detection at a 1:2 dilution, viral examination by fluorescent antibody staining with multiple passes using the test serum at 15% by volume on multiple cell lines susceptible to various viruses (MadinDarby Bovine Kidney, bovine turbinate or primary bovine kidney cells and Vero cells). Negative cultures are frozen and thawed three times, centrifuged and the supernate inoculated on to fresh cultures of the indicator cells. Fluorescent antibody staining, haemadsorption testing and May Greenwald staining to detect viruses and other cytopathogenic agents are conducted. Samples which are negative to the above tests are pooled, and checked for purity from mycoplasmas, bacteria and fungi by prescribed 9CFR tests. These pools are then either heat-treated at 56 C° for thirty minutes or gamma-irradiated at 25-35 kiloGrays at -40C°. Testing of master seeds and cells is done as described in 9CFR 113.55 and 113.52 respectively (Table 2) [10].

Table 2: Testing required in 9CFR 113.47.

All cells	bovine viral diarrhoea virus, reovirus, rabies virus
Bovine, caprine, ovine	bluetongue virus, bovine adenoviruses, bovine parvovirus, bovine respiratory syncytial virus
Canine	canine coronavirus, canine distemper virus, canine parvovirus, (infectious canine hepatitis, bluetongue)
Equine	equine herpesvirus, equine viral arteritis virus, (bluetongue)
Feline	feline infectious peritonitis virus, feline panleukopaenia virus (feline coronavirus, bluetongue)
Porcine	porcine adenovirus, porcine parvovirus, transmissible gastroenteritis virus, porcine haemagglutinating encephalitis virus (bluetongue, porcine rotavirus, porcine reproductive and respiratory syndrome virus, pseudorabies virus, swine encephalitis virus, swine influenza virus)

Testing in parentheses is done in addition to required testing.

RESULTS

Virus or antibody to BVDV was detected in from 38 to 75 % of the 1-3 bovine serum pools tested at the CVB during the period 1985 to 2003 (Table 3). Most were rejected due to BVDV, but some lots were also rejected due to bovine herpesvirus-4, reovirus and bovine leukaemia virus. No master cell lots submitted for testing since 1986 have been found to have a viral contaminant, although before that time this laboratory has detected virus in PK2a, PK-15, CRFK and EBK cell submitted for testing [2]. Vaccine lots have been found to be positive for BVDV and bluetongue viruses [3,12,13], introduced by contaminated serum. One master seed was found contaminated with BVDV [3]. A master seed virus was found to be contaminated with reovirus, introduced in serum.

Table 3: Testing of foetal bovine serum 1985-2003.

	1985-1990	**1990-1997**	**1999**	**2000**	**2001**	**2003**
Number tested	2224	6318	34	483	327	881
Number antibody-positive	853	1320	0	112	36	189
Percent antibody-positive	38	21	0	23	11	21
Number virus-positive	61	1673	13	250	107	354
Percent virus-positive	37	27	38	52	33	40
Total rejected	1097	2993	13	362	143	543
Percent Rejected	49	47	38	75	44	62

Discussion

The high rejection rates for FBS reflect the difficulty in obtaining non-contaminated materials for use in cell culture production. The use of materials of animal origin carries with it both positives and negatives. On the positive side, serum contains growth factors, hormones, vitamins, trace minerals, metabolism modifiers, and other factors, all of which are important in cell attachment, growth and viral replication. Serum reduces cell clumping and mediates damage that may occur during trypsinization. The use of serum reduces the need for extensive media reformulations to optimize growth. Most veterinary products are prepared using serum supplemented media, unlike some of the human pharmaceuticals produced on such cell lines as Chinese hamster ovary (CHO) and baby hamster kidney (BHK-21). Although there are more than forty cell types from over a dozen species used in the production of veterinary vaccines and diagnostic kits (Table 4), there are fewer than ten basic media preparations needed to propagate those cells. While there are some serum-free media available for mammalian cell culture, adapting culture procedures may alter cells and change production results. Thus there is a demand for a limited number of standardized media, rather than a multitude of media adapted to specific cell lines.

There are, however, definite risks involved in having to use materials of animal origin to propagate cells. It is necessary to conduct confirmatory testing of each of the materials used to eliminate the possibility of inducing viral disease in vaccinated animals [12,14,15]. In addition to the fact that the original substrate may have been derived from a contaminated source, serum, trypsin and the viral seed may each be contaminated. Serum is not completely characterized, there is lot-to-lot variability, protein levels vary and may interfere with downstream processing, and sourcing and adulteration are possible drawbacks. Another consideration is the need to check all

Table 4: Licensed cell types for veterinary vaccine production by species.

Bovine MDBK Turbinate Kidney	**Equine** Kidney Dermis	**Porcine** Kidney PK-15 PK-2a Testis	**Feline** Foetal heart Emb. lung CRFK	**Canine** Kidney MDCK NLDK Macrophage Histiocyte Bone marrow
Monkey Vero MA-104 BS C-1 BGM	**Hamster** CHO BHK21	**Mouse** Macrophage Hybridomas McCoy	**Ovine/Caprine** Lamb kidney Goat kidney	**Avian** Hepatic carcinoma Thymocytes RP-19 QT-35
Insect T. ni Tick Sf21 Sf9	**Human** Rectal tumour Lung	**Misc.** CHSE Bat lung Rat kidney		

material used in production thoroughly, to see if they either came from an animal source, were produced using animal source materials or came in contact with animal source materials at some time [16]. This is especially true of seed viruses, which may start in one species, be passaged in the cell culture of another, including eggs and mice, and intended for use in yet another species. This requires testing for extraneous agents for all species of origin, production and usage.

The final area of consideration is that serum from animals other than cattle is capable of sustaining cell growth. Researchers and diagnostic laboratories often use equine, sheep or goat, or porcine serum to supplement culture growth. Although at present little is being used in vaccine production, there is the possibility that they could be sources of extraneous viral contaminants. It is necessary to examine sourcing, and conduct testing on all supplemental sera to ensure that no contamination occurs. Many viruses may cause inapparent illnesses, while still being viraemic. If an animal is used as a blood/serum donor while in this status, there is the possibility that infective virus could be collected. Recent reports on porcine circoviruses indicate that they may be the cause of post-weaning multi-systemic wasting syndrome, either in association with other porcine viruses such as porcine parvovirus and possibly "atypical" porcine reproductive and respiratory syndrome virus (PRRS) [11,17-19]. Japanese B encephalitis has shown a high rate of seroconversion in sentinel pigs, corresponding to an increase in children, where infected mosquitoes are thought to be the vector [20]. The virus has been detected in serum and tissues. PRRS virus may produce a persistent infection despite an active immune response with antigenic variability [21]. African swine fever can be either acute or sub-clinical, with the possibility of spread due to mobility and global trade. Viraemia may be greater than two weeks [22]. A report from Japan indicates that a human hepatitis E isolate has high identity with swine hepatitis isolates [23]. Equine serum is often used to maintain cultures, and as such may need to be tested for West Nile virus, encephalitis viruses, vaccinia and vesicular stomatitis, as well as equine viral arteritis, equine infectious anaemia, equine influenza and equine herpesviruses. Sheep and goat serum, less widely used, may need to be tested for encephalitis, scrapie, and bovine pustular stomatitis and porcine serum for encephalitis, vesicular stomatitis, and swine hepatitis. This is in addition to those viruses which are already described in 9CFR for required testing. Cells and other animal origin ingredients, such as porcine trypsin for parvovirus, may need to be tested for these same diseases, as well as others. The most likely sources for contamination will still remain bovine serum or FBS. In addition to the 9CFR requirements, either sourcing or testing may be required for encephalitis, BSE, vaccinia and paravaccinia, and vesicular stomatitis viruses. A consideration for in vitro fertilization work is the fact that BVDV contamination has occurred in batches of FBS used in transport media. Although no transmission of the virus was reported and none of the embryo recipients seroconverted, it was reported that polymerase chain reaction assays did detect the BVD virus in the medium, which could cause a false-positive result for the embryos [24]. Bovine herpesvirus type 4 has been shown to grow in cell cultures from multiple species including humans, so the possibility of human infection or disease must be considered [25,26].

The United States requires certification that all non-U.S. collected serum will be from areas not having foot-and-mouth disease, rinderpest, or BSE, according to 9CFR 95.4. The FDA defines animal sera and media for culture use as Class I medical devices and requires of regulated manufacturers that they maintain traceability record showing that animal materials from a designated BSE country were not used [27]. Materials from areas that are known to be free of specified diseases may still need to be tested for certain diseases such as contagious bovine pleuropneumonia, Rift Valley fever, vesicular stomatitis, African swine fever, classical swine fever, pseudorabies, swine vesicular disease and Japanese B encephalitis, as well as FMD, BSE and rinderpest. Among the diseases not currently in the United States, FMD and Rift Valley fever are diseases which can cross over to humans (Table 5). FMD, Rift Valley fever and rinderpest, because of their ease of spread, have also been suggested as potential bioterrorism agents. Although FMD causes a mild or often inapparent illness in humans, Rift Valley fever can cause fever, influenza-like symptoms and death [28,29].

Conclusions

The use of serum and other animal ingredients can result in contamination of cell cultures, cell and virus stocks, with possible infection of animals and humans. There is no single solution, as cell and virus stocks will have a history of past exposure to animal origin ingredients, and certain viruses will only grow in certain cell lines. Testing, serum-free media, sourcing, viral inactivation, and improvement of test methods must all be used to ensure that vaccines will protect animals and humans and that accidental infection of inoculated animals and zoonotic disease does not occur.

Table 5: Zoonotic diseases and considerations.

Virus type	Primary host of concern/vector	Disease in humans
Eastern equine encephalitis	Horses/mosquitoes	Encephalomyelitis
Western equine encephalitis	Horses/mosquitoes	Encephalomyelitis
Venezuelan equine encephalitis	Horses/mosquitoes	Encephalitis
Japanese B encephalitis	Horses,cattle,swine	Encephalomyelitis
Tick-borne encephalitis	Cattle,goats/ticks	Encephalitis
California encephalitis	Horses, cattle/mosquitoes	Encephalitis
Rabies	Cattle	Paralysis, death
Prions	Cattle, sheep	
Bovine spongiform encephalopathy agent	Cattle	Atypical Creutzfeldt-Jakob disease
Vaccinia	Cattle, horses	Skin lesions
Bovine pustular stomatitis virus	Sheep, goats	Skin lesions
Vesicular stomatitis	Cattle, swine, horses	Fever, chills, headache
Parainfluenza 3	Cattle	Flu-like symptoms
Rift Valley fever	Cattle, sheep	Fever, influenza-like symptoms, death

References

1 Wessman SJ, Levings RL: Benefits and risks due to animal serum used in cell culture production, in Brown F, Cartwright T, Horaud F, Speiser JM (eds): Animal Sera, Animal Sera Derivatives and Substitutes Used in the Manufacture of Pharmaceuticals: Viral Safety and Regulatory Aspects. Dev Biol Stand. Basel, Karger 1999: 99:3-8.

2 Levings RL, Wessman SJ: Bovine viral diarrhea virus contamination of nutrient serum, cell cultures and viral vaccine, in Horaud F, Brown F (eds): Virological Aspects of the Safety of Biological Products. Dev Biol Stand. Basel, Karger, 1990; 75:177-181.

3 Wessman SJ, Levings RL: Collective experiences of adventitious viruses of animal-derived raw materials and what can be done about them. Cytotechnology 1998;28:43-48.

4 Bolin SR, Mathews PJ, Ridpath JF: Methods for detection and frequency of contamination of fetal bovine serum with bovine viral diarrhea virus and antibodies against bovine viral diarrhea virus. J Vet Diag Invest 1991;3:199-203.

5 Bolin SR, Ridpath JR: Prevalence of bovine viral diarrhea virus genotypes and antibody against those viral genotypes in fetal bovine serum. J Vet Diag Invest 1998; 10:135-139.

6 Horzinek MC: Non-arthropod borne Togaviruses. London, Academic Press, Inc. 1981.

7 Potts BJ, Sawyer M, Shekarchi JC, Wismer T, Huddleston D: Peroxidase labeled primary antibody method for detection of pestivirus contamination in cell cultures. J Virol Meth 1989;26:119-124.

8 Bolin SR, Ridpath JR, Black J: Survey of cell lines in the American Type Culture Collection for bovine viral diarrhea virus. J Virol Meth 1994;48:211-221.

9 Audet SA, Crim RL, Beeler J: Evaluation of vaccines, interferons and cell substrates for pestivirus contamination. Biologicals 2000;28:41-46.

10 Code of Federal Regulations Title 9, Part 1, Sections 113.47, 113.52-113.55. Office of the Federal Register, Washington DC, National Archives and Records Administration, 2004.

11 Allan GM, Ellis JA: Porcine circoviruses: a review. J Vet Diagn Invest 2000;12:3-14.

12 Ianconescu M, Akita GY, Osburn BI: Comparative susceptibility of a canine cell line and bluetongue virus susceptible cell lines to a bluetongue virus isolate pathogenic for dogs. In vitro Cell and Dev Biol-Animal 1996;32:249-254.

13 Everman, JF, McKiernan AJ, Wilbur LA: Canine fatalities associated with the use of a modified live vaccine administered during late stages of pregnancy. J Vet Diag Invest 1994;6:353-357.

14 Falcone E, Tollis M, Conti G: Letter to the Editor. Bovine viral diarrhea disease associated with a contaminated vaccine. Vaccine 2000; 18:387-388.

15 Caramelli M, Ru G, Casalone C, Bozzetta E, Acutis PL, Calella A, Forloni G: Evidence for the transmission of scrapie to sheep and goats from a vaccine against Mycoplasma agalactiae. Vet Record 2001;148:531-536.

16 Collins NF: Animal derived components in cell culture media. Art to Science 1998;17(2):1-3.

17 Calsamiglia M, Segales J, Quintana J, Rosell, C, Domingo M: Detection of porcine circovirus types 1 and 2 in serum and tissue samples of pigs with and without postweaning multisystemic wasting syndrome. J Clin Microbiol 2002;40(5):1848-1850.

18 Ellis JA, Bratanich A, Clark EG, Allan et al: Coinfection by porcine circoviruses and porcine parvovirus in pigs with naturally acquired postweaning multisystemic wasting syndrome. J Vet Diag Invest 2000;12:21-27.

19 Morozov I, Sirinarumitr T, Sorden SD, Halbur PG et al: Detection of a novel strain of porcine circovirus in pigs with postweaning multisystemic wasting syndrome. J Clin Microbiol 1998;36(9):2535-2541.

20 Kabilan, L, Ramesh S, Srisivasan S, Thenmozhi, V, et al: Hospital- and laboratory-based investigations of hospitalized children with central nervous system-related symptoms to assess Japanese Encephalitis virus etiology in Cuddalore district, Tamil Nadu, India. J Clin Microbiol 2004;42(6):2813-2815.

21 Chang CC,Yoon KJ, Zimmerman JJ, Harmon KM et al: Evolution of porcine reproductive and respiratory syndrome virus during sequential passages in pigs. J Virol 2002;76 (10):4750-4763.

22 Leitao A, Cartaxeiro C, Coelho R, Cruz B et al: The non-haemadsorbing African swine fever isolate ASFV/NH/P68 provides a model for defining the protective anti-virus immune response. J Gen Virol 2001; 82:513-523.

23 Takahashi M, Nishizawa T, Okamoto H: Identification of a genotype III swine hepatitis E virus that was isolated from a Japanese pig born in 1990 and that is most closely related to Japanese isolates of human hepatitis E virus. J Clin Microbiol 2003;41(3):1342-1343.

24 Givens MD, Riddell KP, Galik PK, Stringfellow DA, Brock KV, Loskutoff, NM: Diagnostic dilemma encountered when detecting bovine viral diarrhea virus in IVF embryo production. Theriogenology 2002;1399-1407.

25 Egyed L: Replication of bovine herpesvirus type 4 in human cells in vitro. J Clin microbial 1998;36(7):2109-2111.

26 Egyed L: Bovine herpesvirus type 4: a special herpesvirus (review article). Acta Vet Hung 2000;48(4):5012-513.

27 Code of Federal Regulations Title 21, 864.2800. Office of the Federal Register, Washington DC. National Archives and Records Administration, 2003.

28 Durand JP, Bouloy M, Richecoeur L, Peyrefitte CN, Tolou H: Rift Valley fever virus infection among French troops in Chad. Emerg Infect Dis 2003;9(6).

29 Sidwell RW, Smee DF: Viruses of the Bunya- and Togaviridae families: potential as bioterrorism agents and means of control. Antiviral Res 2003;57(1-2):101-111.

S.J. Wessman, USDA, APHIS, VS, Center for Veterinary Biologics, 1800 Dayton Avenue,
Ames, IA 50010, USA.
E-mail: stephen.j.wessman@aphis.usda.gov

Petricciani J, Sheets R (eds): Vaccine Cell Substrates 2004.
Dev Biol (Basel). Basel, Karger, 2006, vol 123, p 281.

Assays for Bovine Viruses

M. Wisher

BioReliance, Innovation Park, Stirling, FK9 4NF, U.K.

Abstract: Regulatory guidance for the detection of bovine viruses in bovine-derived raw materials is given in the U.S. Code of Federal Regulations 9CFR 113.53 and in the EMEA Note for Guidance on the Use of Bovine Serum in the Manufacture of Human Biological Medicinal Products. Both these documents require that the presence of viral contaminants should be screened for using detector cell lines and examining for viral cytopathic effects, haemadsorption, and specific assays for bluetongue, bovine adenovirus, parvovirus, respiratory sycytial virus, bovine viral diarrhoea virus (BVDV), rabies and reovirus 3. The details of these assays has been described. The EMEA document strongly recommends inactivating serum by validated methods. The presence of BVDV should be tested before and after inactivation and, since anti-BVDV antibodies may mask the detection of BVDV in an infectivity assay, a validated test should be used to detect such antibodies. Tests to perform this analysis will be discussed.

Published PCR methods to detect bovine polyoma virus (BPyV) DNA suggest that this virus is a common contaminant of bovine serum. The EMEA document encourages manufacturers to apply infectivity assays for BPyV. The presentation will describe the development of an infectivity assay for BPyV using MDBK cells and measuring replication of BPyV using a Q-PCR endpoint assay. This infectivity assay or a direct Q-PCR assay have been used to screen batches of bovine serum and cell lines grown in FBS. To date no contaminated cell line has been detected. While > 90 percent of FBS batches were positive by Q-PCR (between 0.1 and 2.6×10^4 genome equivalents/ml), no replicating BPyV has been detected using an infectivity assay in PCR positive FBS.

Dr. M. Wisher, BioReliance, Innovation Park, Stirling, FK9 4NF, Scotland, U.K.
E-mail: mwisher@bioreliance.com

Petricciani J, Sheets R (eds): Vaccine Cell Substrates 2004.
Dev Biol (Basel). Basel, Karger, 2006, vol 123, pp 283-290.

Animal-Free Culture Process for Manufacture of Viral Vaccines: Process Validation and Regulatory Perspectives

I. Pierard, J.-C. Mareschal, M.-M. Gonze, M. Duchêne

GlaxoSmithKline Biologicals, Rixensart, Belgium

Introduction

Several biological reagents of animal origin are currently used as constituents of the culture media for the manufacture of vaccine. Among these, the following additives are the most frequently used:

- bovine serum from foetal or young animals
- porcine trypsin
- human albumin
- miscellaneous animal protein hydrolysates

The use of these components imposes numerous difficulties which have an impact on the safety (presence of adventitious agents, endotoxin), consistency (variable performance in terms of cell growth and product manufacturing) and the cost of production of biopharmaceuticals in animal cells.

To reduce the risks associated with the use of the serum or trypsin during production, a number of measures have to be considered and combined to ensure the highest guarantee for the absence of adventitious agents [1,2]:

(i) selection of the source animals;

(ii) thorough testing for adventitious agents;

(iii) specific treatment to inactivate or remove the extraneous agents.

In addition, in the early nineties, CPMP, WHO and FDA issued guidelines on minimising the risk of BSE transmission via medicinal products. The guidelines requested manufacturers of biological products to undertake development work to reduce and at best eliminate the use of substances of animal origin in their manufacturing process.

More recently in 2000, the TSE crisis reinforced the need to minimize the risk associated with the use of additives of animal origin. Therefore, to comply with these recommendations, attempts are being made to identify non-animal derived replacement components for the culture media.

The development of serum-free processes was justified with respect to the product aspects of safety, quality and reproducibility since it would allow the following benefits [3]:

(i) simplified and defined composition of the culture media;

(ii) better batch-to-batch consistency;

(iii) reduction of the range of contaminants with the elimination of the known source of adventitious agents;

(iv) lower cost.

The proof of the concept of serum-free media was established in the eighties and the first new developments of serum-free culture, on an industrial scale, were obtained for different continuous cell lines (CHO, hybridomas, recombinant myelomas) [4,5].

Since the use of a serum-free medium has to be envisaged on an individual basis and requires the expenditure of manufacturing and laboratory resources, it was found more profitable to consider it in the context of the development of a new biotechnology product.

Based on the good results obtained using the continuous cell lines, the development of serum-free media was extended to the propagation of cultured mammalian cells and hence to the improvement of existing vaccines.

Contrary to the adaptation of a cell line in a serum-free condition during the early development of a new vaccine, this cell adaptation at large-scale manufacturing processes is far more challenging when it addresses existing processes, since the biological performance as well as the product quality have to be maintained [6,7].

To reduce the potential introduction of adventitious agents further, the development and use of serum-free media that are also free of animal-derived products are to be considered. The aim is to minimize, as much as possible the risk of inadvertent introduction of contaminant by replacing all additives of animal or human origin by a synthetic substance or components of plant origin. Here again, such development is not straightforward; indeed, besides the high investment in both money and time during the process development, the principal drawback is that no single animal-free medium is satisfactory for the cultivation of the different cell lines.

That is the reason why the animal-free process principles are preferably envisaged early in vaccine development, since the investment in time and resources are judged too high for adapting existing marketed products without sacrificing biological performance.

Like other companies, GSK Biologicals has taken up the challenge and embarked on the development of new culture media to reduce or eliminate the use of substances of animal origin in their manufacturing process. To meet the Regulatory Authorities request, the Company's intent is to develop animal-free media not only for new candidate vaccines but also for existing products. For both applications, the use of medium devoid of animal-derived components gave rise to manufacturing processes that are robust and reliable and vaccines that are consistently safe and of high quality.

In the case of currently registered vaccines with long safety and efficacy experience, the question was how to manage technically this change and how to prove comparability. Due to the fact that regulatory requirements specifically addressing the replacement of animal origin additives in media are not available, the aim of this paper is to share our view and approach for this significant change and to present the validation studies that should sustain such licensing variations.

Having been encouraged by all authorities to reduce or eliminate the use of substances of animal origin in their existing manufacturing processes, GSK Biologicals achieved substantial progress in the development and the optimization of medium devoid of animal-derived components used in the currently registered vaccines.

It should be mentioned that, in addition to the manufacturing investments for the cell culture itself, resources within the laboratory have to be allocated to build up and run a strong validation package to allow further progress on the commercialization of such improved vaccines.

Since the use of animal-free medium concerns all the manufacturing steps from the culture of the cell bank to the growth of the virus on the cell substrate, this validation package can be made up of five distinct groups of laboratory studies:

(i) the testing programme of the new cell banks prepared in animal-free conditions;

(ii) the biological properties of the cells;

(iii) the characterization of the viral vaccine produced;

(iv) the evaluation of the process in animal-free conditions;

(v) the stability of the vaccine produced.

For each study, it is necessary to develop a testing package using sophisticated and sensitive methods that will provide confidence at two different levels: (i) thorough analysis of the vaccine product to verify that it is not altered by the process change and therefore ensuring that the product profile (i.e. safety, efficacy or stability) is comparable, and (ii) verification of the new manufacturing process capacities to ensure a consistent vaccine production.

These extensive characterization studies can support a technical licence variation by assessing and proving the product comparability and the equivalence of performance with the existing manufacturing process without the need of additional clinical studies.

As mentioned in the CPMP/ICH document addressing the comparability of biological products [8], the demonstration of comparability does not necessarily mean that the quality attributes of the new products are identical to the current ones. Indeed, differences in quality attributes may exist if one can assess that they have no adverse impact upon the safety or efficacy of the product.

Testing Programme of the Cell Banks Grown with the Animal-Free Medium

The QC testing programme of the new animal-free cell bank, which includes testing at the MCB, MWCB and EPC stages, is first established based on the testing plan performed on the current cell bank. However, this plan is adapted based on the following considerations:

(i) the new animal-free cell bank is established from the same parent cell seed as the current cell bank. The extensive testing already performed on the latter allows a retrospective validation of the parent cell seed;

(ii) as mentioned in the CPMP/ICH document addressing the comparability of biological products [8], the use of animal-free medium for the preparation of the cell bank allows reduction within the QC testing and more specifically in the testing programme of adventitious agent (i.e. removal of the specific tests for bovine and porcine viruses);

(iii) the QC is adapted to the current version of the following guidelines:
   - Points to Consider for the Characterization of Cell Lines Used to Produce Biologicals – CBER, FDA (1993)
   - ICH Topic Q5A – Quality of biotechnological products: viral safety evaluation of biotechnology products derived from cell lines of human or animal origin CPMP/ICH/295/95 (1997) / CFR, vol. 63, N° 185, 1998
   - ICH Topic Q5D - Quality of biotechnological products: derivation and characterization of the cell substrates used for production of biotechnological/biological products CPMP/ICH/294/95 (1998) / CFR, vol. 63, N° 182, 1998
   - WHO Technical report Series No. 878, 1998 – Annex 1: Requirements for the use of animals cells as in vitro substrates for the production of biologicals.
   - Ph. Eur. 5.2.3. Cell substrates for the production of vaccines for human use.
   - CBER's letter to manufacturer dated 14 December 1998: request for PCR-based reverse transcriptase (PBRT) assay to characterize cell substrates

Table 1 summarizes the QC to be performed at the three stages of the cell bank and highlights changes between the current and the new animal-free cell bank.

Biological Properties of Cells Grown with the Animal-Free Medium

Different cell culture parameters must be studied, at large-scale cultures, to verify the equivalence in performance between the animal-free and the existing processes:

(i) to study the cell morphology and growth characteristics of the cells grown in animal-free conditions, parameters such as general appearance and cell shape, proliferative rate and cell density should be evaluated;

(ii) in an animal-free process, an alternative for trypsin should be found to detach the adherent cells at each subculture. The new protease should be tested for efficacy by verifying: (a) the detachment of the cell sheet as well as the dissociation of the cells from each other; (b) the cell viability of the detached cells; (c) the growth of the subculture after detachment;

Table 1: Overview of testing of cell bank systems: current cell bank versus animal-free cell bank.

Tests	Current cell bank			Animal-free cell bank		
	MCB	MWCB	EPC	MCB	MWCB	EPC
• **Identity**						
– morphology	X	-	X	X	-	X
– iso-enzyme analysis	X	X	X	X	X	X
– karyotype	X	X	-	-	-	X
– purity (DNA fingerprinting)	-	-	-	X	X	X
• **Sterility**						
– mycoplasma	X	X	X	X	X	-
– bacteria	X	X	X	X	X	-
– fungi	X	X	X	X	X	-
– mycobacterium tuberculosis	X	X	X	-	-	-
• **Adventitious agents in cell cultures**						
– human diploid cell	X	X	X	X	X	X
– human amnion cell	X	X	X	X	X	X
– cercopithecus kidney cells	X	X	X	X	X	X
– rabbit kidney cells	X	-	-	-	-	-
• **Adventitious agents in animals and eggs**						
– adult mice	X	X	X	X	-	X
– suckling mice	X	X	X	X	-	X
– guinea-pigs	X	X	X	X	-	X
– embryonated eggs	X	X	X	X	-	X
– rabbits	-	-	X	-	-	-
• **Bovine viruses**	X	-	-	-	-	-
• **Porcine parvoviruses**	X	-	-	-	-	-
• **Retroviruses**						
– detection by electron microscopy	X	-	X	-	-	-
– detection of reverse transcriptase activity	X	-	X	X	-	X
• **Specific human viruses by PCR**	X	-	X	-	-	-
Tumorigenicity						
– in vivo	-	-	X	-	-	-
– in vitro	-	-	X	-	-	-

(iii) diploid cells grown in animal-free medium should be seen to maintain their diploid characteristics;

(iv) in the context of diploid cell lines, the senescence of the cells grown in animal-free conditions should be determined to verify their limit of in-vitro cell age for production;

(v) finally, it will be interesting to verify that the cells grown in animal-free conditions maintain their permissibility to various viruses.

Characterization of the Viral Vaccine Produced on Cells Grown with the Animal-Free Medium

The effect of the new culture technologies on the properties of the virus call for careful examination, including the monitoring of phenotypic, biological and molecular characteristics. These evaluations should be performed in comparison with the existing ones.

Two important investigations should be carried out to assess the comparability between the existing process and the animal-free process:

(i) verification of the ability of the animal-free process to support the manufacture, at substantial titres, of the desired vaccine;

(ii) thorough characterization to detect any drift in the biological properties of the virus produced. Different aspects of the virus can been investigated: (a) genomic analysis; (b) antigenic analysis, and (c) biological analysis to study the behaviour of the virus grown in animal-free conditions further.

In the case of a viral vaccine, the following tests are useful to address these two questions fully and therefore ensure the potency, safety, purity and efficacy of the product:

(i) viral growth kinetic;

(ii) virus titration to assess the viral yield of the bulk vaccine;

(iii) SDS-PAGE;

(iv) Western Blot;

(v) electron microscopy;

(vi) antigenicity (ELISA);

(vii) reactivity of the vaccine product with polyclonal, monoclonal and human antibodies;

(viii) immunogenicity studies in animals;

(ix) ability of the produced virus to grow on different cell substrates;

(x) genotypic characterization (sequencing, hybridization experiments, RFLP,...);

(xi) ratio intact infectious to dead particles;

(xii) for attenuated viral vaccine if markers are available, the attenuation should be verified.

Process Evaluation

Having gathered the above information, it is critical to establish that the animal-free process will consistently produce, on a large scale and at an acceptable yield, a product which meets its specifications in terms of safety, potency and purity. It is also important to run an adequate battery of in-process control tests to verify whether the new product lies within the historical manufacturing trend, thus ensuring that the change is well controlled and maintains the quality of the products.

Stability

Finally, the impact of the process change on the vaccine stability should be evaluated and real-time concurrent stability studies should be conducted.

In addition, accelerated and stress stability studies are useful tools to compare the current and the new vaccine product further

Conclusions

Regulatory concerns have increased regarding the potential contamination of biological products by adventitious agents. Therefore, to reduce the risks associated with the use of biological reagents of animal origin, regulatory authorities strongly encourage the development of industrial-scale serum-free processes that are also animal and human-derived materials-free.

This kind of process development requires significant resources that are more compatible with the development of a new vaccine. Indeed, when addressing existing processes, this cell adaptation should guarantee equivalent biological performances (i.e. productivity, safety, purity and stability).

Therefore, in the case of already marketed vaccines, extensive laboratory testing has to be performed to prepare a solid validation package. This validation work must prove the product comparability with the existing vaccine process and hence should address all the different manufacturing steps from the cell culture of the cell bank to the growth and purification of the virus.

The validation package presented in this document can serve as a basis for future guidelines on the replacement of additives of animal origin in cell culture media. In our view, this testing package, using sophisticated and sensitive methods, provides confidence that the vaccine is not altered by the process change. Finally, since the adaptation of existing vaccine towards animal-free processes was strongly recommended, this thorough testing programme would allow Regulatory Authorities to expedite the approval of such technical licence variations without the need for additional clinical studies.

References

1 Mareschal J-C: Quality control of bovine serum used for vaccine production, in Brown F, Cartwright T, Horaud F, Spieser JM (eds): Animal Sera Derivatives and Substitutes Used in the Manufacture of Pharmaceuticals: Viral Safety and Regulatory Aspects. Dev Biol Stand. Basel, Karger, 1999, vol 99, pp 61-68.

2 Eloit M: Risks of virus transmission associated with animal sera or substitutes and methods of control, in Brown F, Cartwright T, Horaud F, Spieser JM (eds): Animal Sera Derivatives and Substitutes Used in the Manufacture of Pharmaceuticals : Viral Safety and Regulatory Aspects. Dev Biol Stand. Basel, Karger, 1999, vol 99, pp 9-16.

3 Froud SJ: The development, Benefit and disavantages of serum-free media, in Brown F, Cartwright T, Horaud F, Spieser JM (eds): Animal Sera Derivatives and Substitutes Used in the Manufacture of Pharmaceuticals: Viral Safety and Regulatory Aspects. Dev Biol Stand. Basel, Karger, 1999, vol 99, pp 157-166.

4 Mertens O-W: Safety issues of animal products used in serum-free media, in Brown F, Cartwright T, Horaud F, Spieser JM (eds): Animal Sera Derivatives and Substitutes Used in the Manufacture of Pharmaceuticals: Viral Safety and Regulatory Aspects. Dev Biol Stand. Basel, Karger, 1999, vol 99, pp 167-180.

5 Mertens O-W, Litwin J: Serum-free medium for fermentor cultures of hybridomas. Cytotechnology 1991;5:69-82.

6 Jayme DW, Blackman KE.: Culture media for propagation of mammalian cells, viruses, and other biologicals. Adv Biotechnol Processes 1985;5:1-30.

7 Kan M, Yamane I: In vitro proliferation and lifespan of human diploid fibroblasts in serum-free BSA-containing medium. J cellular physiology 1982;111:155-162.

8 Comparability of biotechnological/biological products subject to changes in their manufacturing process (CPMP/ICH/5721/03).

Dr. I. Pierard, GlaxoSmithKline Biologicals, 89, rue de l'Institut, BE-1330 Rixensart Belgium.
E-mail: isabelle.pierard@gskbio.com

Petricciani J, Sheets R (eds): Vaccine Cell Substrates 2004.
Dev Biol (Basel). Basel, Karger, 2006, vol 123, pp 291-308.

PANEL V DISCUSSION

Bovine (and Porcine) Viruses

Panel Members: *Jim Robertson (Panel Chair), Steve Wessman, Martin Wisher, Isabelle Pierard, Mark Abdy, Phillipe Vannier*

Panel Questions and Points for Discussion

1. Which viruses might be present in serum? Which are zoonotic? Which can be amplified in cell culture? Which are potentially pathogenic for humans?
2. If inadequately documented serum was used in the passage history of a cell substrate or viral seed, which could be detected.
3. Testing for BVDV and anti-BVDV antibodies.
4. Bovine polyoma virus.
5. The effects of TSE in the U.S.A. and Canada and the need for alternative sources.
6. The acceptability of contaminated serum.
7. The use of serum-free media

Dr. Robertson: In addition to four presenters from this morning, we have been joined by two panelists, Dr. Mark Abdy from FDA, and Dr. Philippe Vannier from AFSSA, the French food safety agency.

Dr. Lubiniecki: Dr. Wisher, in the system that you discussed with the use of Q-PCR in infectivity tests for bovine polyoma, have you had a chance to look at what happens to materials that were positive and then irradiated? Does the virus propagate in your system in MDBK cells?

Dr. Wisher: We had some inactivated serum, but I do not think we have specifically done the experiment. However, Dr. Onions might have done it.

Dr. Onions: The answer to that is that we have not. We have been conducting validations of irradiation processes using other polyoma viruses such as SV-40 and we are proposing to validate irradiation using bovine polyoma virus but that has not yet been done. What is surprising about SV40 is that it is much more radio-resistant, in our hands at least, than you might expect from its theoretical genome target size.

I would also just comment that it is worth remembering that data were developed by Invitrogen on irradiation of porcine trypsin, but what is relevant to serum is that the porcine circovirus has incredibly high radio-resistance. It is 40 kiloGrays. You only get about a one log drop in infectivity. If bovine circovirus becomes an issue, we do not have an infectivity assay for it. It is also likely to fall into a highly radio-resistant group that may not be inactivated by the kind of standard kiloGrays that we are giving serum at the moment.

Dr. Wisher: I wonder what form the bovine polyoma is in your serum that you are looking at? Is it actually still particulate, for example? And if it is particulate, are you not seeing infectivity because you have it neutralized by antibodies in the serum? If you pellet it, for example, does it come down in the pellet and, if you treat that with acid, can you recover infectivity?

Dr. Onions: We have not done those experiments. We are in the process of purifying the material out of serum. But we do have, in many of our assays, a control in which we will take that serum and spike it with infectious bovine polyoma virus, a positive control, and show that it is not being inhibited.

Dr. Kahn: I also had a question for Dr. Wisher. With regard to your 28 batches of serum that you found positive by PCR but negative by infectivity, can you indicate how many different vendors or sources these came from?

Dr. Wisher: They came from all the major vendors. I cannot remember how many from which, but, in terms of our conventional PCR studies which I also showed you on 80 batches, they came from all the major vendors and all continents. So it seems to be similarly spread, at least the PCR-positive material, in all continents.

Dr. Onions: I want to put this proposal to you which might sound outrageous to most of us at first hearing. We talked yesterday about the way that we could approach testing, if not of individual donations then maybe at a 10-lot donation level for this range of viruses, which would overcome many of these issues about BVDV and the way we screen for it at the moment. But even if we apply those technologies, which means we will not dump so many batches of FBS, we still have a shortage of FBS for certain critical manufacturers. Last year, two very high-profile licensed monoclonal antibodies did not reach their production targets and, therefore, did not reach clinics because there was a physical shortage of FBS.

One of the proposals being considered within the industry but not yet put to regulators is that we look at alternative sourcing outside New Zealand, Australia and the United States. Europe is clearly a "no, no" because of the BSE issue but there are countries such as Argentina which are approaching standards where the OIE might put a reasonable stamp of approval in terms of TSE testing.

The question is if we use the kind of approach I mentioned, that we would look at 10-lot or 10-donation level minipools or even the individual donation, how would the panel feel about using a new source? Clearly, there are new virus issues in Argentina. There are some interesting exotics down there that we would have to consider. How does the panel feel about extending the source of FBS to a country such as Argentina?

Dr. Vannier: To reply about the risk analysis in regard to sources from South America, I would say that, from the veterinary point of view, the problem for South America is that of foot-and-mouth disease. I think the risk is higher for foot-and-mouth disease virus in veterinary vaccines and then its spread, than for BSE. So that is effectively the major problem in comparing the level of risk according to the sources of the serum.

Dr. Onions: I understand that foot-and-mouth is a matter of major concern, but we would be talking here about screening, not just for antibodies, but screening individual donations or the 10-lot donation, which is not currently done. Remember, the minimum testing that is done at the moment is usually at the 100-lot, 100-litre, testing. We are now talking about going down to about 10 litres.

That means we would pick out individual polyoma virus contaminants as well as diseases such as foot-and-mouth. We would also be talking about using irradiated serum which would be irradiated to 30 kiloGrays. So I can understand the concerns and I share them, but I am just trying to articulate, is there not a scientific method of approaching this, that would allow us to widen the sourcing of FBS.

Dr. Robertson: There should be no reason not to use such serum if you are performing qualitative tests, and other aspects of guidelines and regulations are followed. You are saying that if everything is in place, we can use material of that nature. I cannot conceive why not, if all bases are covered.

Dr. Berry: I was wondering how the outside testing laboratories control for the issues of viruses and problems with serum. How do we know that it was our serum-based product that had the virus and it was not something to do with your cell line? How do you actually assure us that it is not you who had the problem?

Dr. Wisher: The detector cells that we use in these assays have all been tested for the panel of viruses against which we test the serum. We have actually prepared the master and working cell banks of the detector cells. So I am very confident that we do not have anything in our cell line.

The other way of controlling it is by the controls in the laboratories in terms of how we handle test articles, how we separate them, and how we run positive and negative controls in all our assays.

Dr. Dalemans: In one of the slides from Steve Wessman on the 9 CFR 113 subpart 47, I understood that the selection of viruses that had to be tested for depended to some extent on the target substrate. For human vaccines, where you use human cell lines, the FDA also refers to this USDA 9 CFR Chapter. What is the selection criteria for which viruses should be screened for and on which basis are they selected, because we could screen for everything which is over there, or is there a panel adapted for the human vaccine situation?

Dr. Wessman: I think that you probably need to get direction on that from someone from FDA because we have the regulations set up in 113.47 for veterinary biologics. FDA has chosen to use at least part of that as their general testing, but I think that you need to determine just exactly what FDA would say about that. I think that you still need to look at the legacy, if you want to call it that, anything that the cell substrate has been exposed to and then the intended host.

Dr. Abdy: I am sure you have heard already that everything I say is my opinion and not the FDA's official policy. The current policy is to follow what is in 9 CFR. I think there are some things that have been said today that would make us look a little harder at polyoma virus.

Dr. Vannier: I can add some words about the rationale involving, on the one part, the sources of the material and, on the other part, the target species for which the vaccine is intended. In Europe, the rationale was founded effectively on the main risk analysis in regard to the target species for the vaccines. But the problem is how can we be sure that there is a real species barrier from infected sources which will be completely blocked according to the target species for which the vaccine is used. You can never say that the species barrier will be strong enough to protect the animal species or humans. So that is the limit; how to be sure the species barrier will be efficient. Nobody can ever say yes, it will be 100 per cent efficient. In Europe, the rationale is a compromise between them both.

Dr. Egan: If you would like a little additional comment to follow up Mark's response to that question; what is being looked for is a panel of things that those cell lines or that product may have been exposed to. So if it is a human cell line, you may be looking for some of those bovine viruses because it may have been exposed to bovine serum or porcine enzymes. Of course, in addition to that, there is the number of human viruses that the cell lines could have been exposed to during their handling, isolation and storage. So I guess legacy is a very nice word; to use the "legacy" of those things that the cells could have been exposed to and, therefore, for testing. Then, of course, for the virus, the seeds, again the same thing, where they came from and their history. So I think that is the rationale for why you would test for bovine viruses in mammalian or other cells.

Dr. Robertson: The viruses of cattle are quite well recognized. There is an extensive list and I think it is impractical that we start attacking every possible virus of concern. So, how do we decide what is a virus of concern? Is it the frequency with which it might turn up in bovine serum, whether it be foetal, newborn, or adult serum? Is it that the serum is for use in human vaccine situations? Is it that it may be a zoonotic agent? What are the issues? If a virus is only turning up now and again, the very fact that it has been found in serum being used for vaccine manufacture, should we then, from a regulatory point of view, insist that all serum is screened for a virus which is not in the current list of specified viruses.

Dr. Onions: The two that concern me most are bovine polyoma because, as Martin pointed out, John Parry's rather nice serological data suggest that that virus is probably zoonotic. The other one that worries me, are things like Cache Valley that sneak through although that frequently just causes a mild influenza-like disease in people. At least one probable fatal case was reported in the New England Journal of Medicine a few years ago.

So the bunyaviruses in general, and some of these other arthropod viruses, do concern me, although we usually pick them up because they are highly cytopathic. But, in addition, bovine herpes viruses 2 and 4 which have been flagged in the European document are worth mentioning.

Bovine herpes virus 2 will replicate very efficiently in human cells. This virus actually infects the teats of cattle. One of the hypotheses, partly based on sequence information, is that this was a human virus at one point or it has evolved from HSV-1, because where you find it is on the teats of cattle and maybe it is associated with domestication. Bovine herpes virus 4 looks like a species of bunyavirus. In fact, there is nice work done at IAH in the U.K. suggesting that the virus might actually have been a buffalo virus originally. But it also appears to have a more diverse host range than some of the other bovine herpes viruses and it belongs to the gamma herpes virus group.

In terms of its assay, Steve may have more experience with this virus, but in our hands it does not come up on the usual cell substrates quite so easily. It is not very permissive and it requires mitotically active cells to pick it up. So the assay does need rather careful adjustment to pick up infectious virus.

Dr. Robertson: This is a difficult area and, for those involved from the regulatory side with viral safety, whether it is bovine, equine, porcine, or any other animal species, there has always been a question mark about exactly which viral agents should be looked for. We have the difficulty of deciding what to impose on industry.

I would like to get Martin Wisher to say a little more on polyoma. One of the big issues seems to be PCR, i.e. the detection of DNA versus infectivity.

What I find extremely curious from Martin's data is that so much serum has polyoma DNA but zero infectivity. Do you have any thoughts or explanations on that result from the 28 batches of serum?

Dr. Wisher: Yes, I do find it rather curious; that assay clearly supports the growth of bovine polyoma virus, and there are no high levels of neutralizing antibody inhibiting infectivity.

Dr. Robertson: Does it make you think that there is a limitation in your infectivity assay for picking up alternative strains of bovine polyoma?

Dr. Wisher: I was just going to say, with the inactivation data, – data for bovine polyoma virus are incomplete; but for SV-40 inactivation, both gamma irradiation and U.V. irradiation are not effective. You will see about 2 logs of inactivation. It is a double-stranded DNA virus. Perhaps that may be the problem.

Dr. Egan: For all these assays for infectivity, what are the levels of antibodies that are present in the foetal bovine serum? I do not know what they are, how they affect these assays, and how variable are they?

Dr. Wisher: We have not looked at antibody levels. They do not seem to be significant in foetal bovine serum. There does not seem to be significant levels; if we spike the test serum with infectious virus, we do not see a significant decrease in infectivity.

Dr. Onions: On this point, we now have a validated assay but this crucial question is simply not resolved. We are putting quite high levels of infectious virus into that challenge so we really do need to do proper kinetics here to understand what is happening. One possibility is that the assay is not sensitive enough or is not sensitive enough to variants, which I do not think is likely. I do not think there are going to be wide-range variants that we are not picking up. I think a more likely explanation is that we are probably getting some antibody interference. The reason we do not see it in our challenge is that we are using quite a large amount of virus. We have to do proper kinetics to demonstrate that. What I do not understand is why Schuurman's work showed such a high infectivity using a system that, at least in our hands, is not so sensitive. So I do not understand that dichotomy, and it may be just incompetence on our part.

Dr. Peden: I did not think we had an answer for what was asked about the state of the DNA in the serum. Is it known whether the DNA is full-length and is it worth trying to isolate the DNA from the batches of serum to see whether it is infectious in your cells, just to have some idea if the genome is infectious itself, not just in particles.

Dr. Wisher: We are looking at that, and trying to isolate the genomic material, purify it and characterize it more fully.

Dr. Lubiniecki: The role of antibody in these assays is going to be problematic. One of the reasons is that all bovine sera are not created equal. If you look at a number of lots from a number of manufacturers, you see levels of gamma globulin that range from 100 micrograms per ml to 5 milligrams per ml.

Clearly, these materials are, in many ways, different and so this has to be taken into account in some way. Further, I believe there is old literature on the irradiation of polyoma viruses that has, perhaps, slightly larger values of inactivation. I also know that, from my own work with SV40 years ago, that SV40 preparations are frequently quite clumpy and this can also affect the apparent target size. So I would urge that that be controlled in such studies.

Dr. Dalemans: Looking at it from the other side, my understanding of the whole story of polyoma is that you find it in batches all over the world. It is also quite infectious when you look at the data on seroconversion in veterinary people. Are there any studies being done looking at children who have received live or inactivated vaccines, where serum is used and where there is little purification, so you would imagine that polyoma virus, or at least the polyoma signal, would be at least quite high and you have millions of children who have received it. Are there any data available on that?

Dr. Robertson: I am not aware of such data, only the data by Schuurman on veterinary vaccines and interferons.

Dr. Onions: You might have noticed from John Parry's data, which I think form the only dataset on this, that John picked up this case in a rabies vaccine. He did not know whether that seropositive person was an agricultural or veterinary worker. Given that that is the group of people who are most likely to be vaccinated, that is probably just because they have been exposed to cattle. At least on that limited dataset, there is no evidence that polyoma in serum and vaccines has caused infection. There is the Muntz group as well who also tested it. There is another group of childhood vaccinees who were also tested by John. But they were all seronegative.

Dr. Wisher: There were fairly small numbers of people tested.

Dr. Onions: You are absolutely right.

Participant: Naked DNA has proven to be a poor immunogen in a vaccine context. My assumption is that the DNA, just on its own, without some form of electroporation or lipofection is poor in getting into the cells. It seems to me that the issue is a very small amount of the DNA present in the serum and, therefore, the likelihood of anything like this occurring is almost zero.

Dr. Wisher: Unfortunately, I do not think you have data to make that assumption. Neither do we. The experiments have to be performed to be that clear. I gather that you would like it to be just naked DNA and not infectious virus, but we do not have those data.

Dr. Robertson: I find it strange that you have a great deal of DNA there without infectivity. It is difficult to develop a guidance document with words beyond "encourage". But the fact that there is a signal that the virus may be present very frequently in reasonable titre if the DNA does translate into infectivity, that it does appear to be zoonotic, that there is no apparent recognized disease syndrome in those persons who seroconverted, I am tempted to change "encourage" to "recommend". We seem to have a virus here which is zoonotic, and which you cannot inactivate readily. I guess we will wait for more data coming through.

I am very appreciative of Martin and his colleagues for generating that kind of data. I am hoping other people are generating similar data to help us try to answer these questions and resolve this one way or the other.

Dr. Murphy: Is anyone taking one of these supposedly contaminated lots and putting it back into an animal which is seronegative and then examine what happens, and is this known to infect foetal bovines, because you are observing this in foetal bovine serum, right? Is this naturally transmitted to the foetus? Is any of the biology of this virus known and how would answers to those questions help us?

If you really want information on that, putting it into tissue culture is good, but putting it back into a seronegative animal and seeing whether there is any infectivity there and then looking at the natural pathogenesis of this in its natural host, looking at what happens to infected foetuses, would be interesting as well.

Dr. Abdy: It does not cause clinical disease in cattle, from my understanding, so at least looking for clinical disease would probably not be an option. But replication is a different issue. You can do that.

Dr. Onions: In John Parry's Ph.D. thesis which was really the only body of work that is substantial on bovine polyoma virus, he looked for disease associations for this virus. The only thing that they could come up with is a slight association with reproduction, not abortion. But if you look at cattle reproductive rates, you have failure to conceive. There was a mild association with serocoversion on that event in young heifers. But doing those animal experiments is completely meaningless.

First, if you look at seropositivity in the U.S., and it is probably true worldwide, when you have cattle at about six years of age, which we do not have any more in the U.K. because nearly all are dead from foot-and-mouth disease or BSE, but when we used to have them, the seropositive rate was about 90 percent. This is a ubiquitous virus.

Clearly, in veterinary medicine, it is of no importance whatsoever. But the whole issue about polyoma viruses is that we know that when they are put into heterologous hosts and they undergo abortive replication, that is when the danger is.

So it is rather irrelevant to the human sector what they do in cattle. What is relevant is what they might do in cells. We know from Schuurman's data that at least the T-antigen is transforming in rodent cells.

Dr. Vannier: Considering polyoma virus from the veterinary side, we did not see any practical solution for the use of polyoma-free bovine serum as it is impossible to achieve, in vaccines. Also the data were showing that these viruses did not cause any impact on bovine or other species.

According to the data given by Martin Wisher about the role of transformation of cells, it was only in vitro when the cells were transfected with plasmid containing the early region of bovine polyoma under the transcriptional control of rous sarcoma virus. The cells cannot be transformed by the early genes of polyoma virus under the control of bovine polyoma virus's own transcriptional regulatory sequences.

According to that, no polyoma virus, including bovine polyoma virus, has been detected in a wide range of tumours in man, reported in a publication of Volter in 1997, nor in vaccines associated with sarcomas in cats. This research was published by Kidney in 2001.

According to that, from the veterinary side, we decided not to include bovine polyoma virus in the tests to be done for the bovine serum.

Dr. Minor: I do not understand the pathogenesis of bovine polyomavirus. If you look at the SV40 data we have from our cynomolgus macaque colony, I think there is clear evidence that you do not have intra-uterine vertical transmission. You may observe transmission from the mother to the baby, but it is not vertical transmission.

If you look at things such as JC and BK, it seems to me that there is no vertical transmission. You have post-natal infection. What you are describing here is foetal bovine serum where almost everything is actually stiff with bovine polyoma. Again, going back to what David was saying about when we had six-year-old cattle, it was 90 percent positive. Why were they not all positive at birth if you have so much in the foetal tissue? So I do not understand the basic virology and pathogenesis here either.

Dr. Robertson: Let us move on to a slightly different topic that was raised earlier, the issue of BVDV testing for the virus and the influence of antibodies, and the tests which are being recommended in the European guidelines. I would like to ask Philippe Vannier to make some comments on this.

Dr. Vannier: First, I should say some words about the biology of the virus which is very particular compared to all the other viruses in bovines, and the same occurs in ovines and porcines. Why is there such a big problem with BVD virus? It is because foetal animals which are infected before the onset of immunocompetence become immunotolerant and persistently infected. That means, they are persistently infected by the virus and they are permanent shedders and never produce antibodies because they become immunotolerant.

These animals can maintain a high level of virus up to 10^7 or 10^6 per millilitre of serum, and that is creating the major problem. About BVD antibodies, when you consider the data we have now, there is a big problem because you cannot find any batch of serum without antibodies. For example, for foetal bovine serum, with the data I have, of 990 batches which were controlled, only 19 per cent are free of BVD antibodies.

For newborn calves, out of 40 batches, none is without BVD antibody. For donor bovine, from 63 batches, no batches are without antibodies. For general bovine serum, in 17 batches, no batch is without antibody.

The European guidelines shift very clearly to the possibility of a high level of antibodies, so now we have deleted the comparative titration test which was mentioned by Martin Wisher. Why? Because it was very confusing for the manufacturers and it was quite impossible to have files for marketing authorization in compliance with the European guidelines.

For that reason, we completely deleted the test and reached the conclusion that maybe, if you have bovine serum with high level of antibodies, it is better than having serum without antibodies. Why? Because, in fact, if you have a batch with high levels of antibodies, if this batch is composed of one or several foetuses which are persistently infected, probably the virus will be neutralized by the antibodies you have in other animals when you merge the different sources. So, this is acting in addition with the compulsory inactivation treatment, and that is the reason why we request titration before inactivation, to be sure that the level of virus you could find is below the level of virus which is used for validation of the gamma inactivation treatment.

The antibodies are a marker of Infection. But, the major problem is the one of persistently infected calves. The only request we now have in the latest version, which was finalized three weeks ago at the CVMP level, is if you find a high level of antibodies in your batch of serum, the vaccine manufacturer has to have a procedure not to use that batch for the production of pestivirus vaccines because effectively, if you use serum with a high level of antibodies, you cannot produce classical swine fever or BVD vaccine because they interfere with the multiplication of such pestiviruses.

That is the only request we have at the moment at the European level with regard to BVD antibodies.

Dr. Robertson: What about the EU CHMP's note for guidance? At this time, the BWP advising the CHMP has not re-addressed these issues following on from what the vets have concluded, but it is certainly something that will be raised in the very near future.

The CHMP and the CVMP expert groups are quite willing and prepared to listen to what is happening. If we have erred in our guidance, then we are quite willing and able to reverse that guidance or alter the guidance accordingly. I will certainly take this message back to the BWP and I am sure other European members will be aware of what the veterinary expert group has now recommended which, potentially, will be reflected also in the equivalent human guideline.

As for the draft European Pharmacopoeia Monograph, which also states that you shall do this comparative titration test, that is up to the group responsible for the monograph to listen to and, if necessary, to adopt what other people are saying and doing.

Dr. Wisher: It is a significant problem for some vaccine manufacturers who are not using foetal bovine serum but newborn calf serum.

Participant: Is the majority of the high rate of rejection due to BVD or a range of other viruses?

Dr. Wessman: By far it is BVD. Much of it we reject because of the diagnostic uses where any antibody could interfere with a virus that is present in a diagnostic sample or in a biological sample at a very low level.

Dr. Onions: Although the data are not so consistent, if you look at the titre range in the viraemic calves, it is probable that with every batch, at some level, there is some viraemia there. It is simply that statistically we do not test enough. It does not normally interfere with our assays because, again, in diagnostic assays we are using relatively small amounts of serum so the chance is that, when we have tested it as being negative, its interference is low.

But it is exactly the same argument as for Cache Valley virus. With Cache Valley virus, you may have an infected donation going into a thousand litres, but you do not normally pick that up. You will never pick that up in testing FBS. But you put 20 litres into a fermenter and you pick it up. We do not do that with BVDV, but I think it is there in every batch of FBS.

Dr. Robertson: Is it possible that we could think about using human serum in place of bovine serum?

Participant: I think a manufacturer would have to be deranged to use human serum rather than calf serum. You know very well that over the last few years, every so often the requirements for human serum are increased again and again. One of the things that I think regulators have been trying to do is persuade manufacturers not even to use human serum albumin in their products because, every time the rules change, you have to change the whole production.

Dr. Robertson: It would be rather tricky. But can anybody provide information regarding to what extent one is moving on from the use of foetal bovine to newborn calf or to the use of equine from bovine? Has anybody moved over to equine? Has anybody moved away from bovine to porcine?

Dr. Sheets: Regarding human serum, there is testing for specific viruses, but not all human viruses, so there might be other things in the serum. So there are still going to be human viruses in serum that the donors are not screened for.

Dr. Vyas: The safety of transfusion is now so well established that this community should consider using type AB serum. Type AB is one that does not have anti-A and anti-B and, therefore, could be used as a nutrient in tissue-culture medium.

We have tested this and it is as good or slightly better than foetal bovine serum in growing wild-type HIV, that is, clinical isolates in human CD4 cell substrates. So I think the time has come for the scientific community to raise the issue of whether AB serum can be used or not.

There is no evidence for any other viruses that are transmitted to humans. So, at least for human vaccines, AB serum may be a better alternative than foetal bovine serum. This needs to be debated.

Dr. Onions: Every time you give a blood donation, a whole blood donation, you transmit TTV from one person to another. We now know that TTV is ubiquitous. All of us in the room are probably viraemic with TTV viruses. It is probably like the papilloma story, with so many genotypes that many of these are completely harmless, non-pathogenic.

What we do not know is whether there are pathogenic genotypes out there. So, to say that you do not transmit virus is palpably incorrect. Whether any of those could do any harm, I do not know, but I certainly would not want to take that risk. It is a rather like saying it took fifteen or twenty years to discover that HPV 16, 18 were associated with cervical carcinoma. How are you going to determine quickly out of about 80 genotypes of TTV that one of those genotypes is pathogenic?

Dr. Vyas: TTV produces no disease. But if that is a concern, then there are conserved genomic sequences of TTV that can be tested for.

Dr. Lubiniecki: For those who think it is only a theoretical concern, though, I would urge them to remember one of the clinical trials of cell therapy that was done about twenty years ago at the NIH in which human lymphocytes, autologous human lymphocytes, were grown on human plasma or serum and the result was, I believe, about a dozen cases of hepatitis A.

Dr. Robertson: Given the concern of virus infection of this kind of reagent, clearly there is much effort put into inactivation procedures to inactivate infectious agents. I would like to highlight here for discussion the acceptability of contaminated serum.

Most of the guidelines, if not all of them, accept a certain degree of contamination in the starting material and that really involves the presence of BVDV. Potentially, it also involves bovine polyoma, except there is still a question mark as to whether the signals there represent infectivity or not.

Should the serum be retested after an inactivation procedure on material previously shown to have infectious virus?

In many cases, it is assumed that inactivation will destroy the infectivity that has previously been detected but can that be assumed when the serum is being used for preparing a human biological? Should it be used at all if there were a virus present, and I am referring to viruses other than BVDV, and what effect might any virus contamination have on the cell line being used?

If you can demonstrate that a particular virus does not propagate on the cell line that you are using for your vaccine manufacture, is it then of no concern that the virus may be present in serum?

Dr. Vannier: In Europe, there was a very big problem of BVDV Type 2 in the field in cattle that was coming from the use of a live vaccine, contaminated by BVDV Type 2.

This vaccine was tested and no BVD virus was detected by carrying out the control tests. The source of contamination was identified: it was a bovine serum used for the production of the vaccine. This serum had been gamma-irradiated. Despite that, the controls failed to detect the virus and the gamma radiation failed to inactivate the virus and, for that reason, we decided in Europe to implement the guideline with relatively strong requirements in regard to BVD virus.

In 1994, we had to change the serum batch for our research and diagnostic purposes. We had six serum suppliers, seven different batches of serum to be tested and we found five batches which were BVD-virus positive and all the batches were certified BVD-free for viruses and one was gamma-irradiated at 29 kiloGray. So the problem is, first, BVDV is very difficult to detect and people have to be very competent to handle that virus. That is not so evident to all vaccine manufacturers, serum suppliers and also to the subcontractors who are in charge of BVD testing first. Second, the gamma-irradiation treatment has to be validated in a proper way because the serum supplier has to be sure that the proper dose reached the mass of the products.

That is not so easy and is the reason why we request a test before inactivation to be sure that the validation technique process is taking into consideration a certain level of contamination and to retest to be sure that the gamma-irradiation treatment was really effective through the mass of the product.

Dr. Wisher: I would say that much of the data on gamma irradiation of serum would not be acceptable now in terms of the type of controls and data that we have to produce for human viral-clearance studies.

Very often, there has been no statistical analysis, nor any basis of the sample volume nor of Poisson distribution and they have said they have not detected anything. They put six logs in; therefore, they have six logs inactivation. That would not be acceptable data for viral clearance studies.

Dr. Lewis: I am not aware of any systematic attempts to inactivate various viruses with gamma-irradiation. Maybe there is some information in the literature, but I have not seen it.

Back in the 1960s when people were doing a lot of work on the sensitivity of viruses to various physical and chemical things, there was work done on U.V. inactivation, especially on adenoviruses and SV40. But I am not aware of gamma-irradiation studies. I am also not aware of any systematic studies showing how sensitive, or resistant, various viruses are. However SV40 appears to be very resistant.

Dr. Onions: There is not a large body of data, but we do have considerable in-house data now on irradiation. Part of that information also comes from Mark Plavsic when he was at Invitrogen where he did studies on the circovirus which have this extraordinary radio-resistance.

There is a nice Canadian study which goes through a number of viruses including bunyaviruses and also includes data on polyoma virus on irradiation kinetics.

There is an interesting study on retroviruses that came from Howard Temin's laboratory, if I remember correctly. It is pertinent because we all know from the early studies with retroviruses (HIV) that they have an extraordinary high radio-resistance because they have diploid genomes and they will do this strand repair. So, atypically, they have extremely high resistance. I think you can go to 45 kilo-Grays with, say, 6 logs of HIV, and you will have infectious virus at the end.

Dr. Vannier: I think there are a lot of data, but the problem is that they are not published. In Europe, we had a lot of data from manufacturers and from serum suppliers about the efficiency of gamma irradiation and I can give you some information about that. The most resistant ones are the naked viruses such as adenovirus and parvovirus, which are the most resistant.

You need about 19 kiloGray to reduce one log of these viruses. So that means, according to the dose you apply to the product, you start to have a significant reduction of titre, at least at 25 or 30 kiloGray. The more sensitive viruses are the pestiviruses and the coronaviruses which are enveloped RNA viruses.

These data have been collected by the same suppliers, but they do not publish them. They were very useful for revising the guidelines in Europe, because the first version set too high a requirement which was completely impossible for the serum suppliers or the vaccine manufacturers to implement or to comply with.

For successive versions, we had more and more data. So it allowed us to adjust the requirements to what was really practicable for the serum supplier and vaccine manufacturers. So, the data are there, but they are not published.

Dr. Robertson: Should there be a greater drive to change from serum to serum-free? What are the regulatory issues? What happens when cells have prior exposure to serum or if your cells are frozen and serum has been used in cryopreservation, what are the issues?

Dr. Sheets: If you cannot go serum-free, then what about trying to develop closed herds that would be SPF and BSE-free.

Dr. Egan: We were discussing that at a recent TSE advisory committee. I think we were talking more in terms of using some of the higher-risk material from certified herds. I think, for foetal bovine serum, given the several hundred thousand litres that are used each year by the biotech industry, that having closed documented herds may not be feasible just because of the size of the herds that would be necessary. It is certainly worth looking into further.

Dr. Onions: Just to be very clear in case anyone is misunderstanding this, you cannot have a closed-herd system to generate FBS. That is just not feasible, for obvious reasons when you think about it.

It is theoretically possible to do for donor serum, but the economics do not look very sensible. But those kinds of herds have been set up for other purposes; for instance, for the production of collagen where we have defined closed herds set up here in the United States for that purpose.

The mechanisms of doing it are possible, but it comes down to a purely economic one because the only product you would be talking about is newborn calf serum.

Dr. Petricciani: Many years ago, Connaught did have a closed herd that they used for their vaccine production. I am not sure what happened to it, but it was used by at least one manufacturer for many of the same reasons that we are discussing now.

Dr. Lubiniecki: Perhaps it would be useful to consider encouraging manufacturers to think about and to study why the cell substrate that they have chosen requires serum.

My guess is that you will find different answers for different cell substrates in different products. Sometimes, it is just a source of hormones, in which case, there may be other alternatives that will eventually get you away from using animals, such as recombinant hormones. But if it is also attachment factors or unknown factors, then you have a different problem.

But I think the options for going serum-free are different depending on the specific cell line and the specifics of the biology in terms of what it needs from serum.

Dr. Robertson: The importance of the kinetics of growth of the virus was stressed by Dr. Pierard.

Are you measuring the kinetics subsequently in cells that have been grown with serum or in cells that have been grown without?

Dr. Pierard: In fact, when we want to compare the existing process versus the new process. For the existing process that contains serum, we already have all that kind of kinetic data. We do not expect to have, for instance, the same yield of virus without serum. The important thing to know is if it is feasible in terms of the manufacturing process. We want to have the same purity, the same efficacy.

Dr. Vannier: It is necessary to have a good certification or qualification of what we call a serum-free medium because I remember that some years ago we were approached by some manufacturers proposing a serum-free medium. In fact, we tested it and found BVD virus or BVD antibody.

Dr. Onions: Do you have different glycosylation patterns on your enveloped viruses and are they of any consequence?

Dr. Pierard: All the studies are continuing; we do not have all the results yet.

Dr. Peden: Do the cells senesce?

Dr. Pierard: All these studies are continuing. But, up to now, what the results say is that we do not see a very big difference between the animal-free cells and the current processes.

Dr. Onions: Have you tested some of the lines that you have adapted to the serum-free for tumorigenicity?

Dr. Pierard: Everything is continuing.

Dr. Robertson: Can anybody comment on the extent to which there is a drive to try to grow cells in serum-free conditions?

Dr. Vyas: Most of the primary human lymphocytes do not grow well in vitro without some serum supplement in the tissue-culture medium. Serum-free medium is not so desired for the propagation of primary CD4 cells, for example.

Dr. Wisher: Well over 50 percent of the cell banks that we are now testing are grown in serum-free or animal-free media. Interestingly, it is certainly the larger companies that have now invested very heavily in this. Although 50 percent are still using serum, these are smaller companies or academic groups.

Dr. Sheets: I want to make sure that all of us understand that some of the viruses identified in 9CFR are infectious in people, for instance, bovine PIV3. But are all of them infectious in humans? I do not know the answer to that. Are all those viruses infectious as shown in vitro in human cells or in vivo based on serology? We have heard some information but not for each of the viruses. Also, are they all present in foetal serum?

Dr. Vannier: In the first guidelines we set up, we considered the risk linked to the viraemic viruses, the zoonotic viruses and the health status of the countries which could be the origin of serum batches and possible contaminants.

We tried to compose the guidelines with all these aspects and to advise the manufacturers on detecting viruses with regard to the risk. In fact, it was quite impossible to implement because, for example, we looked at the possibility of certification of the serum with regard to the health status declared or notified at the OIE level in regard to OIE lists A and B of animal diseases. It was quite impossible to restrict the number of tests to be done by the manufacturers and so that means it was a wide range of testing according to the risk.

So we preferred, at the end, to give general tests and to focus on the highest risk for the finished products of the vaccines, i.e. BVD virus, and to delete the rest because it was quite impossible to check for compliance.

Dr Abdy: Bluetongue is spread by Culicoides midges. They have to be infected with that virus. In order to find a seronegative cow, I needed to find that calf before its first summer. I could do that and find seronegative animals. As soon as it had been through a summer and a Culicoides season, they were all positive.

Dr. Vannier: We prefer to focus the effort on the validation process of gamma irradiation. That means to have a variety of viruses to be validated as inactivated by the process which is proposed by the serum supplier or the vaccine manufacturer and to be sure that all these viruses are really inactivated. The tests are done in such a way that we have assurance that it is well inactivated.

Dr. Sheets: The problem is that there is this legacy for cell substrates and viral seeds that may have been propagated in serum as far back as the 1950s and 1960s, and you have no idea where it came from and whether it was irradiated, heat inactivated, or treated at all.

So, for cell banks, you still have to do testing. Even if you are growing in serum-free, you may have to do testing because of the unknown history. So, we should focus on the viruses that are likely to have been in foetal serum or likely to be in these cell substrates, in that list of 9CFR, and ask which of them are relevant?

Dr. Vannier: Effectively, this problem has to be taken into consideration in regard to the test to be done on the master seed cells or working seed cells. What was the origin of the cells but, also, what is the target species where the vaccine is to be used? That is the reason why, if the vaccine is to be used in bovines, it is requested to test for bovine viruses because we do not know exactly some parts of the history of the cells in the 1950s or the 1960s.

Dr. Sheets: That is what I am trying to do also for humans. Given that the products are going to be used in humans, not in cows or horses, are we doing the right thing in following the USDA guidelines when making human vaccines? Is this really appropriate?

Dr. Moore: One of the things which is interesting is that there was some effort put in by both the FDA and, I believe, the USDA, to address viral concerns associated with the use of transgenic animals. The key in those discussions was the identification of zoonotic viruses which would be of specific concern to products derived from transgenic animals.

So, somewhere within FDA, within some of those working group notes, there is probably some very relevant information as to the bovine agents of concern.

If there is a significant change in the geographical origin of serum, we will need to remember to go back and check some of the components of our cell characterization batteries because there is definitely a geographical bias.

There are classes of viral agents that are not accommodated in some of the screening assays simply because, at the time they were put together, there were geographical constraints, as mentioned, primarily related to animal disease in various countries, with foot-and-mouth disease being a primary example.

Summary Statement

Virus testing involves general screening combined with tests for specified viruses and is based on USDA requirements as specified by 9CFR 113.

Manufacturers of human biological products, in the absence of alternative guidance to 9CFR, also validate their serum according to 9CFR.

Recent EU guidance to a large degree is comparable to 9CFR with respect to specified viruses, although the note for guidance issued by the CVMP encourages studies on the prevalence of infectious bovine polyoma virus.

Although other bovine serum viral contaminants have occasionally occurred, there was no consensus on expanding the current list of specified viruses. Indeed, the need for the current seven specified viruses for the use of serum in human vaccine production was questioned since they may not all be zoonotic.

BVDV is recognized as a common contaminant of bovine serum. Some controversy has arisen from the new EU guidelines, both veterinary and human, on recommended tests for the effects of BVDV antibodies on virus assays. This has arisen because of the high titres of BVDV antibodies frequently being found in serum, especially non-foetal bovine serum.

The conference was informed that the veterinary expert group of the EU's CVMP has recently modified their guidance in response to the requests by serum manufacturers and users. The CHMP expert group will be informed of this development so that appropriate follow-up action may be taken.

Bovine polyoma virus is an area of concern. It has been known for some time that foetal bovine serum is often PCR-positive for bovine polyoma DNA. One recent study found 90 percent of batches to be PCR-positive.

Polyoma viruses are often oncogenic in cells and animal species of alternative host origin. Bovine polyoma virus antibodies are prevalent in humans with close contact with bovine species. Consequently, there is concern regarding this virus. However, in the above recent study, none of the batches of serum was positive in an infectivity assay although at least one earlier study had detected infectious virus in serum.

No data are available regarding the nature of the DNA detected by the PCR assay; for example, its size and particulate nature. Clearly, more data are required to expand our knowledge of this virus and its prevalence in serum as soon as possible.

Foetal bovine serum is in very heavy demand and alternatives were discussed. New geographical areas could be tapped, such as South America, although concerns were expressed regarding local infectious agents such as foot-and-mouth disease.

However, assuming appropriate testing of such serum and compliance with all aspects of current regulations and recommendations, there is no apparent reason why such sera could not be used.

There is an increased use of newborn and calf sera in lieu of foetal bovine serum. No concern regarding this beyond the increased levels of BVDV antibodies was expressed. The use of serum from alternative species was considered; for example, equine, porcine or human.

Little comment was made on the use of serum from non-human species. The idea of the use of human serum was not popular.

There is a strong recommendation in EU guidelines to use serum that has been inactivated, for example, by gamma irradiation. It was noted that many viruses, including bovine polyoma virus, are highly resistant to irradiation. Of the few studies performed to investigate the ability of gamma irradiation to inactivate specific viruses, many of these were performed some time ago and would not be deemed of adequate quality if performed today. And many are not published.

Overall, the presence of a viral inactivation step is believed to contribute considerably to the safe use of bovine serum.

A final major point for discussion was that of moving from the use of serum-containing to serum-free media. In this way, and as long as the serum-free medium is animal-component-free, then all the ensuing issues become irrelevant.

Changing cells to growth in serum-free media requires considerable effort and, although the process is conceived to be difficult, it may well be that all cell lines can eventually be adapted to serum-free growth.

Beyond the adaptation process, from a regulatory point of view, considerable work is required to establish comparability of the viral vaccine being produced in serum-free media with the vaccine produced in serum.

It is also important to recognize that cells currently in storage may have been exposed to serum and its potential contaminants at some point in their history.

It was encouraging to hear that 50 % of cell lines currently being submitted to a contract laboratory for testing purposes have been grown in serum-free conditions.

Petricciani J, Sheets R (eds): Vaccine Cell Substrates 2004.
Dev Biol (Basel). Basel, Karger, 2006, vol 123, p 309.

SESSION VI

Bovine Spongiform Encephalopathy

Petricciani J, Sheets R (eds): Vaccine Cell Substrates 2004.
Dev Biol (Basel). Basel, Karger, 2006, vol 123, p 311.

Susceptibility of Common Laboratory Fibroblasts to TSE Infection

S.A. Priola

Laboratory of Persistent Viral Diseases, Rocky Mountain Laboratories, National Institute of Allergy and Infectious Diseases, National Institutes of Health, Hamilton, MT, USA

Abstract: Cell lines that are susceptible to infection with transmissible spongiform encephalopathy (TSE) agents tend to be brain-derived and/or express high levels of prion protein (PrP), a host protein that is essential for TSE infection. Because many cell lines used in medical applications are not of neuronal origin and express relatively low levels of PrP, it has long been assumed that they are at very low risk for infection by TSE agents, such as bovine spongiform encephalopathy (BSE), that may contaminate cell culture medium products.
Contrary to these assumptions, we have recently shown that the commonly used laboratory fibroblast cell lines NIH/3T3 and L929 are highly susceptible to infection with mouse-adapted TSE agents. Both these cell types express low levels of mouse PrP and multiple cell passages were sometimes necessary before infection became apparent. Neither the cell type, the level of PrP expression, nor the TSE strain used were predictive of the susceptibility of a particular cell culture to TSE infection. Our results suggest that the testing of cells for TSE susceptibility might be necessary for all cell lines that are routinely used in vaccine production and other medical applications.

Dr. A. Priola, Laboratory of Persistent Viral Diseases, Rocky Mountain Laboratories, 903 South 4th Street, Hamilton, MT 59840, USA.
E-mail: spriola@niaid.nih.gov

Petricciani J, Sheets R (eds): Vaccine Cell Substrates 2004.
Dev Biol (Basel). Basel, Karger, 2006, vol 123, pp 313-323.

Diagnostics for TSE Agents

A.J. Raeber, B. Oesch

Prionics AG, Schlieren, Switzerland

Abstract: Bovine Spongiform Encephalopathy (BSE) is a fatal acquired neuro-degenerative disease in cattle, belonging to the group of transmissible spongiform encephalopathies (TSEs) or prion diseases. Since its first recognition in the U.K. in 1986, BSE has raised great public health concerns because the BSE agent has been shown to cause variant Creutzfeldt Jakob Disease (vCJD) in humans. With the introduction of mandatory active surveillance programmes in the European Union the need to develop rapid tests to diagnose BSE has become a high priority. Up to now, the European Union has approved twelve rapid tests for BSE monitoring in cattle, and approval for two new tests which have been evaluated in 2004 is pending. These rapid screening tests have been used in active surveillance of BSE and have greatly improved the detection of infected cattle before their entry into the human food chain. At present, no diagnostic test exists for the detection of prions in live animals or humans. New diagnostic techniques aimed at increasing the sensitivity and specificity of PrP^{Sc} detection in body fluids and at identifying novel surrogate markers are under development.

TSEs and the Prion Protein

Transmissible spongiform encephalopathies (TSEs, or prion diseases) such as bovine spongiform encephalopathy (BSE) in cattle, scrapie in sheep and goats, chronic wasting disease (CWD) in deer and elk and Creutzfeldt-Jakob disease in man are caused by a conformational variant (PrP^{Sc}) of the normal cellular form (PrP^{C}) of the prion protein (for review see [1]). The normal form of the prion protein (PrP^{C}) can be distinguished biochemically from the disease-associated form PrP^{Sc} by the different sensitivities to protease digestion: whereas PrP^{C} is totally digested, PrP^{Sc} loses only its N-terminus, giving rise to the protease-resistant core designated PrP27-30 [2]. PrP^{Sc} forms large aggregates which can be separated from PrP^{C} by centrifugation [3,4]. Both forms of PrP have the same amino acid sequences but differ in their tertiary structure. While PrP^{C} is predominantly α-helical (42 %) with low β-sheet content (3 %), PrP^{Sc} has considerably more β-sheet structure (43 %) and a similar α-helical content (30 %). Most of the antibodies raised against PrP recognize linear sequences on PrP and are therefore unable to differentiate between the two forms of PrP. Several antibodies have been published which recognize a conformational epitope on PrP^{Sc} and are able to discriminate between the two forms of PrP. Some of these antibodies bind exclusively to PrP^{Sc} without cross-reacting to PrP^{C} and thereby allow for a direct one-step detection of PrP^{Sc} [5].

Rapid BSE Tests Evaluated by the European Authorities

Most diagnostic tests approved by the EU authorities exploit the relative protease resistance of PrP^{Sc} in brain samples to discriminate between PrP^{C} and PrP^{Sc}, in combination with immunological detection of the protease-resistant part of PrP^{Sc} (PrP27-30). Two tests which underwent evaluation in 2004 do not include a protease digestion step: one uses a specific monoclonal antibody and the other a chemical polymer for a selective PrP^{Sc} capture. Sample preparation is a crucial part of the diagnostic tests and influences the diagnostic sensitivity and specificity as well as the throughput. Because of the membrane attachment of PrP^{C} and PrP^{Sc}, tissue solubilization requires detergent extraction. In addition, conditions for Proteinase K digestion were optimised to allow complete digestion of PrP^{C} but only removing the N-terminus of PrP^{Sc}, thereby leaving intact the protease-resistant core, PrP27-30, for the subsequent detection by anti-PrP antibodies. For non-size discriminating immuno-assays, such as ELISAs this issue is not of direct importance, since size shifts are not a part of the result. In Western blot assays a size shift is directly visible and constitutes an independent criterion for a positive assay result and thereby greatly increases specificity [6]. This has the additional advantage of providing a direct in-process control for the performance of the protease. For ELISA based or similar assays it is of paramount importance that the digestion conditions are solid as even slightly incomplete digestion of PrP^{C} can lead to false positive results.

In 1999, the European Commission carried out a first evaluation of rapid post mortem tests for BSE [7]. Four tests were evaluated on brain tissue from clinical BSE cases and the EU commission approved the following three rapid tests for BSE monitoring in cattle according to regulation (EC) 999/2001 (Table 1).

- **Prionics®-Check WESTERN (Western blot, Prionics AG, Switzerland):** simple automated one-step sample preparation procedure followed by protease treatment. Detection occurs after the separation of the treated sample by denaturing polyacrylamide gel electrophoresis and transfer to a membrane using a PrP specific antibody and an alkaline phosphatase-coupled secondary antibody detection system generating chemi-luminescence. The presence of a PrP-immuno-reactive signal with the additional two criteria of a reduced molecular weight (due to digestion of the N-terminus of PrP^{Sc}) and a typical three-band pattern (due to different glycosylation forms of PrP) result in a TSE-positive diagnosis.
- **TeSeE test (ELISA, Bio-Rad, USA):** sample preparation involving protease treatment followed by precipitation and a centrifugation step for enrichment of the analyte. Detection by sandwich ELISA using a colour-converting enzyme-coupled detection antibody. Cut-off setting with grey zone (leading to repetition of the sample).
- **Enfer test (ELISA, Enfer Scientific, Ireland)**: one-step sample preparation involving sample transfer into and out of a plastic bag followed by protease treatment. The analyte is coated individually on to ELISA plates followed by detection via polyclonal antibodies and an enzyme-coupled secondary antibody. Detection is based on chemi-luminescence.

Table 1 : Evaluation of rapid post mortem tests for BSE.

	Laboratory evaluation			Field trial				
				Healthy slaughtered			Poor quality	
	Sensitivity in % (95 % CI in %)	Specificity in % (95 % CI in %)	Detection limit	Sensitivity in %	Specificity in % (95 % CI in %)	IR (95 % CI in %)	Specificity in % (95 % CI in %)	IR
1999								
Prionics®-Check WESTERN	100 (99)	100 (99.7)	1 : 100[a]	nd	nd	nd	nd	nd
Bio-Rad TeSeE	100 (99)	100 (99.7)	≥ 1 : 200	nd	nd	nd	nd	nd
Enfer TSE kit	100 (99)	100 (99.7)	1:30	nd	nd	nd	nd	nd
2003								
Prionics®-Check LIA	100 (93.8)[b]	100 (98)	≥ 1 : 200	100 (98.7)	100 (99.97)	0.86	100 (98.5)	
CDI-5	100 (93.8)	100 (98)	≥ 1 : 200	100 (98.6)	100 (99.97)	1.45	100 (98.5)	
2004								
CediTect® BSE test	100 (94)	100 (98)	≥ 1 : 200	99.5 (97.7)[c]	100 (99.97)	0	100 (98.6)	0
FRELISA BSE	100 (94)	100 (98)	≥ 1 : 200	ip	ip	ip	ip	ip
IDEXX HerdChek® BSE test	100 (94)	100 (98)	≥ 1 : 200	100 (98.5)	99.99 (99.95)[d]	0.39	100 (98.5)	0
Institut Pourquier Speed'it BSE	100 (94)	100 (98)	1 : 64	100 (98.5)	100 (99.97)	1.88	100 (98.5)	0
PDL Rapid BSE test	96 (87.4)[e]	100 (98)	1:10					
Prionics®-Check PrioSTRIP	100 (94)	100 (98)	1 : 100	100 (98.5)	100 (99.97)	0.37	99.6 (98.2)	1.34
Priontype® post mortem ELISA	100 (94)	100 (98)	1:25	ip	ip	ip	ip	ip
Roboscreen Beta Prion BSE test	100 (94)	100 (98)	≥ 1 : 200	100 (98.5)	100 (99.97)	0.67	100 (98.5)	0
Roche PrionScreen	100 (94)	99.3 (96.2)[f]	1 : 100	100 (98.5)	100 (99.97)	0.2	100 (98.5)	27
Enfer TSE kit v2	100 (94)	100 (98)	≥ 1 : 200	100 (98.5)	100 (99.97)	5.81	100 (98.5)	0

Table 1: continued.

2004	CediTect® BSE Test	2.5-3 h
	FRELISA BSE	3.5 h
	IDEXX HerdChek® BSE Test	4 h
	Institut Pourquier Speed'it BSE	3 h
	PDL Rapid BSE test	
	Prionics®-Check PrioSTRIP	2 h
	Priontype post mortem	3.5 h
	Roboscreen Beta Prion BSE test	3.5-4 h
	Roche PrionScreen	2.5-3 h
	Enfer TSE kit v2	

95 % CI in %: 95 % confidence interval in %.
IR: initial reactive rate in ‰.
nd: not done.
ip: in progress.
[a] Detection limit in the 1999 EU evaluation was 1:10; the EU evaluation in 2002 showed a detection limit between 1:81-1:243
[b] One sample which was initially a false negative was found positive after retesting of additional tissue and must be considered as a sample with low and/or heterogeneous PrP^{Sc}
[c] Three samples were initially classified as false negative. A change in the cut-off setting reduced this number to one false negative
[d] Four samples were classified initial reactive; three were negative after retesting and one re-tested positive.
[e] Three positive samples were not correctly classified initially and re-tested again negative in a duplicate analysis. Re-testing of these samples starting with new tissue slices were positive for one sample but remained negative for two samples. Based on this weakness in diagnostic sensitivity and the low performance in the analytical sensitivity, the test was not allowed to enter the field trial.
[f] One negative sample out of three initial reactives was positive in re-testing.

A second call for expression of interest to participate in an evaluation for rapid BSE test was issued in 2002. This exercise was split into a laboratory evaluation and a field trial to assess the performance of the new tests under field conditions and allow a comparison with already approved tests [8]. The following two tests passed the laboratory evaluation and the field trial and were approved by the EU commission in 2003 for BSE monitoring in cattle (Table 1).

- **Prionics®-Check LIA (ELISA, Prionics AG, Switzerland):** simple one-step sample preparation procedure followed by protease treatment. Detection occurs via sandwich ELISA using monoclonal antibodies, one of them enzyme-labelled. Detection is based on chemi-luminescence using a plate-specific cut-off setting.
- **Conformation Dependent Immuno-assay CDI-5 (ELISA, Inpro, USA):** in addition to a mild protease treatment, the CDI uses the differential binding of antibodies to native or denatured PrP^{Sc}. The detection antibody recognizes a conformation-dependent epitope which, while always exposed in the non-infectious form (PrP^{C}), becomes exposed in the infectious form of PrP (PrP^{Sc}) only upon denaturation. Quantification of the binding events leads to a signal difference (signal "total denatured PrP" minus signal "native PrP^{C}" = PrP^{Sc}) which is used as a diagnostic criterion.

A third evaluation was carried out in 2004 following a call for expression of interest in January 2003. This evaluation exercise was managed jointly by the European Commission's Directorate General Joint Research Centre, Institute for Reference Materials and Measurements (IRMM) and the Directorate General Health and Consumer Protection in collaboration with a working group of the European Food Safety Agency (EFSA). This evaluation was again carried out in two phases: a laboratory evaluation [10] and a field trial [11]. Of a total of 10 tests that were assessed in the laboratory evaluation, nine were allowed to enter the field trial. Based on a low diagnostic sensitivity, the PDL rapid BSE test (Prion Developmental Laboratories, Inc., U.S.A.) was not allowed to undergo the field trial. Until September 2005, the following seven tests have successfully passed the field trial and have received for approval by the EU commission (Table 1).

- **CediTect® BSE Test (ELISA, CEDI Diagnostics, the Netherlands):** protease digested samples are fixed to an activated PVDF membrane by filtration in a 96-well format. One replicate of each sample is further treated with a chaotropic agent resulting in the T-sample; the second replicate (N-sample) as a control is not exposed to that reagent. Detection of the protease-resistant PrP is based on a chemi-luminescent read-out using an enzyme-conjugated monoclonal antibody. Calculation of the ratio of the T- and N-value is used to determine the status of the homogenate.
- **IDEXX HerdChek® BSE Antigen Test Kit, EIA (ELISA, IDEXX Laboratories Inc., U.S.A.):** sample homogenisation is carried out with a ribolyser. Detection of PrP^{Sc} is performed by a classical two-sided immuno-assay using a chemical polymer (technology licensed from Microsense Biotechnologies, London, UK) for selective capture of PrP^{Sc} – without the need for a protease digestion step – and detection by an enzyme-conjugated monoclonal antibody using a colorimetric substrate. A plate-specific cut-off setting is used in the diagnosis of the sample.
- **Institut Pourquier Speed'it BSE (ELISA, Institut Pourquier, France):** tissues are homogenised using a ribolyser followed by treatment of the samples with Proteinase K. Detection of PrP27-30 uses a classical two-sided immuno-assay with two monoclonal antibodies. Capture and detection of the protease-resistant core of PrP^{Sc} is performed in a single step in a microplate. Detection is based on chemi-luminescence using a plate-specific cut-off setting.
- **Prionics®-Check PrioSTRIP® (Lateral flow device, Prionics AG, Switzerland)**: rapid microplate-based chromatographic immuno-assay using two monoclonal antibodies to detect the protease resistant core of PrP^{Sc}. The homogenate is subjected to treatment with Proteinase K and then incubated with the conjugate consisting of a coloured latex bead-labelled antibody. The PrioSTRIP® – a comb-shaped cartridge holding eight strips – is inserted into the sample conjugate mixture in a microplate, and the immuno-complexes are captured by a monoclonal antibody sprayed as a line perpendicular to the long axis of a strip (Fig. 1). The resulting immuno-"sandwich" appears as a coloured line and the reading can be done either visually or automated allowing a quantitative analysis using a specially designed cartridge holder for a maximum of 12 cartridges (= 96 strips), a flat-bed scanner and a dedicated software (PrioSCAN).

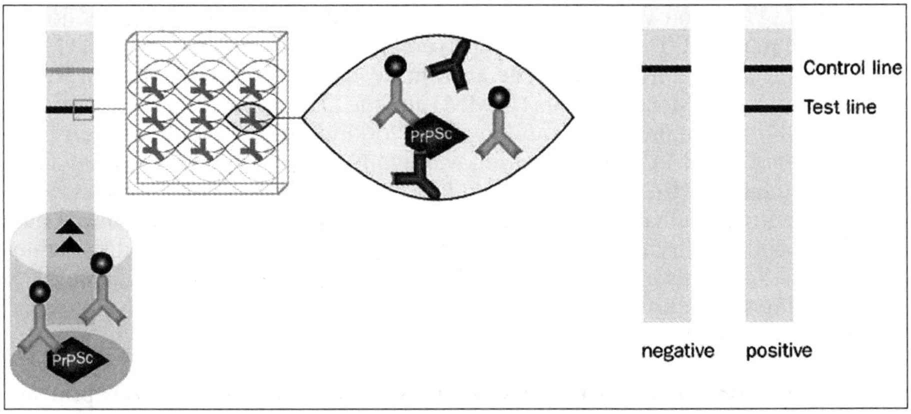

Fig. 1: **Detection of protease treated PrP^{Sc} in the PrioSTRIP®.** Medulla oblongata samples are homogenized and reacted with Proteinase K to digest cellular PrP. Samples of digested homogenates are then mixed with the PrioSTRIP® conjugate consisting of anti-PrP mAb-bound colored latex particles. Cassettes consisting of eight individual strips are then inserted into the sample/conjugate mixtures causing the analytes to chromatograph the strip towards an absorbent waste pad. During chromatography, PrP binds to the latex-bound mAb, thereby forming immuno-complexes. These complexes are captured at the test line position resulting in a visible line across the strip. Excess conjugate is bound at the control line. PrioSTRIP® forces analyte and antibodies into close proximity and concentrates the analyte at the test line. This increases assay sensitivity, accelerates reaction kinetics and minimizes incubation times.

- **Roboscreen Beta Prion BSE EIA Test Kit (ELISA, Roboscreen GmbH, Germany):** tissues are homogenised using a ribolyser followed by treatment of the samples with Proteinase K. After protease digestion, the homogenate is subjected to a protein precipitation and delipidation step. The test is a classical two-sided immuno-assay using two monoclonal antibodies. Detection is based on a colorimetric enzyme reaction using a fixed cut-off setting with grey zone (leading to repetition of the sample).
- **Roche Applied Science PrionScreen (ELISA, Roche Diagnostics GmbH, Germany):** homogenisation and protease digestion of the tissue samples is carried out in a 96-well plate homogeniser. Samples are then analysed in a classical two-sided immuno-assay using two monoclonal antibodies to detect the protease-resistant core of PrP^{Sc}. Capture and detection of PrP27-30 is performed in a single step in a streptavidin-coated microplate using a biotin-conjugated capture and an enzyme-conjugated detection antibody. Detection is based on a colorimetric enzyme reaction using a plate-specific cut-off setting.
- **Enfer TSE Kit version 2.0, automated sample preparation (ELISA, Enfer Scientific Ltd., Ireland):** due to a change in the homogenisation system, the EC approved Enfer TSE test was subjected to a full re-evaluation in accordance with the rules defined by the Community Reference Laboratory for TSE. The new homogenisation procedure uses a disposable plastic tube in which a tissue sample is ground between two rough surfaces. The process of homogenisation is automated and the samples are then transferred to a liquid-handling robot for the subsequent ELISA.

By November 2004, field trials for the following two tests were still continuing:

- **FRELISA BSE (ELISA, Fujirebio, Japan):** tissue homogenisation with beads followed by a treatment with DNAse I, collagenase and protease. The homogenate is then subjected to protein precipitation, protein unfolding and sonication steps. Detection of protease resistant PrP^{Sc} is performed by a one-step ELISA using two monoclonal antibodies, one of which is an enzyme conjugate allowing a colorimetric detection with a plate-specific cut-off setting.
- **Priontype® post mortem ELISA (Labor Diagnostik Leipzig GmbH, Germany):** homogenisation is performed with ceramic beads followed by protein denaturation, protein precipitation, resolubilisation of the protein pellet and a stringent protein unfolding step. PrP^{Sc} is captured by a monoclonal antibody that differentiates between PrP^{C} and PrP^{Sc}. Detection with an enzyme-coupled second monoclonal antibody is based on a colorimetric reaction using a plate-specific cut-off setting.

The hallmark of any diagnostic test is its diagnostic sensitivity (ability to identify infected reference samples as positive) and specificity (ability to identify uninfected reference samples as negative). This will determine the usefulness of a test in the field, i.e. how many false and true positives it will produce. The diagnostic sensitivity and specificity of the TSE tests were evaluated by the EU commission as well as in individual countries by using 50 positive and 150 negative samples from animals previously diagnosed for TSE with a reference method such as histopathology or immunohistochemistry [6-9].

In addition, the evaluation protocol included an assessment of the detection limit using serial dilutions of a BSE-positive brain homogenate diluted in negative homogenates. A comparison of three ELISA tests showed very similar analytical sensitivities (Fig. 2). This rather technical parameter should reflect the level of antigen that can be detected by a particular assay and should give an indication of the capability of the tests to detect preclinical cases of BSE. However, the analytical sensitivity may or may not correlate with the diagnostic sensitivity depending on the appearance and concentration of the analyte during the course of the disease. In the case of protease-resistant PrP as an analyte of TSE diagnosis, the two parameters differ because PrP^{Sc} does not accumulate linearly but exponentially in the brain of infected animals [12]. Furthermore PrP^{Sc} may be present in different aggregation states during the incubation phase of the disease. It has been shown that small PrP^{Sc} aggregates which may be present in preclinical animals can not be easily precipitated [13]. Therefore, PrP^{Sc} in a dilution of a brain sample of a clinical animal does not necessarily reflect PrP^{Sc} in a preclinical animal. For this reason, slight variations in analytical sensitivity do not necessarily reflect differences in diagnostic sensitivity. This is illustrated by a comparative study on the surveillance data from France (Table 2) which show equal diagnostic sensitivity for two of the most used BSE tests in Europe despite some claims of higher analytical sensitivity for one of the tests.

Field Assessment of Rapid Test

The rapid tests were further evaluated in a field trial to determine their performance under normal routine conditions. It further enables a thorough assessment of

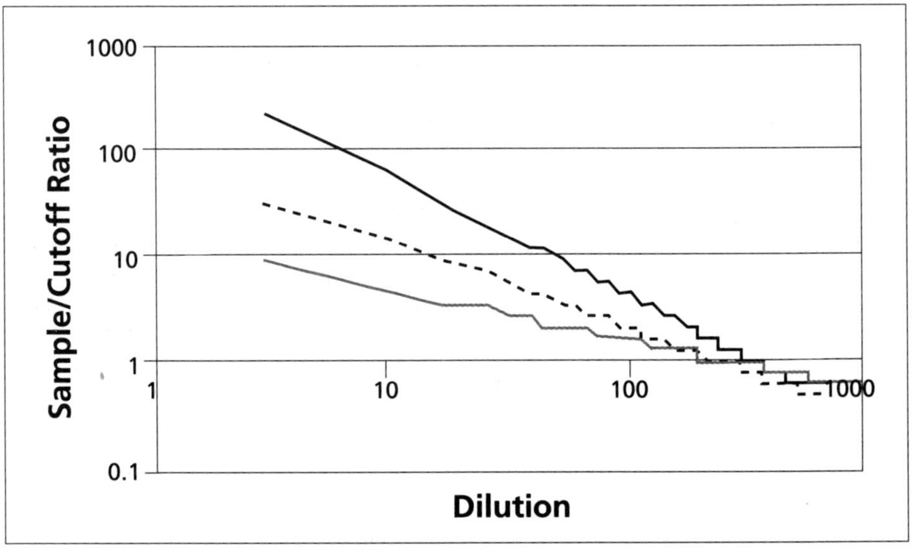

Fig. 2: **Analytical sensitivity of rapid TSE tests.**
BSE positive brain homogenate was diluted in a negative homogenate in dilutions down to 10^{-3} and analysed with the Prionics®-Check LIA (——), the CDI (- - - -) and the Platelia (——) ELISA. Measurements are expressed as a ratio of mean signals to the cut-off value for the LIA and the Platelia. For the CDI the measurements are expressed as a ratio of mean signals to negative controls (Data taken from http://europa.eu.int/comm/food/fs/bse/bse42_en.pdf).

Table 2: **Results of BSE testing in France.** Numbers of positively and negatively tested cattle from two different populations (normal slaughter and cattle from the risk category) in France using two different assay systems (Prionics Western and BioRad). The low χ^2 values (compared to $\chi^2\alpha = 3.8$) of a χ^2 test indicate no significant difference in the relative populations of positively tested animals between the two assay systems.

Normal slaughter		
	Prionics	***Bio-Rad***
neg	2'356'421	753'124
pos	79	26
tot	2'356'500	753'150
χ^2	0.02	

Risk category		
	Prionics	***Bio-Rad***
neg	161'308	37'336
pos	92	24
tot	161'400	37'360
χ^2	0.27	

Source: French Food Agency (Direction Générale de l'Alimentation (DGAL)).

parameters such as robustness, applicability in different laboratories, expected rate of false initial reactives under high throughput field conditions and necessary cut-off adjustment before market introduction. For this evaluation, over 10,000 tissue samples from healthy slaughtered cattle that were tested negative with an already approved rapid BSE test were analysed to assess the diagnostic specificity. The diagnostic sensitivity of a new test was likewise assessed using a set of 200 true positive samples that were tested positive using a reference test. In addition the tests were assessed for their performance on 200 poor quality samples (http ://europa.eu.int/comm/food/fs/sc/ssc/out316_en.pdf). This is important because field samples often deteriorate due to poor storage conditions of carcass samples over long periods of time. Such samples show extended autolysis and a robust diagnostic test has to give accurate results under such conditions. Previous studies have shown that PrP^{Sc} can be readily detected in autolyzed samples with Western blot and IHC [14-16].

Whereas it has been extensively documented that sample autolysis does not influence PrP^{Sc} detection by Western blot, a recent study carried out by the National Institute of Animal Health in Japan found that some ELISAs might be unsuitable for screening of fallen stock samples [17]. Hayashi et al. found that one of the commonly used ELISA rapid tests showed a significant reduction in OD values on deteriorated samples.

Current and Future Testing for BSE

By the middle of 2004 more than 33 million rapid tests have been performed in Europe within the mandatory rapid testing of all slaughtered cattle above the age of 30 months. Currently, the Prionics®-Check tests are the most used BSE-tests in the world followed by the TeSeE test. In total, about 10 million BSE tests per year are performed in Europe and about one million in Japan. Following the 2004 EU evaluation, it can be expected that in 2005, a total of 12 rapid tests based on different detection principles will be approved for BSE monitoring in cattle: one Western blot, 10 ELISA, and one Strip. While Western blot-based BSE tests have been shown to be the most reliable and accurate rapid BSE tests, giving no false positives – or initial reactives – ELISA tests have been widely used because they can be automated for high throughput screening. However, future BSE testing needs might be tackled by a new generation of rapid BSE tests based on an immuno-chromatographic principle similar to lateral flow devices commonly used in e.g. pregnancy tests. While two such tests passed the EU laboratory evaluation in 2004, only one – the Prionics®-Check PrioSTRIP – has successfully completed the EU field trial. This rapid test combines high-testing speed (100 minutes from brain to result) with simple handling and at the same time delivers results with 100 % sensitivity and specificity.

The future extent of BSE testing is currently the subject of intense discussions among food safety and public health authorities in Europe, Japan and the Americas. The U.K. government, advised by the Food Standards Agency (FSA) has recently announced that testing for BSE could replace the current Over Thirty Months (OTM) Rule which prevents cattle over thirty months old from entering the human food chain. Since its introduction as a public health protection measure in 1996, the OTM scheme has cost more than £360 million a year or £3 billion in total, compared to around £60 million a year for the proposed BSE monitoring programme, which could start in the second half of 2005 (http ://www.defra.gov.uk/news/2004/041201a.htm).

In contrast, Japanese food safety officials have recently given the green light for reviewing the test-all approach which was adopted by Japan in 2001 after the discovery of the first BSE case and which requires testing for BSE of all cattle that enter the human food chain. Going back to a less stringent testing regime including only cattle older than 20 months could eventually help reviving bilateral talks between policy makers in Japan and U.S. over lifting the ban on U.S. beef imports.

The ban on U.S. beef came as a result of the first BSE case in the U.S. in December 2003. In June 2004, the USDA implemented a targeted surveillance programme to estimate the BSE prevalence in the U.S. cattle population. Within this testing regime, 268,500 animals, corresponding to 0.6 % of the high-risk population, will be screened. This is a rather restricted number of cattle when compared to BSE test programmes in the EU where many countries are screening up to 10 % of the high-risk cattle population and in addition a significant proportion of the healthy slaughtered cattle.

Ante-Mortem TSE Test Development

PrP^{Sc} is used as a marker for the post-mortem diagnosis of TSEs because high concentrations are found in the CNS of infected animals and humans. In contrast, PrP^{Sc} concentrations in the blood are extremely low, if present at all, and methods to enrich for PrP^{Sc} become mandatory for ante-mortem test development. Antibodies that specifically recognize the disease-associated form of the prion protein have been generated and shown to be capable of detecting PrP^{Sc} in brain homogenates without the need for Proteinase K digestion [5]. Enrichment for PrP^{Sc} in the blood of TSE-infected organisms using 15B3 should pave the way for a new live test for TSEs.

Other disease-indicating markers other than PrP^{Sc} from easily available body fluids are needed. Using differential RNA display technologies, it was found that a novel transcript coding for erythroid differentiation-related factor (EDRF) showed a progressive decrease in the expression during the course of the disease in scrapie-infected mice [18]. In a similar approach, cellular nucleic acids were identified in the serum of BSE-infected cattle that appeared to correlate with the development of the disease [19]. If such nucleic acids were indeed present in blood, this might form the basis of a simple PCR-based diagnosis of TSE and related disorders from live organisms.

Another simple, non-invasive test to diagnose TSEs uses a technique known as high-resolution electrocardiogram (ECG) measuring a unique heart rate variability signature observed at the early stages of infection [20]. The test was used on 150 cows experimentally infected with the BSE agent and detected the disease based on increased levels of RSA (respiratory sinus arrhythmia) in two animals that died eight months later.

References

1 Weissmann C: The state of the prion. Nat Rev Microbiol 2004; 2 (11): 861-871.

2 Oesch B, Westaway D, Walchli M, McKinley MP, Kent SB, Aebersold R et al: A cellular gene encodes scrapie PrP 27-30 protein. Cell 1985; 40: 735-746.

3 Bolton DC, McKinley MP, Prusiner SB: Identification of a protein that purifies with the scrapie prion. Science 1982; 218 (4579): 1309-1311.

4 Hope J, Morton LJ, Farquhar CF, Multhaup G, Beyreuther K, Kimberlin RH: The major polypeptide of scrapie-associated fibrils (SAF) has the same size, charge distribution and N-terminal protein sequence as predicted for the normal brain protein (PrP). EMBO J 1986; 5: 2591-2597.

5 Korth C, Stierli B, Streit P, Moser M, Schaller O, Fischer R et al: Prion (PrPSc)-specific epitope defined by a monoclonal antibody. Nature 1997; 390 (6655): 74-77.

6 Schaller O, Fatzer R, Stack M, Clark J, Cooley W, Biffiger K et al: Validation of a western immunoblotting procedure for bovine PrP (Sc) detection and its use as a rapid surveillance method for the diagnosis of bovine spongiform encephalopathy (BSE). Acta Neuropathol (Berl) 1999; 98 (5): 437-443.

7 Moynagh J, Schimmel H: The evaluation of tests for the diagnosis of transmissible spongiform encephalopathy in bovines. http://europa.eu.int/comm/food/fs/bse/bse12_en.html. 1999.

8 Schimmel H, Catalani P, Le Guern L, Prokisch J, Philipp W, Trapmann S, Zeleny R: The evaluation of five tests for the diagnosis of transmissible spongiform encephalopathy in bovines (2nd study) http://europa.eu.int/comm/food/food/biosafety/bse/bse42_en.pdf

9 Biffiger K, Zwald D, Kaufmann L, Briner A, Nayki I, Purro M et al: Validation of a luminescence immuno-assay for the detection of PrP (Sc) in brain homogenate. J Virol Methods 2002; 101 (1-2): 79-84.

10 Philipp W, van Iwaarden P, Goll M, Kollmorgen N, Schimmel H, Vodrazka P: The evaluation of 10 rapid post mortem tests for the diagnosis of transmissible spongiform encephalopathy in bovines. http://www.irmm.jrc.be 2004.

11 Philipp W, Vodrazka P: The field trial of seven new rapid post mortem tests for the diagnosis of bovine spongiform encephalopathy in bovines http://www.irmm.jrc.be 2004.

12 Jendroska K, Heinzel FP, Torchia M, Stowring L, Kretzschmar HA, Kon A et al: Proteinase-resistant prion protein accumulation in Syrian hamster brain correlates with regional pathology and scrapie infectivity. Neurology 1991; 41 (9): 1482-1490.

13 Gabizon R, McKinley MP, Groth DF, Kenaga L, Prusiner SB: Properties of scrapie prion protein liposomes. J Biol Chem 1988; 263: 4950-4955.

14 Oesch B, Doherr M, Heim D, Fischer K, Egli S, Bolliger S et al: Application of Prionics Western blotting procedure to screen for BSE in cattle regularly slaughtered at Swiss abattoirs. Arch Virol Suppl 2000; (16): 189-195.

15 Debeer SO, Baron TG, Bencsik AA: Immunohistochemistry of PrPsc within bovine spongiform encephalopathy brain samples with graded autolysis. J Histochem Cytochem 2001; 49 (12): 1519-1524.

16 Chaplin MJ, Barlow N, Ryder S, Simmons MM, Spencer Y, Hughes R et al: Evaluation of the effects of controlled autolysis on the immunodetection of PrP (Sc) by immunoblotting and immunohistochemistry from natural cases of scrapie and BSE. Res Vet Sci 2002; 72 (1): 37-43.

17 Hayashi H, Takata M, Iwamaru Y, Ushiki Y, Kimura KM, Tagawa Y et al: Effect of tissue deterioration on postmortem BSE diagnosis by immunobiochemical detection of an abnormal isoform of prion protein. Journal of veterinary medical science/the Japanese Society of Veterinary Science 2004; 66 (5): 515-520.

18 Miele G, Manson J, Clinton M: A novel erythroid-specific marker of transmissible spongiform encephalopathies. Nat Med 2001; 7 (3): 361-364.

19 Brenig B, Schutz E, Urnovitz H: [Cellular nucleic acids in serum and plasma as new diagnostic tools]. Berl Munch Tierarztl Wochenschr 2002; 115 (3-4): 122-124.

20 Pomfrett CJD, Glover DG, Bollen BG, Pollard BJ: Perturbation of heart rate variability in cattle fed BSE-infected material. Veterinary Record 2004; 154 (22): 687-691.

Dr. A.J. Raeber, Prionics AG, Wagistrasse 27A, CH-8952 Schlieren, Switzerland.
E-mail: alex.raeber@prionics.ch

Petricciani J, Sheets R (eds): Vaccine Cell Substrates 2004.
Dev Biol (Basel). Basel, Karger, 2006, vol 123, pp 325-333.

Production Cell Substrates and vCJD

P.D. Minor, J. Cooper

National Institute for Biological Standards and Control, Potters Bar, Herts., U.K.

Introduction

The outbreak of bovine spongiform encephalopathy in the United Kingdom caused great concern about the safety of materials derived from cows. While the primary issues were to do with food, biological medicinal products where manufacture involved ingredients such as calf serum were also subjected to scrutiny. In 1996 a new form of Creutzfeld Jakob Disease (vCJD) was identified in the U.K. in young people and linked to exposure to BSE [1-3]. As human blood and blood products such as clotting factors from U.K. sources incubating the disease could be contaminated with the infectious agent, the U.K. government stopped the fractionation of U.K.-sourced plasma for the manufacture of blood products. This caused difficulties world-wide as varying exclusions were introduced in other countries for individuals who had spent some time in the U.K. during the period of highest risk.

The transmissible spongiform encephalopathies are insidious diseases where neurodegeneration follows a long silent incubation period. While they are infectious their ability to transmit between species depends on the strain and the species concerned. There is good reason to believe that there is negligible risk of transmission of scrapie of sheep to humans [4,5]; transmission of CJD itself between humans has been shown by the use of contaminated biological medicines including growth hormone. It remains to be determined how transmissible the agent of BSE is to humans. At the time of writing there are fewer than 150 cases of vCJD world wide (www.cjd.ed.ac.uk), and some believe that the incidence is falling. However it is still not clear how many individuals are incubating vCJD, and it is possible that a larger epidemic is incubating, or that completely silent infections occur [6]. The BSE epidemic in the United Kingdom is well past its peak, although infected animals are still identified, and other countries including the U.S.A. have also reported cases. Current and future exposure to BSE is thus likely to diminish, although it may not stop altogether, and the extent of the human infection is still not known.

Exposures to materials of bovine origin, particularly if those materials could not be convincingly shown to be of non-U.K. origin, have been a major regulatory concern for biological medicinal products for some time. Where there is a gap in the record of passage of a cell line, so that it might have been exposed to serum of U.K.

origin, the acceptability of the line for production has been questioned. This is based on the possibility that the serum was contaminated with the agent of BSE, and that the cells might have become infected as a result. The infectability of cell lines is of interest from two practical points of view, therefore. Firstly, if it can be shown that cells used in production are not easily infected under optimum conditions, it provides some reassurance that they have not been infected by accidental exposure to serum contaminated with the BSE agent. Secondly, if cells can be infected with the agents of either BSE or vCJD, they could provide a possible in vitro infectivity assay for BSE or vCJD for production cells or for blood or other tissues. Current infectivity assays involve the use of animals, frequently transgenic mice, and take a very long time to complete. Certain strains of TSE have been shown to be able to infect cells in culture [7, Vorberg, Raines and Priola, Journal of Biological Chemistry (in press), Wessmann this volume]. As an aid to the development of assays for CJD in general we have developed reference reagents from tissues from infected patients. This paper describes the features of the reagents and their use in attempting to infect a number of human diploid cell lines, including several which have been used in the production of biological medicines. The reagents are available for use in assay development (www.nibsc.ac.uk/divisions/virology/cjd/). The attempts at infection have so far been unsuccessful.

The Reference Reagent Bank

Samples of four brains were provided by the CJD Surveillance Centre, Edinburgh. Two were from cases of sporadic CJD, one from a case of vCJD, and one from a patient who had died of other causes. The prion protein (PrP) is involved in the pathogenesis of the transmissible spongiform encephalopathies. It is a normal component of the surface of many cells but is deposited in an abnormal, less biodegradable form, termed PrP^{sc} in the disease; the most commonly held view at present is that PrP^{Sc} is the infectious agent itself, catalysing the deposition of the normal form (PrP^{c}) in the abnormal conformation. The sequence of the protein is a crucial element in susceptibility to disease, including transmission from one species to another [8] and in humans there are two allelic forms at codon 129, which influence disease susceptibility, such that methionine/methionine homozygotes are heavily over-represented in cases of sporadic CJD, and represent all vCJD cases to date [9,10]. The alternative allele is valine. All four brains selected were of the methionine/methionine genotype; a fifth brain sample, of the methionine/valine genotype was provided by Professor H. Budka, Vienna. In addition a spleen sample from a case of vCJD, again of the methionine/methionine genotype was supplied by the CJD surveillance Unit in Edinburgh. The tissues were prepared as 10 % homogenates in unbuffered 25mM sucrose, each distributed in approximately 2000 vials containing 0.5ml and stored at – 80^{0} C as described in detail elsewhere [11]. The first four brains were used in a collaborative study under the auspices of the World Health Organisation, in which in vitro assays were used to estimate the amount of PrPsc, the abnormal isoform of PrP. The assays mainly used immunoblotting after treatment with proteinase K to remove the normal isoform, PrPc. Participants were requested to assess serial dilutions of the preparations to gain an idea of the amount of PrPsc present, and to compare the effects of slight differences in technique on the analytical sensitivity achieved. The minimum detectable volumes are shown in Table 1 in microlitres of 10 % homogenate for six immunoblot-based assays (laboratories 1 to 6). The detectable volume of the sporadic CJD sample Sp1CJD varied from 0.067 to 0.33 microlitres; of sp2CJD from 0.35 to 1.0 and of vCJD from 0.067 to 0.6. The ranges were thus of about three- to ten-fold similar to the dilution series used. This implies that all the methods examined were of similar sensitivity despite the differences in technique used, which included dilution in normal brain or buffer before proteinase K treatment, and a variety of concentration steps, reflected in the volume loaded.

Table 1: Minimum detectable volume of homogenates.

Laboratory	Volume loaded[1]	Sp1 CJD[2]	Sp2 CJD	VCJD
1	5	0.22	0.50	0.091
2	20	0.20	2.0	0.20
3	3	0.25	1.0	0.60
4	8	0.067	0.35	0.067
5	5	0.33	1.25	0.25
6	100	0.18	0.44	0.18
7a	167	0.076	0.13	0.076
7b	167	0.0017	0.0073	0.00098

1 Equivalent volume (µl) of 10 % brain homogenate in undiluted sample.

2 Equivalent volume (µl) of 10 % brain homogenate in loading last track containing detectable PrP.

Laboratories 1-6 used immunoblots. Laboratory 7 used CDI in two formats, a and b.

The seventh laboratory used the Conformation-Dependent Immunoassay as initially described by Safar et al [12] and modified by Bellon et al [13]. This method detects PrP using an antibody that reacts only with native or denatured PrPc but not PrPsc. The sample is assayed initially to determine the amount of PrPc, then reassayed after denaturation with guanidinium hydrochloride to measure total PrP. Detectable PrP is measured by attaching the protein to an ELISA plate, adding the antibody and measuring the amount bound by DELFIA (dissociation enhanced lanthanide fluorescence assay). The samples are pre-treated with Proteinase K to reduce the background and the ratio of the signals after and before denaturation of the sample is a measure of the amount of PrPsc. The sample is considered positive if the ratio is higher than the background. In the first format of the technique the samples were attached to the plate using glutaraldehyde, and the minimum detectable volume was of the same order as that found for the immunoblot methods as shown in Table 1 under 7a. In the second format, which is that currently used, the PrP is attached to the plate by use of a specific antibody so that it is partly purified from the sample. This reduces the background further and increases the sensitivity by about two orders of magnitude, as shown by the results for (7b) in Table 1.

The CDI method has been used at NIBSC to examine the 10 % spleen homogenate for the presence of PrPsc, and the results are shown in Figure 1, where it can be seen that the amount present is approximately one hundred-fold less than in the vCJD brain preparation. This is consistent with published infectivity studies [14]. The spleen preparation gave a barely detectable signal by the immunoblot method used at NIBSC.

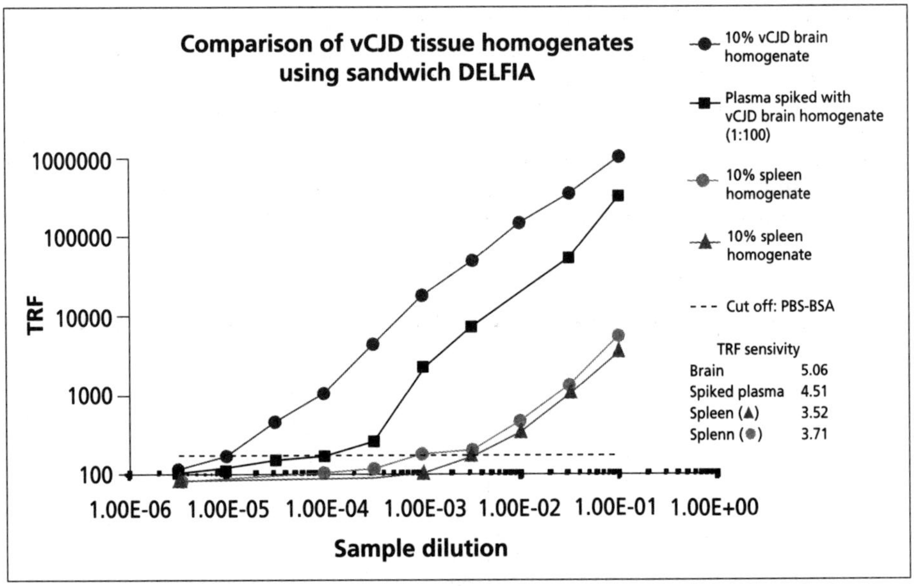

Fig. 1: CDI assay of vCJD brain homogenate and vCJD spleen homogenate in different diluents.

The preparations are available from the CJD Resource Centre at NIBSC for assessment of detection methods (www.nibsc.ac.uk/divisions/virology/cjd/); the sporadic CJD preparations have been examined for infectivity by Dr L. Chernevokova in transgenic mice carrying the human prion gene and have a titre of about 10000 in this system (L. Chernevokova, pers. comm. It is intended to assay the infectivity in the vCJD brain and spleen preparations in mice.

Expression and Genotype of Prion Gene in Human Diploid Cells

A number of human diploid cells were developed in the late1960s from embryonic tissue. They are characterised by a normal karyotype, limited life span in culture and lack of tumorigenicity, and some such as MRC5 [15] and WI38 [16] have been used very extensively in the production of biological products, particularly vaccines. Other cells including MRC9 have been used on a smaller scale, and other cells are available if needed, although they have been less widely used and characterised. Cells in the stocks at NIBSC were recovered and tested by STR analysis. This established their unique DNA fingerprint, and where patterns were available demonstrated their identity with cells of the same name from other sources such as ATCC. The prion gene was sequenced to establish the codon at 129, and the results are shown in Table 2. No cells were found to have mutations associated with familial Creutzfeld Jakob Disease or GSS, the other hereditary form of CJD, and the representation of allelic forms was similar to that in the general population, with a slight excess of methionine/valine heterozygotes over methionine methionine homozygotes. MRC5 cells were found to be from a methionine/valine heterozygote, and WI38 from a methionine/methionine homozygote; no valine/valine homozygotes were found, as might have been expected from the number of cell types examined. Samples of the cells containing the same amount of protein were resolved on a polyacrylamide gel, transferred to nitrocellulose and probed with the antibody 3F4 [17], which will react with the human prion protein. Figure 2 shows that all cells tested expressed the protein in readily detectable amounts although some, such as MRC1, expressed less than others. The null cell is a continuous cell line derived at NIBSC from a mouse in which the prion gene is deleted [18] and was included as a negative control; 3F4 does not react with the murine prion protein, which in any case is absent from the cell line.

Table 2.

Cell line	Genotype
MRC1	MV
MRC5	MV
MRC7	MM
MRC9	MV
IMR90	MM
Wi38	MM
HEK-293T	MM

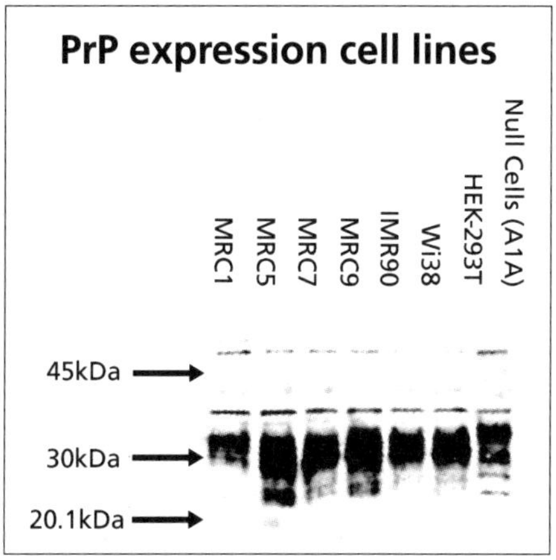

Fig. 2: Immunoblot of human diploid cells and null mouse cell line with 3F4 using nine micrograms of protein from each cell type to assess the amount of human prion protein.

EXPOSURE OF CELLS TO vCJD MATERIAL

Cell monolayers were incubated with a one to ten dilution of the 10 % vCJD brain homogenate in culture medium for 96 hours at 37°, washed in fresh medium and allowed to recover for 24 hours. They were then sub-cultured at a ratio of 1:3. The cultures were maintained for a number of passages using a split ratio of 1:3 as they became confluent, usually every four to five days. Two thirds of the cell harvest was retained and assayed for the presence of PrPsc by CDI. A number of parameters were investigated. Some workers treat infectious suspensions by boiling them before inoculation into animals in infectivity assays to remove toxins or bacterial or other contamination, and cells were exposed to heated and unheated samples. Sub-culture of cells can involve the use of trypsin, which could remove PrPsc from the cell surface, even if it does not digest it; cells were therefore sub-cultured using either trypsin or other reagents. It was found that sub-culture using trypsin retained more material than with the other options, possibly because of the poorer quality of the cells that resulted from the use of reagents such as versene. Typical results are shown in Figure 3 for six cell strains: MRC5, MRC7, MRC9, MRC1, IMR90 and WI38. Readily detectable amounts of PrPsc were adsorbed to all cell types. The material detected represents the residue remaining attached to the cells after the monolayers have been harvested. Thus the first column in each case is the signal found in two thirds of the cells when the monolayers were harvested after the 96 hours exposure followed by the twenty four hour recovery period. Material may have been removed by the harvesting process. Similarly the second column, labelled pass one, is the signal found in two thirds of the cells at the next sub-culture, when material would have been diluted one to three by the initial sub-culture step. In the absence of any removal or dilution of PrPsc by washing, the signal at the first pass should therefore be thirty three percent of that of the initial signal. In fact it is between 5 % and16 % as shown in Figures 3 and 4. It appears that about 50 to 80 % of the adsorbed material is removed by washing or treatments used in sub-culture on passage, as opposed to dilution by cell growth and splitting. The material thus appears to be fairly firmly attached to the cells. Cells from subsequent passages did not give a detectable signal. The fraction of the PrPsc applied which adhered to the cells was measured, and varied from experiment to experiment for reasons that are not yet understood. The range is illustrated in Figure 5. For the unheated preparation the proportion adsorbed was 1 to 5 %. The heated preparations tended to adsorb more completely, as shown in Figure 5, up to 29 % in one experiment with WI38. There was no convincing difference between cells in the amount adsorbed; any differences might have been masked by the variation between experiments. There is no evidence of infection of the six cell types when following the protocol described, which has been shown to be effective in other systems [Vorberg, Raines and Priola Journal of Biological Chemistry (in press)]. The cells will ultimately be tested for infectivity. The infectivity in the vCJD brain homogenate used to treat the cells is not yet known.

Exposure of Cells to vCJD by Modified Procedures

It has been reported that TSE agents adsorb strongly to metal surfaces, particularly steel. CJD has been transmitted by the use of surgical instruments in brain surgery, and the instruments proved impossible to sterilise [19]. Wires dipped into homogenates and then implanted in the brains of animals have efficiently transmitted infection, and it has been reported that certain cells may be infected by growing them in the presence of a wire mesh dipped in infectious homogenate [8]. This could be an effective method for infection in vitro. Wire mesh was immersed in dilutions of the vCJD brain homogenate and after rinsing applied to the surface of confluent cell sheets of MRC5 or WI38 cells for 72 hours. The cells were then sub-cultured using either trypsin or versene and grown on before the sheets were assayed by CDI for the presence of PrPsc. All samples were negative. It is not known how much material adsorbed to the metal, but there was no evidence of transfer.

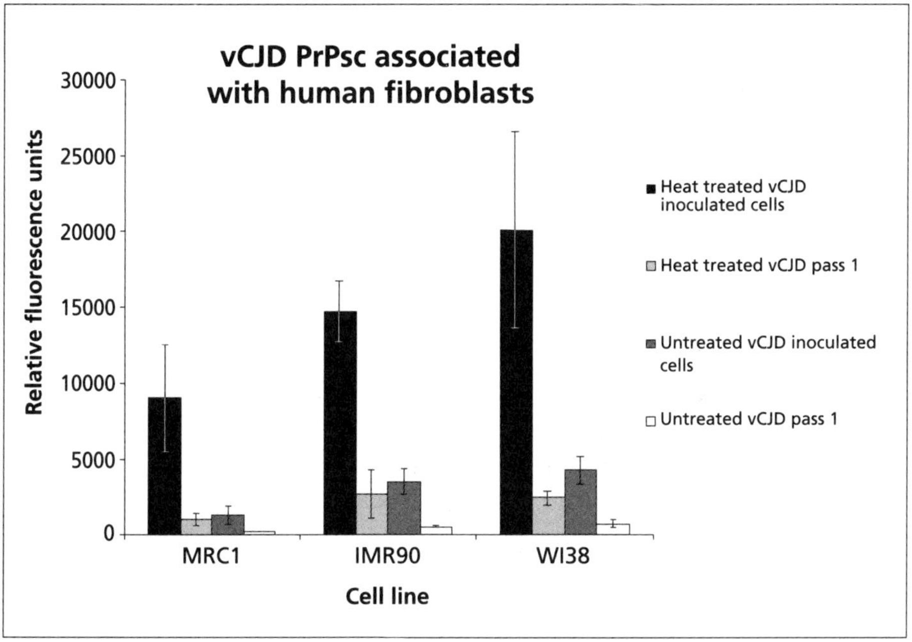

Fig. 3: Binding of PrPsc to cells and loss on passage as measured by CDI. MRC1, IMR90, and WI38 cells and the cells sub-cultured as they became confluent.

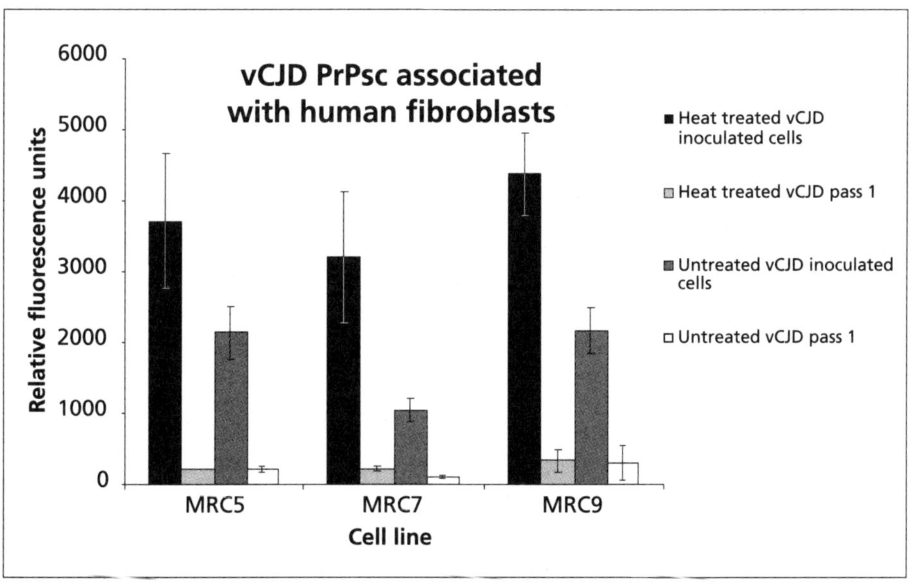

Fig. 4: Binding of PrPsc to cells and loss on passage as measured by CDI. Heated and unheated preparations of vCJD brain homogenate were bound to MRC5, MRC7, and MRC9 cells.

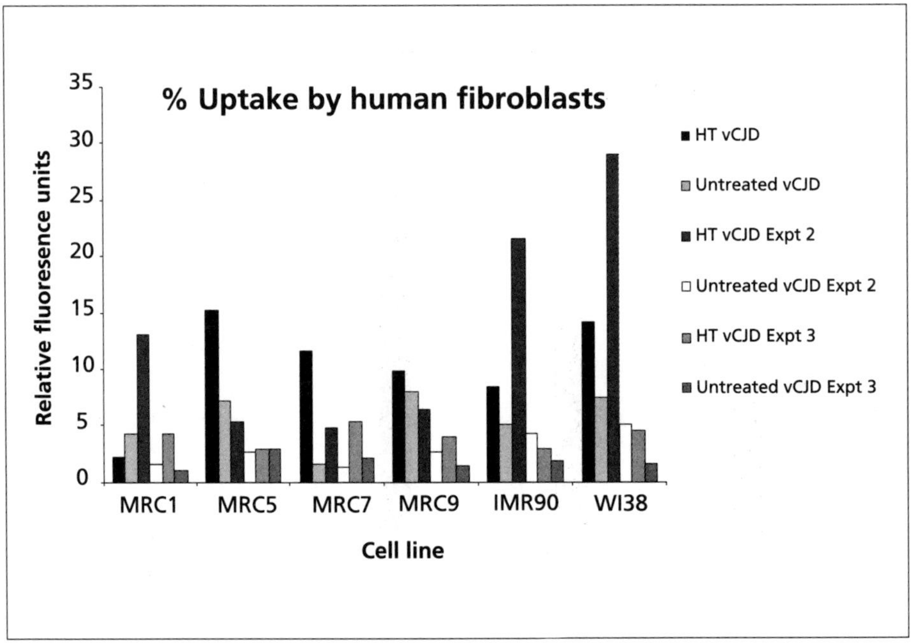

Fig. 5: Amount of input vCJD brain bound to different cell types.

CONCLUSIONS

Reagents have been prepared for the assessment of assays for CJD. The amount of PrPsc present in brain preparations has been assessed by International Collaborative study using in vitro assays, of which the most sensitive CDI has been applied as a detection system for other purposes, such as assessing spleen homogenates for PrPsc. Attempts to infect human diploid cells of various genotypes have been unsuccessful so far, although no positive control infection has yet been attempted. Future work will include assessment of the cells for infectivity and the exploration of other cell types for their infectability with the agent of vCJD.

REFERENCES

1 Anonymous: Tenth Annual report 2001. Creutzfeld-Jakob disease Surveillance in the UK. 2002, www.cjd.ed.ac.uk.

2 Collinge J, Sidle KCL, Meads J, Ironside J, Hill A: Molecular analysis of prion strain variation and the aetiology of 'new variant' CJD. Nature 1996; 389: 685-692.

3 Bruce ME, Will RG, Ironside JW, McConell I, Drummond D, Suttie A, McCandle L, Chree A, Hope J, Birkett C, Cousens S, Frazer H, Bostock CJ: Transmissions to mice indicate that 'new variant' CJD is caused by the BSE agent. Nature 1997; 389: 498-501.

4 Davanipour Z, Alter M, Sobel E, Callahan M: Sheep consumption: a possible source of spongiform encephalopathy in humans. Neuroepidemiology 1979; 4: 240-249.

5 Brown P, Cathala F, Raubertas RF, Gajdusek DC, Castaigne P: The epidemiology of Creutzfeld Jakob disease: conclusion of a 15 year investigation in France and review of the world literature. Neurology 1987; 37: 895-904.

6 Race R, Raines A, Raymond GJ, Caughey B, Chesebro B: Long term subclinical carrier state precedes scrapie replication and adaptation in a resistant species: analogies to bovine spongiform encephalopathy and variant Creutzfeld Jakob disease in humans. J Virol 2001; 75: 10106-10112.

7 Weissmann C, Enari M, Kohn P-C, Rossi D, Flechsig E: Transmission of prions. J Inf Dis 2002; 186 (supp 2): S157-S165.

8 Scott M, Foster D, Mirenda C, Serban D, Coufal F, Walchi M, Torchia M, Groth D, Carlson G, DeArmond SJ, Westaway D, Prusiner SB: Transgenic mice expressing hamster prion protein produce species specific scrapie infectivity and amyloid plaques. Cell 1989; 59: 847-857.

9 Palmer MS, Dryden AJ, Hughes JT, Collinge J: Homozygous prion protein genotype predisposes to sporadic Creutzfeld-Jakob disease. Nature 1991; 352: 340-342.

10 Collinge J, Palmer MS, Dryden AJ: Genetic predisposition to iatrogenic Creutzfeld Jakob disease. Lancet 1991; 337: 1441-1442.

11 Minor PD, Newham J, Jones N, Bergeron C, Gregori L, Ashre D, van Engelenberg F, Stroebel T, Vey M, Barnard G, Head M, the WHO Working Group on International Marterials for the Diagnosis and Study of Transmissible Spongiform Encephalopathies: Standards for the assay of Creutzfeld-Jakob disease specimens. J Gen Virol 2004; 85: 1777-1784.

12 Safar J, Wille H, Itri V, Groth D, Serban H, Torchia M, Cohen FE, Prusiner SB: Eight prion strains have PrP (Sc) molecules with different conformations. Nature Medicine 1998; 8: 1157-1165.

13 Bellon A, Seyfort-Brandt W, Lang H, Baron H, Vey M: Improved conformation dependent immunoassay: suitability for human prion detection with enhanced sensitivity. J Gen Virol 2003; 84: 1921-1925.

14 Bruce ME, McConell I, Will RG, Ironside JW: Detection of variant Creutzfeld Jakob disease infectivity in extraneural tissues. Lancet 2001; 358: 208-209.

15 Jacobs JP, Jones CM, Bailie JP: Characteristics of a human diploid cell designated MRC-5. Nature 1970; 227: 168-170.

16 Hayflick L, Moorhead PS: The serial cultivation of human diploid cell. Res 1961; 25: 585-621.

17 Kovacs GG, Head MW, Hegyi MD, Buhn TJ, Hicker H, Hainfellner JA, McCardle L, Laszlo J, Jarius C, Ironside JW, Budka H: Immuohistochemistry for the prion protein: comparison of different monoclonal antibodies in human prion disease subtypes. Brain Pathology 2002; 12:1-11.

18 Bueler H, Aguzzi A, Sailer A, Greiner RA, Autenrie P, Aguet M, Weissman C: Mice devoid of PrP are resistant to scrapie. Cell 1993; 73: 1339-1347.

19 Will RG, Matthews WB: Evidence for case to case transmission of Creutzfeld-Jakob disease. J Neurol Neurosurg Psychiatry 1982; 45: 235-238.

Dr. P. Minor, NIBSC, Blanche Lane, South Mimms, Potters Bar, Hertfordshire, EN6 3QG, U.K.
E-mail: pminor@nibsc.ac.uk

Petricciani J, Sheets R (eds): Vaccine Cell Substrates 2004.
Dev Biol (Basel). Basel, Karger, 2006, vol 123, pp 335-345.

TSE Risk Assessment for Starting Materials Used During, or in the Manufacture of Vaccines for Human Use

A consultant's view of the commercial approach

R. Bradley

Private BSE Consultant, Guildford, U.K.

Key words: Risk assessment, vaccines, starting materials, bovine spongiform encephalopathy (BSE), transmissible spongiform encephalopathy (TSE)

Abstract: Vaccination is an essential tool to protect humans from many infectious diseases. Materials derived from bovine tissues are used in, or during the manufacture of, most vaccines. Following the occurrence of BSE in cattle in three continents since 1986, the need to ensure the TSE safety of vaccines is vital. Various actions have been taken to ensure safety targets are reached. This paper shows that selecting starting tissues used in vaccine manufacture on the basis of their inherent TSE risk is of prime importance and is more secure than using geography as the primary criterion of a safe origin. Other approaches can also contribute to safety assurance. These include avoiding, where possible, tissues from animals, risk tissues and tissues of unknown or uncertain provenance. It is essential to ensure tissues are collected and processed appropriately and detailed records are kept. The risk assessment approach, based on the main safety factors of Source, Process and Use, coupled with these additional features, combine to establish that the TSE safety of vaccines is ensured, despite the challenge resulting from the occurrence of BSE.

INTRODUCTION

Vaccination is the most effective way of protecting humans and animals from many infectious diseases. A high proportion of the world population of animals and humans is protected from these diseases by single dose or multiple dose vaccination. The safety of vaccines used is therefore paramount and this (among other parameters) is regulated by national and international regulatory agencies. The safety of vaccines in regard to conventional bacteria, viruses, fungi and the like is relatively easy to achieve as most of these agents are removable, or susceptible to inactivation or destruction by conventional physical and chemical processes. With regard to conventional bacteria, viruses, fungi and the like, the means by which the safety and quality of vaccines is ensured, is well established.

Materials of bovine origin are used in, or during the manufacture of, a high proportion of vaccines. Until the 1980s there were few bovine materials used for these purposes that posed infection risks that could not be effectively controlled. However, in 1986 a new disease of cattle, bovine spongiform encephalopathy (BSE), was identified in dairy cattle in Great Britain and subsequently elsewhere in Europe, the Middle East, Asia and North America. BSE is a transmissible spongiform encephalopathy (TSE) akin to scrapie of sheep and goats and Creutzfeldt-Jakob disease of man. In 1986 it was not known if the BSE agent posed a risk to species other than cattle, though it was considered a possibility. A number of measures were taken by the U.K. authorities to eliminate BSE from the cattle population and to protect public health. This was followed by the introduction of similar measures by the European Commission (EC) and the Swiss government. These included actions to protect vaccines from bovine starting materials that might carry BSE infectivity.

The main measures were assorted feed bans to protect animal health, identification and complete destruction of clinically BSE-suspect animals and removal and complete destruction of bovine (and later some small ruminant) tissues that are considered a real or potential risk of carrying the BSE agent – the so-called specified risk materials (SRM). Identification of BSE-infected animals was improved from 1999 onwards by the introduction of active surveillance using newly developed 'Rapid' tests on brain material. These enabled detection of the misfolded form of protease-resistant protein (PrP-res), a molecular marker of BSE-infectivity. Rapid tests are regarded as effective only towards the end of the incubation period (about three to six months before clinical onset) and into the clinical phase of disease so it is not possible to detect infected cattle during the bulk of the incubation period which is on average around five years and even then it is only possible post mortem.

In the context of this presentation the hazard is the BSE agent. In regard to the safety of biological products, including vaccines, methods of assessing the risk from this hazard in bovine starting materials have fallen into two broad classes: assessment of the TSE risk in each source/starting material (adopted by the author [1]) or basing the assessment primarily upon the BSE risk in the country of origin of the cattle providing the source/starting material – a geographical BSE risk (GBR) assessment.

This paper seeks to demonstrate and confirm that the TSE safety of vaccines prepared for human use is most securely determined by a broad process of TSE risk assessment of the animal materials used in the course of manufacture, rather than relying upon their geographical origin only.

Factors Determining the TSE Risk

The factors determining risk can be considered under three headings: **Source, Process** and **Use.**

In regard to **Source,** the main points to consider are the species and tissues of origin. The geographical origin of the source animal should also be considered but not as the primary determinant of risk.

In regard to **Process,** this is important for collection procedures, to eliminate or reduce risks of contamination of source materials with tissues of higher risk. It is also important to consider the processing parameters of derived materials such as gelatin and tallow derivatives.

In regard to **Use,** the regime and route of administration of vaccines is fixed; there is little or no flexibility and this factor will not be discussed further.

Of all the factors source has the highest influence and use the least. Within the source factor the species and tissue of origin have the highest influence and geographical origin the lowest.

The Basis of the Risk Analysis

First it is necessary to distinguish the difference between a **starting material** (any tissue or substance derived whole or in part from an animal (including man), whether processed or not, and used during vaccine manufacture e.g. amino acids) and a **source material** i.e. tissues taken directly from a live animal e.g. blood or milk.

The procedure is to identify the original animal source of the material, determine the TSE risk in the **source material** and assess the impact of any processing.

Only when a TSE risk assessment for all starting materials is complete can an effective risk management strategy be developed.

All ingredients of starting materials must be traced back to the species and tissues of origin before a TSE risk assessment can be initiated. This is relatively easy for blood, milk and their products but more difficult for gelatin (multiple animal sources) and tallow derivatives (multiple species, animal and tissue sources).

Regulatory Considerations

In the European Union (EU), the Committee for Human Medicinal Products (CHMP), previously the CPMP (Committee for Proprietary Medicinal Products) has, for more than a decade, maintained a "Note for Guidance" on minimising the risk of transmitting animal spongiform encephalopathies via human and veterinary medicinal products that has been periodically updated in the light of developing knowledge [2]. The Note for Guidance is published by the European Commission and has legal status within the EU. It is also widely used in countries outside the EU. The latest revision is effective from 1 July 2004. This classifies tissues into one of three categories (A, B and C) based on Guidelines issued by the World Health Organisation (WHO) in February 2003. The classification is based in turn upon bioassay and/or presence of PrP-res in natural TSE of animals and man and experimental BSE in cattle, as follows:

Category A – High infectivity tissues

High infectivity tissues e.g. central nervous system (CNS) tissues and tissues anatomically associated with CNS e.g. brain, spinal cord and associated ganglia and eye.

Category B – Lower infectivity tissues

Lower infectivity tissues are peripheral tissues that are PrP-res or bioassay-positive in at least one natural form of TSE or experimental BSE e.g. lymphoreticular tissues, nerve, blood.

Category C – Tissues with no detectable infectivity

Bio-assayed tissues with no detectable infectivity and/or are PrP-res negative e.g. milk, skeletal muscle.

Source

Table 1 lists some starting materials that may be used during commercial vaccine manufacture, the species and tissue source and the WHO/CHMP Category to which the source tissue belongs.

Table 1: A selection of starting materials used for vaccine manufacture, the species and tissues from which derived, and the WHO/CHMP category ascribed.

Starting material	**Species/Tissue source**	**WHO/CHMP Risk category**
Beef heart	Bovine heart muscle	C
Bovine meat extract	Bovine skeletal muscle	C
Haemoglobin	Bovine blood	B*
Haematin	Bovine blood	B*
Donor calf serum	Bovine blood from live cattle	B*
Foetal calf serum	(Bovine fetal blood)	B/C ?
Sheep blood	Ovine blood from live sheep	B
Skimmed milk	Bovine milk from live cattle	C
Casamino acids, casein, casein hydrolysate, casein peptone, galactose, hycase, lactalbumin hydrolysate, lactose+	Bovine milk from live cattle	C

* **Bovine** blood and/or components have been bio-assayed with negative result (see Table 2).

? Blood is classified as category B, based on transfusion studies in sheep. Foetal bovine blood has not been classified separately. The nearest equivalent is 'Cord blood' which is Category C.

\+ This table does not consider risks from collection and processing procedures that should be considered under **Process**.

Bovine blood

No detectable infectivity has been found in bovine blood, its components and source tissues, nor in foetal calf blood derived from naturally, clinically-affected cattle with confirmed BSE. Neither has detectable infectivity been found in buffy coat, spleen or lymph nodes or in bone marrow of cattle experimentally challenged orally with BSE during incubation (Table 2).

Cell lines, banks and seeds

In regard to cell lines, banks and seeds, none contains bovine or ovine material for which TSE infectivity has been demonstrated.

Table 2: Results of bioassay in mice and cattle of bovine blood and its components and related tissues derived from natural or experimental cases of BSE.

Tissue	**Natural BSE**		**Experimental BSE**	
	Tested in cattle	**Tested in mice**	**Tested in cattle**	**Tested in mice**
Blood clot		NDI		
Serum		NDI		
Buffy coat		NDI	NDI*	NDI
Foetal calf blood		NDI		
Spleen	NDI	NDI		NDI
Lymph nodes	NDI	NDI		NDI
Bone marrow		NDI	NDI+	NDI++

NDI: No detectable infectivity.

* From 32 months incubating donor cattle, seven and a half years after challenge of recipient cattle. At earlier stages of incubation buffy coat still shows NDI > five years after challenge. Experiments still in progress.

\+ In progress.

++ NDI during incubation.

"Tested in cattle" data provided courtesy of GAH Wells and SAC Hawkins by personal communication.

In particular, cell lines are not derived from neural or lymphatic tissues or cells. No cell line used has been shown to replicate any naturally occurring TSE agent. However, in view of the recent publication by Vorberg et al [3], and the presentation by Dr. Priola showing the susceptibility of common fibroblast cell lines to murine scrapie strains, there may be a need to conduct further studies. However, this is likely to be most effective as an academic, rather than a commercial exercise, since both cells and agent strains can change over time. A multi-factorial risk assessment based largely on the TSE safety of animal source materials is a sound way to proceed. It is important to note that epidemiological study of humans and animals vaccinated using commercially prepared vaccines shows no evidence for a vaccination-associated increase in TSE incidence.

Process

The TSE risk assessment of **source materials** used to make starting materials requires knowledge of:

(i) the specification of the source animal and herd health status;

(ii) if source materials are collected post mortem, the method of stunning and appropriate veterinary inspections;

(iii) the method of tissue collection;

(iv) the processing details and any TSE infectivity clearance factors;

(v) any dilution or concentration factors.

In regard to the vaccine manufacturing process the following are required:

(i) the processing details and any TSE infectivity clearance factors;

(ii) any dilution or concentration factors.

Cross-contamination

It is important to recognize that, in regard to tissue collection procedures from cattle slaughtered and passed fit for human consumption, brain penetrative stunning and pithing create brain emboli and dissemination [4-6]. Even conventional captive-bolt stunning without pithing or use of injected compressed air may induce widespread and significant dissemination of brain material in beef carcases [7].

It should be noted that no research has been reported on the effect of non-penetrative stunning on the dissemination of brain tissue. In the EU and in Switzerland all brains from healthy slaughter cattle over 30 months old (and in some countries over 24 months old) are "Rapid" tested for PrP-res by EC-approved tests conducted in approved laboratories. If any test is positive all such carcases and other parts are completely destroyed as well as the same materials from the one before and the two after on the slaughter line.

TSE infectivity clearance factors and dilution are important TSE safety factors. Large production volumes, numerous cell passages, and extensive processing of vaccines result in large dilution. However, dilution should not be relied upon as the sole safety criterion. TSE agents are "sticky" and therefore in a worst case scenario, calculations should assume that they will remain associated with the active principle unless there is positive evidence to the contrary. This is another reason why animal source materials must be carefully selected, collected and processed to ensure they are safe in regard to TSE risks.

To facilitate risk reduction during processing it is necessary to avoid adulteration and cross-contamination, to develop quality assurance procedures, to document the materials used and to detail the processing methods. This applies during collection and processing of source materials, the manufacture of starting materials and during vaccine production.

Other Ways to Reduce Risk

Where possible avoid the use of tissues:

(i) from animals;

(ii) from ruminant animals and other species susceptible to TSE;

(iii) of unknown or uncertain provenance.

Always avoid the use of tissues known to be a risk, tissues inappropriately collected or ineffectively processed.

Future Developments

Continue the global programme away from materials of animal origin whenever possible. Regularly review recommendations in the light of new knowledge and ensure that the precautions taken are commensurate with the risk. Continue risk/benefit analysis because a zero risk is not possible to prove.

Changes over Time

It is important to recognize that the real risk in source materials has not changed over time but, as a result of new knowledge, there have been changes in the interpretation of earlier information with resulting changes in guidance and legislation. None of these changes has in practice required the risk management to be changed. The outcome might have been different if geography had been used as the main criterion for safe sourcing.

Conclusions on Species/Tissue Risk Assessment

The assessment of TSE risk in the starting materials of ruminant origin used for the manufacture of vaccines for human use has revealed no evidence for a degree of risk that is higher than negligible.

Using Geography as the Main Determinator of TSE Risk

Occurrence of BSE

Only three countries remain that have reported BSE in imported animals only. These are the Falkland Islands, the Sultanate of Oman and the U.S.A.

In regard to BSE in native-born animals these have now been reported from Europe, the Middle East, Asia and North America. Europe has had by far the greatest number of cases that developed between 1986 and 2002 (Fig. 1).

A statement

In June 2001 the Office International des Epizooties (OIE) hosted a joint technical consultation with the Food and Agriculture Organisation (FAO) and the WHO in which it was stated that "Materials potentially infected with BSE have been distributed throughout the world through trade in cattle and certain cattle products and by-products. These products include rendered animal proteins and compound feed containing meat-and-bone-meal (MBM)".

Origin of BSE

BSE can arise from exogenous or endogenous sources. Exogenous sources so far identified are the importation of infected cattle or import of infected MBM or feed containing infected MBM. Endogenous sources could result from genesis of TSE in cattle from any species and recycling of the cattle infection via MBM in feed.

Introduction of BSE into other countries

Following on from the statement above, it is clear that the precise geographical destination of cattle and mammalian MBM exported from countries with BSE, or

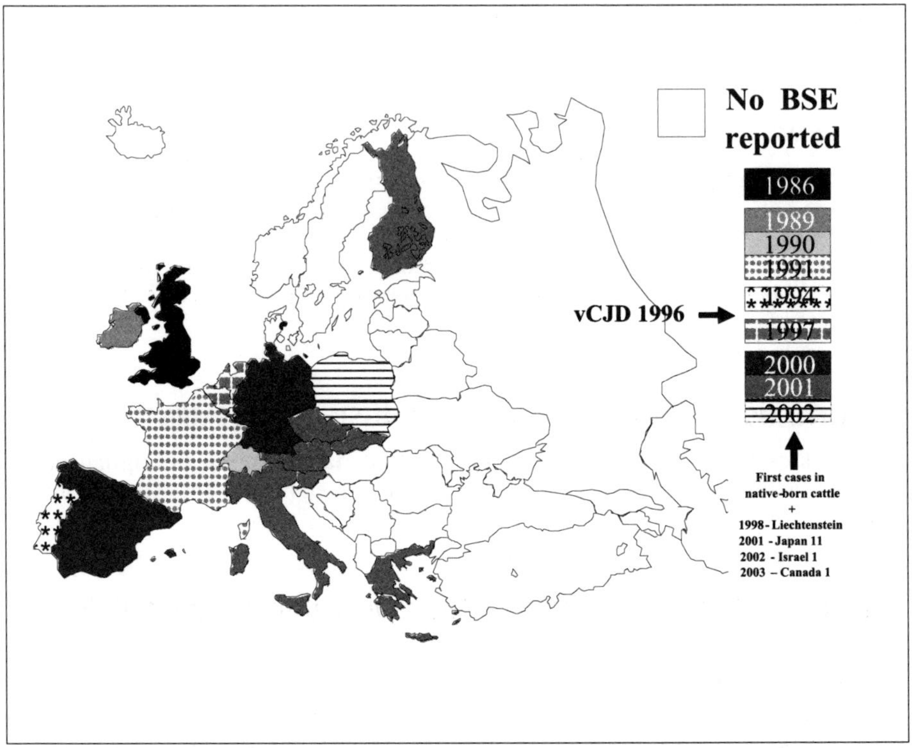

Fig. 1 : Distribution of reported BSE in native-born cattle in Europe by year of first report.

from countries in which cattle are incubating the disease, is uncertain. Thus the analysis of the risk of the TSE infectivity by type of tissue is of fundamental importance.

From 2000 onwards 15 countries have reported BSE in native-born cattle for the first time:

2000 – Denmark, Germany and Spain.

2001 – Austria, Czech Republic, Finland, Italy, Greece, Japan, Slovakia, Slovenia and Japan.

2002 – Israel and Poland.

2003 – Canada.

2004 Onwards?

The identification of BSE in the above countries increases the risk that BSE may have been imported into even more countries in the period before (and after) the disease was discovered if live cattle incubating BSE, infected MBM, feed containing infected MBM or infected cattle products like SRM were exported other than for complete destruction. Geographical BSE risk (GBR) assessments for countries in the EU and countries wishing to trade with the EU had been completed by the Scientific Steering Committee of the EC for most of these countries and indicated that there was indeed a TSE risk, even though cases had not at the time been reported. Although the GBR assessments have not changed, the actual risk is now a reality.

BSE knows no bounds and has affected new regions, continents and countries since 2001. The number of cases so far reported in these regions has been small: 2001 – Asia (Japan 11 cases); 2002 – Middle East (Israel, 1 case); 2003 – North America (Canada 1 case and one case exported from Canada and reported in the U.S.A.). By contrast, to date there have been 183,496 cases of confirmed BSE in the U.K. and 4,278 confirmed cases in the rest of Europe.

Geographical risks

Geographical risks should not be the primary or only determinant of BSE risk because:

(i) no country has a zero risk;

(ii) the distribution of BSE in the world is dynamic and currently uncertain due to the inconsistency of worldwide surveillance (especially targeted active surveillance);

(iii) there is incomplete agreement between different agencies on the countries at risk and not at risk;

(iv) when BSE is reported in a native-born animal, the exposure, on average, would have been five years earlier and exposed feed and birth cohorts may well already have been recycled through MBM;

(v) a low clinical case rate is not necessarily consistent with a low infection rate and is unknown in the absence of active surveillance.

Contrasting the assessment of TSE risks in ruminant tissues with GBR assessment as a primary criterion for assessing the TSE risks in starting materials indicates that:

(i) the former are constant but our knowledge of them changes, usually slowly, and the changes refine and reinforce previous knowledge whereas,

(ii) the latter are dynamic, changes occur rapidly ("overnight") and therefore may demand a rapid response if effective risk management procedures are not already in place.

Consequences of using geography as the primary criterion

If geography has been used as a primary criterion for the assessment of TSE risk in source/starting materials, it does not mean that vaccines prepared before a time when BSE has been discovered in a source country, have a TSE risk. However, risk reassessment is advisable for all source/starting materials that are not in WHO/CHMP Category C and targeted active surveillance for BSE should be undertaken to establish the prevalence of infection in the native national cattle herd and thus determine the extent of any risk there may be.

Conclusions

Vaccination is the most effective way of protecting humans and animals from many infectious diseases. In this regard the safety of starting materials is paramount. Geography should not be the primary factor in deciding the BSE risk or vaccine safety. It is more important to determine the generic BSE risk in source/starting materials and to ensure a risk analysis for all starting materials is completed and regularly reviewed.

Finally

It is very important that international agreement should be achieved on all aspects of TSE risk for vaccines. Wherever possible eliminate animal/ruminant materials in vaccines and in the meantime further develop the global approach to safety and continue risk/benefit analysis since a zero risk cannot be proved. Public confidence in the TSE safety of vaccines is extremely important and must be continuously maintained where appropriate by sound scientifically-based risk communication.

The Message

Selecting starting materials on the basis of the inherent TSE risk in the tissues of origin is an essential primary criterion.

The geographical origin of the host animal may be of value as a secondary criterion for the TSE safety of starting materials.

All countries used as a source of ruminant materials for vaccines with a GBR > 1 (highly unlikely) must conduct active surveillance.

Acknowledgements

The author thanks the organisers for their invitation to participate in the meeting, GlaxoSmithKline Biologicals for their support, Mr. GAH Wells and Mr. SAC Hawkins for information on transmission study progress, DEFRA, OIE and the EC for access to data and Mr. Stuart Woods for helpful collaboration over the years.

References

1 Bradley R: A brief overview of bovine spongiform encephalopathy and related diseases including a TSE risk analysis of bovine starting materials used during the manufacture of vaccines for use in humans. Przegl Epidemiol 2001; 55: 387-405.

2 CPMP: Note for guidance on minimizing the risk of transmitting animal spongiform encephalopathy agents via human and veterinary medicinal products (EMEA/410/01 Rev2 – October 2003) adopted by the Committee for Proprietary Medicinal Products (CPMP) and by the Committee for Veterinary Medicinal Products (CVMP). (2004/C 24/03). Off J EU C24/6-19, 28 January 2004.

3 Vorberg I, Raines A, Story B, Priola SA: Susceptibility of common fibroblast cell lines to transmissible spongiform encephalopathy agents. Journal of Infectious Diseases 2004; 189: 431-439.

4 Garland T, Bauer N, Bailey M: Brain emboli in the lungs of cattle after stunning. Lancet 1996; 348: 610.

5 Anil MH, Love S, Williams S, Shand A, McKinstry JL, Helps A, Waterman-Pearson A, Seghatchian J, Harbour DA: Potential contamination of beef carcases with brain tissue at slaughter. Vet Rec 1999; 145: 460-462.

6 Ramantis SB: Cattle slaughtering and BSE risks Part 1. Potential dissemination of CNS tissue during slaughtering. Vet Bull 2004; 74: 1N-13N.

7 Prendergast DM, Sheridan JJ, Daly DJ, McDowell DA, Blair IS: The use of a marked strain of *Pseudomonas fluorescens* to model the spread of brain tissue to the musculature of cattle after shooting with a captive bolt gun. J Appl Microbiol 2004; 96: 437-446.

Dr. R. Bradley, Private BSE Consultant, 41 Marlyns Dr, Burpham, Guildford, GU4 7 LT U.K.
E-mail: raybradley@btinternet.com

Petricciani J, Sheets R (eds): Vaccine Cell Substrates 2004.
Dev Biol (Basel). Basel, Karger, 2006, vol 123, p 347.

Bovine Spongiform Encephalopathy (BSE) in the United States: an Overview and Critique of the Response with Suggestions for Additional Actions

L.A. Detwiler

Red Bank, NJ, USA

Abstract: On May 20, 2003, the Canadian government announced the detection of BSE in a native-born cow. With this diagnosis it was clear that the BSE agent had been introduced into the North American cattle production system. In December 2003, the BSE case detected in a cow living in Washington State reaffirmed the risk in the United States. The U.S. Department of Agriculture (USDA) responded with an attempt to trace the cow and other cattle at risk. USDA also put in place a number of emergency measures to protect the food supply. Some of these will indirectly protect pharmaceuticals and biologicals. The Food and Drug Administration (FDA) announced plans for certain rule changes, but has yet to publish any specifics.
This presentation will summarize and evaluate the actions taken. It will also provide epidemiological and scientific justifications for future actions.

Dr. L. Detwiler, 225 Highway 35, Red Bank, NJ 07701, USA.
E-mail: lavet22@aol.com

Petricciani J, Sheets R (eds): Vaccine Cell Substrates 2004.
Dev Biol (Basel). Basel, Karger, 2006, vol 123, pp 349-354.

PANEL VI DISCUSSION

Bovine Spongiform Encephalopathy

Panel Members: *Anthony Lubiniecki (Panel Chair), Bob Rohwer, Sue Priola, Alex Raeber, Phil Minor, Ray Bradley, Linda Detwiler, David Asher, Marie-Sol Ruiz, Philippe Monteyne*

Panel Questions and Points for Discussion

Question 1:
Can the BSE agent propagate in cell culture?

Question 2:
Can it be passaged into vaccine products from vaccine cell substrates?

Question 3:
What assay methods are available or in development and are adequate (sensitive and validated) for screening cell substrates and raw materials used in vaccine production?

Question 4:
When there is uncertainty in raw material traceability for a seed material or cell substrate's passage history, and a risk assessment is performed, what degree of residual risk is acceptable?

Dr. Stokes: Regarding Dr. Minor's work on MRC5 cells and their exposure to CJD, do you think it is necessary to have a look at the susceptibility to BSE as well?

Dr. Minor: Yes, the reason for starting off with the CJD was twofold. Firstly, we had the samples available so that was easy. Secondly, human-to-human transmission might be easy. But we certainly wanted to do BSE because it would be a specific risk for the industry, and this is in the pipeline.

Dr. Lubiniecki: Can the BSE agent propagate in cell culture?

Dr. Minor: I think the work that Sue presented seemed fairly reasonable to me and suggests that BSE could propagate. I think there is no evidence for it at the moment, but if you find the right cell culture, I would be surprised if you could not do it.

Dr. Lubiniecki: Can BSE be passaged into vaccine products from vaccine cell substrates?

Dr. Asher: If it were there in the first place for a minimally processed vaccine, one would fear that it could in fact. Fortunately, there has never been a human case of BSE attributed to exposure to a vaccine. But it must not be forgotten that the first observation that attracted attention of the peculiar resistance of the agent to chemical treatment was, in fact, a veterinary vaccine that had been accidentally contaminated with the scrapie agent.

It was a tissue-based vaccine and comparable vaccines are no longer produced for human use, at least in developed countries. But it demonstrates a kind of worst-case scenario, e.g. what could happen if a TSE agent did get into a raw material.

Dr. Rohwer: I think the other point we should remember is that one of the abiding mysteries in variant CJD is why it seems to attack young people instead of older people. We have no definite answer for this but, among the speculations, is that young people are being exposed to something that older people are not and that is why we are seeing it in younger people.

Two of the things that are suspect, in my mind, are milk and vaccines. There have been risk assessments to look at this and they have generally been dismissive of these ideas but we are still left with the phenomenon without an explanation. I think we should bear some residual discomfort with that.

Dr. Minor: I think there are many mysteries surrounding where vCJD comes from, but, certainly, about six years ago, we did an analysis about the year of birth and the coincidence with BSE, and so on. If you assume that there was only one major peak of BSE in 1989 and 1992, or thereabouts, the distribution of the birthdates of the children at the time they would have been vaccinated really did not fit with when the BSE would have been in the vaccines.

I think, from that point of view, it did not add up. That does not necessarily mean that it is wrong. I think that more or less convinced me, at least for those children, that vaccines were not the issue.

With respect to this particular question, it seems to me that if your cell substrates can be infected and your cell substrates are infected, then I do not see why it should be passaged into vaccine products. But I think those are two very big "ifs" because the question of whether they can be infected is something that is not definitively proved. I think the results to date are reasonably encouraging, if not exactly definitive.

The issue of whether they could be infected is that most of these things would most probably be infected by some gap in the serum history. I am not sure that serum is a very effective way of infecting anything. So I think there are at least a couple of big "ifs" about whether they could be infected at all. But if cell substrates were infected, then I suspect the answer to that was yes. But I think they are not.

Dr. Asher: If there is a significant milk risk, you would think that the epidemic curve of BSE in cattle in the U.K. would not be dropping off as promptly as it does considering that there is no cow-to-cow species barrier and one assumes that the consumption of milk, including colostrum by calves, must be substantial.

Dr. Rohwer: Well, the transmission rate could be highly inefficient. But I think the significant factor with milk is that you could have a situation where you had very inefficient transmission but you have very large exposure over a large population and you have a small number of people affected by that exposure.

It is just something that I do not feel we should dismiss entirely until we have some alternative explanation that actually gives us some insight as to why young people are affected preferentially in the case of this disease.

Dr. Bradley: I would like to go back to one thing that David Asher said about the iatrogenic transmissions to sheep with vaccines. Neither of these transmissions which occurred were commercial vaccines. They were done in local laboratories and they committed the cardinal sin of using tissues to prepare the vaccine which were known to replicate the agent; namely brain and spleen. I do not think those things apply in the vaccines we are speaking of, or through commercial companies.

Regarding milk, although the numbers are not high, BSE in the U.K. and in Europe is a dairy-herd disease so the cattle that have all come down have rarely consumed milk from their dams. The whole point of keeping the cows is to produce milk for human consumption. It is the beef herd that would receive milk from the dam directly, often for several weeks or months, as you might perhaps find in the U.S. in pure beef herds. John Wallsmith did an analysis of this and he found zero cases in the offspring of beef cattle with BSE.

I have to say that the numbers are relatively small. We are talking about a few hundred because the incident was very low and, to be able to follow beef herds for the minimum period of five years to see whether BSE occurred was difficult.

I accept that one has to be cautious and we should not close the door on this. As we understand it at the moment, the risk from milk would be extraordinarily low but, as always, perhaps never zero.

Dr. Monteyne: A further comment on the initial question about passage into vaccine products, the way we address that within GlaxoSmithKline is to address the various "ifs" that Dr. Phil Minor was mentioning. Indeed, we translate these "ifs" into risk factors.

We run risk assessments for each of our products, and we have discussed that in detail with all of the regulatory agencies. We consider the source, the process and then the use of each product. For each of these categories of factors, we try to translate into numbers the risk that is actually attributable to each factor, taking into taking into account the worst-case scenario.

For the human vaccines, there is no epidemiological evidence of transmission of vaccine-variant CJD.

One additional point is that, in the past, we also produced veterinary vaccines including bovine vaccines that contained a significant amount of U.K. serum per dose. There was no epidemiological evidence at all of any transmission of BSE.

Dr. Detwiler: When you say no evidence at all, were the vaccines distributed around the world or primarily in Europe?

Dr. Monteyne: As far as I know, it was primarily in Europe.

Dr. Detwiler: If it is a very, very low rate of infectivity in the serum, how would you know that there was no evidence at all, given the continual cases and the unexplained cases born after these intense feed bans, that there was not a small leakage in vaccine?

Dr. Bradley: You cannot absolutely eliminate this. It is just not possible. We have to live with that residual risk. But it is going to be very small, and I should say, actually, undetectable. But it may be there, nevertheless. It is one of the hazards of life.

Dr. Lubiniecki: What assay methods are available or under development that might be adequate, meaning adequately sensitive and validated, for screening cell substrates and raw materials used in vaccine production? In other words, are such methods available for this purpose?

Dr. Raeber: All the methods that are now used have been validated on brain tissue, and we all know that the highest concentration of PrPsc is in the tissue.

There is currently an E.U. evaluation going on in sheep on lymph node and spleen. But, so far, there is clearly no validated method available. There are a number of methods in development. So far, the question is, are they sensitive enough to pick up one infectious unit? To my knowledge, a bioassay is the gold standard for infectivity and to show the transmissible nature of prion diseases.

We all know that it is probably not feasible to carry out a bio-assay. However, a method such as the one presented by Charles Wessman may be practical. But those cells are still not susceptible to any BSE infectivity. They are only adaptable for mouse infectivity.

Dr. Minor: I am sure there is no suitable method for raw materials at the moment, but I do not see why you could not do something about working cell banks. Could you not inoculate those cells into animals and wait to see if they acquire disease in a couple of years?

Dr. Rohwer: I think we have been overfocused on screening and diagnostics as a solution to the TSE risk-management problems. There are other technical ways to go about this. One that we have been working on in our laboratory in collaboration with PRDT, a joint venture of the Red Cross and Promedics Laboratory, is to develop specific absorbance that will remove these very low levels of infectivity that may be associated with raw materials, especially blood and serum.

Diagnosis is so problematical because a lot may consist of a half ton of serum with only a few infectious doses, but yet possibly enough to propagate an infection in a cell line. This may be much more successfully dealt with by just applying methods to make sure you remove that infectivity before it ever sees the cell line.

In the best of all possible worlds, you would have all these tools deployed at the same time so that you have multiple ways of orthogonally minimizing risk. But we should not overlook these other possibilities.

Dr. Lubiniecki : I would agree completely. It is not possible to test quality in any product and this particular issue is certainly no exception to that. Many of the cell lines that we are thinking about for future products were derived in the 1950s, 1960s and 1970s.

These cells originated and were derived at a time when no-one really knew that BSE existed. So it is impossible to do anything about the historical potential exposure that the cell line may have experienced before we knew anything about BSE.

Did the U.S. Department of Agriculture, which regulates the importation of animal materials from outside the U.S., have controls in place over materials imported to the U.S., such as animal serum ?

Specifically, many of these cell lines were derived in academic institutions which typically do not keep track of their purchase orders so it is not possible to know where they acquired their serum.

But, on the chance that they bought it from a reputable vendor, would there have been controls in place on the importation of foetal bovine serum from other countries in the 1950s, 1960s and 1970s, to your knowledge ?

Dr. Detwiler : Going back to the 1950s and 1960s, I do not know.

Dr. Asher : The FDA has adopted a public policy on master seeds and cell banks produced in the past. In general, we have expressed no concern about materials derived before 1980 because of the probable onset of the epidemic in the U.K. after that point. One could say that it might have gone back some years before 1980. But it certainly did not go back as far as the 1950s or 1960s.

Also, since 2000, our policy has been that there will not be concern about master seeds and we will not request rederivation of master cell banks because the risk of rederivation would clearly outweigh the remote theoretical risk associated with the BSE agent exposure.

Dr. Lubiniecki : With that as background, the question is : when there is uncertainty in raw material traceability for seed material or substrate passage history, and a risk assessment is performed, what degree of residual risk is acceptable ?

Dr. Asher : Even if a master seed or cell bank were derived after 1980, the benefits of some of these master seeds and cell banks would be sufficient so that we would not require rederivation from the beginning.

Dr. Ruiz : Around 2000, after the note for guidance became a directive, we conducted a review of vaccine products in the E.U., and we had the same risk assessment as Dr. Asher mentioned.

Dr. Egan: When we dealt with these issues, we did ask for the working cell banks to be rederived and the working seed banks to be rederived, again using serum of known source, free of risk for BSE.

We considered the limitations of the experiments that have been done and the limits of detection for infectivity that could possibly have been in the foetal-bovine serum. For example, we considered that the BSE agent might have been transmitted to at most 10 percent of the donor bovine foetuses. We could estimate what the worst possible risk might be, and even then, those risks were extraordinarily small.

We were dealing with very, very small theoretical risks. So that leads us to the issue of what degree of uncertainty is acceptable versus the enormous benefits of the vaccines. Rederiving those master cell banks and master seeds would be tantamount to deriving new vaccines.

Participant: Could you please clarify whether you mean licensed products or investigational products?

Participant: At that time we were dealing with all the licensed products that were already on the market with proven efficacy.

Dr. Lubiniecki: Would you feel similarly about investigational products?

Dr. Egan: Investigational products are a little more difficult and we dealt with them separately on a case-by-basis, in part depending on what it was possible to do, where they were in the developmental process, etc.

Dr. Lubiniecki: From what you said, I would understand that it is not automatically unacceptable to use cell banks that might carry a very small theoretical risk of BSE.

Dr. Egan: Yes, but there is a wide spectrum of situations that go from evaluating a licence application to doing preclinical testing in a new development programme. We would probably look at those situations differently.

List of Participants

ABDY Mark, Food and Drug Administration/CBER, 1401 Rockville Pike, HFM-475, Rockville, MD 20852, USA.

ACKLAND Jim, Biologics Consulting Group, 1840 Gateway Drive, #200, San Mateo, CA 94404, USA.

ADAMS Amey, Center for Veterinary Medicine, Food and Drug Administration, 7500 Standish Place, Rockville, MD 20855, USA.

ANDREWS Charla, Vaccine Research Center/NIAID/NIH, 40 Convent Drive, Building 40, Room 5508, Bethesda, MD 20892, USA.

ARNOLD Debra, CancerVax Corporation, 4503 Glencoe Avenue, Marina del Rey, CA 90292, USA.

ASHER David M., Veterinary Medical Officer, Food and Drug Administration/CBER, 1401 Rockville Pike, HFM-313, Rockville, MD 20852, USA.

BANSAL Geetha, NIAID/National Institute of Health, 6700-B Rockledge Drive, Bethesda, MD 20892, USA.

BARA-ESTRADA Maria, Health Canada, BREC, Health Canada, Tunney's Pasture, Locator #0603C, Ottawa, CA K1A 0L2, Canada.

BAYLOR Norman, Office of Vaccines Research and Review, 1401 Rockville Pike, HFM-405 Room 355N, Rockville, MD 20852, USA.

BERRY J., Medimmune Vaccines, 3055 Patrick Henry Drive, Santa Clara, CA 95054, USA.

BOUQUET Jean-François, Cell Culture and Virology, Aventis Pasteur, 1514 Avenue Marcel Mérieux, T1, FR-69280 Marcy l'Etoile, France.

BRADLEY R., 41 Marlyns Drive, Burpham, Guildford, GU4 7LT, United Kingdom.

BRANTSCHEN Stefan, Early Process Development, Berna Biotech Ltd, Rehhagstrasse 79, CH-3048 Bern, Switzerland.

BROWN Valerie, McKesson BioServices, 9210 Corporate Boulevard, Suite 210, Rockville, MD 20850, USA.

BUTMAN Bryan, GenVec, Inc., 65 West Watkins Mill Road , Gaithersburg, MD 20878, USA.

CARBONE Michele, Loyola University Medical Center, Department of Pathology, Building 112, Room 205, 2160 South First Avenue, Maywood, IL 60153, USA.

CARDINALI Massimo, HJF, 4541 Chesapeake Street, Washington, DC 20016, USA.

CASADEI Jan, National Institutes of Health, 6130 Executive Boulevard, EPN/7111 MSC 7428, Rockville, MD 20852, USA.

CATANZARO Andrew, National Institutes of Health, 40 Convent Drive, Building 40, Room 2504, Bethesda, MD 20892, USA.

CHANDLER Donna, Division of Vaccinnes and Related Products, 1401 Rockville Pike, Rockville, MD 28052, USA.

CHANG Audrey, Biologics and LADS, BioReliance, 4920 Broschart Road, Rockville, MD 20850, USA.

CHIU Yuan-yuan, Strategic Operations, Genentech, Inc., 1 DNA Way, MS 242, San Francisco, CA 94080, USA.

CHO David, Food and Drug Administration/CBER/OCBQ, 1401 Rockville Pike, Rockville, MD 20852, USA.

CHRISTMAN David, E.M. Sergeant Pulp & Chemical Co., Inc., 6 Chelsea Road, Clifton, NJ 07012, USA.

CHUMAKOV Konstantin, Food and Drug Administration/CBER, 1401 Rockville Pike, HFM-470, Rockville, MD 20852, USA.

CICHUTEK Klaus, Paul Ehrlich Institut, Paul Ehrlich Street 51-59, DE-63225 Langen, Germany.

COFFIN John, Tufts University, 136 Harrison Avenue, Boston, MA 02111, USA

COLLAZZA Carmen, Division of Vaccines and Related Products, 1401 Rockville Pike, Rockville, MD 20852, USA.

COWELL James, Biologics Consulting Group, 5001 Weston Parkway, Suite 103, Cary, NC 27513, USA.

CRISE Bruce, SAIC-Frederick, Inc., NCI-Frederick, Post Office Box B, Building 535, 5th Floor, Frederick, MD 21702, USA.

DALEMANS Wilfred, GlaxoSmithKline Biologicals, 89 rue de l'Institut, BE-1330 Rixensart, Belgium.

DETWILER Linda, University of MD, 225 Highway 35, Red Bank, NJ 07701, USA

DOEHMER Johannes, CEO, GenPharmTox BioTech AG, Fraunhoferstrasse 9, DE-82152 Planegg/Martinsried, Germany.

DRAGUNSKY Eugenia, Food and Drug Administration, 1401 Rockville Pike, HFM -470, Rockville, MD 20852, USA.

D'SOUZA Patricia, Division of AIDS/NIAID, 6700-B Rockledge Drive, Room 5130, Bethesda, MD 20892, USA.

DUBOIS Doria, Division of AIDS, 6700 B Rockledge Drive, Bethesda, MD 20892, USA.

DUCHENE Michel, GlaxoSmithKline Biologicals, 89 rue de l' Institut, BE-1330 Rixensart, Belgium.

DUNCAN Paul, Merck Research Laboratories, WP17-101, Sumneytown Pike, West Point, PA 19486, USA.

ECKELS Kenneth, Department of Biologics Research, 503 Robert Grant Avenue, Silver Spring, MD 20910-7500, USA.

EGAN William, OVRR, Food and Drug Administration, 1401 Rockville Pike-HFM 400, Building 29B, Room 1H16, Rockville, MD 20852, USA.

ELTERMAN John A. Food and Drug Administration/CBER/OCBQ, 1401 Rockville Pike, HFM-670, Rockville, MD 20852, USA.

ENAMA Mary, Vaccine Research Center, NIAID, NIH, 40 Convent Drive, MSC3017, Room 2504, Bethesda, MD 20892, USA.

ERNEST Isabelle, GlaxosmithKline Biologicals, 89 rue de l'Institut, BE-1330 Rixensart, Belgium.

ESHKOL Aliza, International Association of Biologicals, Chemin des Aulx 12, CH-1228 Plan-les-Ouates, Swizerland.

ESPESETH David, Espeseth Consulting, 404 Long Leaf Drive, Perkasie, PA 18944, USA.

FABRY Luc, GlaxoSmithKline Biologicals, 89 rue de l'Institut, BE-1330 Rixensart, Belgium.

FANG Meng, The Institute for Control of Pharmaceutical and Biological Products, Tantan Xili #2, Beijing, CN 100050, China.

FIX Alan, National Institutes of Health, 6700 B Rockledge Drive, Bethesda, MD 20817, USA.

GAGNETEN Sarah, Division of Vaccinnes and Related Products, 1401 Rockville Pike, Rockville, MD 20852, USA.

GALINSKI Mark, Merck & Co., Inc. Sumneytown Pike, West Point, PA 19486, USA.

GALLANT Jackie, Gallant Custome Laboratories, Inc., 1425 Bishop Street North, Units 10 and 11, Cambridge, ON NIR 6J9, Canada.

GALVIN Teresa, Food and Drug Administration, CBER/OVRR, 8800 Rockville Pike, Building 29B, Room 3G15, Bethesda, MD 20892, USA.

GAUDRY Daniel, International Association for Biologicals, 46 Inlets Blvd, Nokomis, FL 34275, USA.

GAVIN Denise, Food and Drug Administration, 1401 Rockville Pike, Room 251N, Rockville, MD 20852, USA.

GEIGERT John, Biopharmaceutical Quality Solutions, 3533 Corte Esperanza, Carlsbad, CA 92009, USA.

GOLDENTHAL Karen, Division of Vaccines and Related Products Applications, Food and Drug Administration, 1401 Rockville Pike , Rockville, MD 20852, USA.

GOLDING Hana, Lab of Retrovirus Research, Food and Drug Administration/CBER, 8800 Rockville Pike, Building 29B Room 1A21, Bethesda, MD 20892, USA.

GOLDMAN Mark, Chiron Corporation, 4560 Horton Street, M/S X-100A, Emeryville, CA 94608, USA.

GOLDMAN Neil, National Cancer Institute, 31 Center Drive, Building 31, Room 3A11, MSC 2440, Bethesda, MD 20892-2440, USA.

GOMEZ Phillip, Vaccine Research Center/NIAID/NIH, 40 Convent Drive, MSC-3011, Building 40, Room 5502, Bethesda, MD 20892, USA.

GOMEZ-Yafal Alicia, Therion Biologics Corporation, 83 Rogers Street, Cambridge, MA 2142, USA.

GOOVAERTS Danny, Intervet, Inc., 29160 Intervet Lane, Millsboro, DE 19966, USA.

GORZIGILLA Mario, Merck & Company, Inc., WP28-70, Post Office Box 4, West Point, PA 19466, USA.

GRAMZINSKI Robert, NIAID Divsion of AIDS, 6700 B Rockledge Drive, Room 5256, Bethesda, MD 20892-7626, USA.

GRIFFITHS Elwyn, Biologics and Genetic Therapies Directorate, Health Canada, HPC Building 7, Tunnney's Pasture, Ottawa, KIA-0L2, Ontario Canada.

GRUBER Jack, National Cancer Institute, Executive Plaza North, Suite 5000 Room 5012, Bethesda, MD 20892, USA.

GUINETMORLOT Francoise, Aventis Pasteur, Campus Merieux, 1541 Av. Marcel Merieux, FR-69280 Marcy l'Etoile, France.

HAGIT David, NAID/DEA/SPP, Food and Drug Administration, 6658 Windsor Court #6, Columbia, MD 21044, USA.

HALPERN Jane, ID Biomedical, Inc., 6996 Columbia Gateway Drive, Suite 103, Columbia, MD 21045, USA.

HAN Jing, CBER/Food and Drug Administration, 29 Lincoln Drive, Building 29B, Room 2NN16, HFM 735, Bethesda, MD 20892, USA.

HARMON Maurice, Chiron Corporation, 4560 Horton Street, M/S X-100A, Emeryville, CA 94608, USA.

HARRIS Trish, Johns Hopkins University, 1650 Orleans Street, CRB HM61, Baltimore, MD 21231, USA.

HAYAKAWA Takao, National Institute of Health Sciences, 1-18-1 Kamiyouga, Setagaya-ku, Tokyo 158-8501, Japan.

HENCHAL Laraine, Department of Biologics Reseach, 1401 Rockville Pike, Rockville, MD 20852, USA.

HILL Jack, Henry Jackson Foundation, DAIDS/NIAID/NIH, 6700 B Rockledge Drive, Room 5245, Bethesda, MD 20892, USA.

HIREMATH Meenaxi, National Institutes of Health, 6001 Executive Boulevard, Suite 2141, Bethesda, MD 20892, USA.

HORTON Rex, Solvay Pharmaceuticals, Inc., 901 Sawyer Road, Marietta, GA 30062, USA.

HOSHINO Yasutaka, National Institutes of Health, 50 South Drive, Building 50, Room 6308, Bethesda, MD 20892-8026, USA.

HU-PRIMMER Jean, National Institute of Allergy and Infectious Diseases/DMID, 6610 Rockledge Drive, Bethesda, MD 20892-6603, USA.

JANCZAK Christin, Food and Drug Administration, 7500 Standish Place, HFV-128, Rockville, MD 20855, USA.

JARRETT Ruth, LRF Virus Centre, Univ. of Glasgow, Bearsden Road, Glasgow, G61 1QH, United Kingdom

JAYARE Geetha, Food and Drug Administration, 29 Lincoln Drive, HFM-457, Bethesda, MD 20892, USA.

JOHNSON Eric, ID Biomedical of Maryland, 6996 Columbia Gateway Drive, Suite 103, Columbia, MD 21046-3303, USA.

JOHNSON Robert, 13406 Oriental Court, Rockville, MD 20853, USA.

JOHNSTON Margaret, National Institute of Allergy and Infectious Diseases/Division of AIDS, 6700 B Rockledge Drive, Room 5142, Bethesda, MD 20892-7628, USA.

KAPIDIAN Albert, National Institute of Allergy and Infectious Diseases, NIH, 50 South Drive, Building 50 Room 6306, Bethesda, MD 20854, USA.

KEARNS Daniel, Food and Drug Administration, 1401 Rockville Pike, Rockville, MD 20852, USA.

KERSTEN Alex, Solvay Pharmaceuticals, C.J. Van Houtenlaan 31, NL-1241 AP Weesp, The Netherlands.

KHAN Arifa, Food and Drug Administration/CBER, 8800 Rockville Pike, HFM-454, Building 29B, Room 4NN10, Bethesda, MD 20892, USA.

KIM Sonnie, National Institutes of Allergy and Infectious Diseases, 6610 Rockledge Drive, MSC 6604, Bethesda, MD 20892, USA.

KRAUSE Philip, Food and Drug Administration/CBER, 29 Lincoln Drive 29A/1C16, HFM-457, Bethesda, MD 20892, USA.

KUMAR Dhanya, Food and Drug Administration/CBER, 8800 Rockville Pike, Building 29B, Room 3G11, Bethesda, MD 20892, USA.

LACHENBRUCH Peter, Food and Drug Administration/CBER, 1401 Rockville Pike, HFM-215, Rockville, MD 20852, USA.

LAMBERT Linda, National Institutes of Allergy and Infectious Diseases, 6610 Rockledge Drive, MSC 6604, Bethesda, MD 20892, USA.

LAPPS William, Aventis Pasteur, Discovery Drive, Swiftwater, PA 18370, USA.

LEBRON Jose, Merck Research Laboratories, 770 Sumneytown Pike, Post Office Box 4, West Point, PA 19486, USA.

LEDWITH Brian, Merck Research Labs, WP 45-215, 770 Sumneytown Pike, Post Office Box 4, West Point, PA 19446, USA.

LeMAY Jamie, Acambis, Inc., 90 Shawmut Road, Canton, MA 02346, USA.

LEONHART Christoph, Baxter Vaccine AG, Biomedical Reasearch Center, Uferstrasse 15, AT-2304 Orth/Donau, Austria.

LEON-MONZON Martha, NIH/OD/OAR, Building 2, Room 4W25, Bethesda, MD 20892, USA.

LEWIS Andrew, Food and Drug Administration/CBER, 29 Lincoln Drive, Building 29A IB13, Bethesda, MD 20892, USA.

LEWIS John, Merck & Co., Inc., 770 Sumneytown Pike, Mailstop: WP17-301, West Point, PA 19486, USA.

LI Sean (Xuguang), Health Canada, Tunney's Pasture, Address Locator: 2201C, Ottawa, Ont. K1A 0K9, Canada.

LUBET Martha, National Cancer Institute, NIH, 6120 Executive Boulevard, Suite 450, Rockville, MD 20852, USA.

LUBINIECKI Anthony, Centocor, 145 King of Prussia Rd, Radnor, PA 19087, USA.

LUEBBER Halger, Chiron Vaccines, Emil-Von-Behring Str. 76, DE-35041 Marburg, Germany.

LUM Melissa, Intervet, Inc., 29160 Intervet Lane, Post Office Box 318, Millsboro, DE 19966, USA.

MA Yunkun, Food and Drug Administration, 8800 Rockville Pike, Room 3G11, Building 29B, Bethesda, MD 20892, USA.

MALLET Laurent, Microbiology Platform, Aventis Pasteur, 1541 Av. Marcel Mérieux, FR-69280 Marcy l'E-toile, France.

MALONEY Carol, Food and Drug Administration, 1401 Rockville Pike, Rockville, MD 20852, USA.

MARCUS-SEKURA Carol, Biotechnology Assessment Services, Inc., 7413 Ottenbrook Terrace, Rockville, MD 20855, USA.

MASON Robin, National Institutes of Allergy and Infectious Diseases, 6610 Rockledge Drive, MSC 6604, Bethesda, MD 20892, USA.

MATHIESON Bonnie, National Insitutes of Health/OD/OAR, Building 2, Room 4W27, Bethesda, MD 20892, USA.

MAX Steven, Wyeth BioPharma, One Burrt Road, A2001, Andover, MA 01810, USA.

McGOWAN John, National Institute of Allergy and Infectious Diseases, 6700-B Rockledge Drive, Room 2142 MSC 7610, Bethesda, MD 20892, USA.

McGUINNES Charlotte, National Cancer Institute, Executive Plaza South, 6120 Executive Boulevard, Suite 450 Rockville, MD 20852, USA.

McVITTIE Loris, Division of Vaccines and Related Products Application, 1401 Rockville Pike, Rockville, MD 20892, USA.

McWATTERS Bernard, Food and Drug Administration, 29 Lincoln Drive, Building 29A IL18, Bethesda, MD 20892, USA.

MEDEMA Jereon, Solvay Pharmaceuticals, CJ Van Houtenlaan 36, Post Office Box 900, NL-1380 DA Weesp, The Netherlands.

MIDTHUN Karen, DHHS/Food and Drug Administration/CBER/OD, 1401 Rockville Pike, Suite 200N, HFM-6, Rockville, MD 20852, USA.

MINOR Philip, NIBSC, Potters Bar, Blanche Lane, South Mimms, Potters Bar, Herts EN6 3QG, United Kingdom.

MITCHELL David, The Immune Response Corporation, 5931 Darwin Court, Carlsbad, CA 92008, USA.

MODRIC Tom, Food and Drug Administration, 7500 Standish Place, HFV-128, Rockville, MD 20855, USA.

MONTEYNE Philippe, GlaxoSmithKline Biologicals, Rue de l'Institut 89, BE-1330 Rixensart, Belgium.

MOORE William, Althea Technologies, Inc., 11040 Roselle Street, San Diego, CA 92121, USA.

MOUTERDE Benedicte Aventis Pasteur, 1541 Av. Marcel Merieux, FR-69280 Marcy l' Etoile, France.

MURANO Gene, Genentech, Inc., 1 DNA Way, MS 242, San Francisco, CA 94080, USA.

MURPHY Brian, National Institute of Allergy and Infectious Diseases/LID, 50 South Drive, MSC 800N, Bethesda, MD 20892, USA.

NELLE Tim, Division of Vaccines and Related Products Applications, 1401 Rockville Pike, Rockville, MD 20852, USA.

NICHOLS Amy, Regulatory Affairs, Wyeth Vaccines Research, 401 North Middletown Road, BH 21-129E, Pearl River, NY 10965, USA.

NIMS Raymond, BioReliance, 9900 Blackwell Road, Rockville, MD 20850, USA.

NISHIMURA Tomohiro, The Chemo-Sero-Therapeutic Research Institute, Kyokushi, Kikuchi, Kumamoto 860-0864, Japan.

NORWOOD Laurie, Food and Drug Administration/CBER/OCBQ, 1401 Rockville Pike, Rockville MD 20852, USA.

OBIRI Nicholas, Food and Drug Administration, 1401 Rockville Pike, Rockville, MD 20878, USA.

OKOYE Zebedee, McKesson BioServices, 9210 Corporate Boulevard, Suite 210, Rockville, MD 20850, USA.

ONIONS David, 14, rua Palmeiras, Quinta do lago, PT-8135-024 Almancil, Algarve, Portugal.

OSTERGAARD Erik, Statens Serum Institut, ArtilleriveJ 5, DK-2300 Copenhagan S, Denmark.

PALANSKY Ijay, Wilmer, Cutler, Pickering, Hale and Dorr, LLC, 2445 M Street, NW, Washington. DC 20037, USA.

PANDEY Rakesh, Division of Vaccines and Related Products Applications, 1401 Rockville Pike, Rockville, MD 20850, USA.

PEDEN Keith, Food and Drug Administration/CBER, Building 29A, Room 3D08, 29 Lincoln Drive, Bethesda, MD 20892, USA.

PENSIERO Michael, NIAID/DAIDS/VPRP/PRDB, 6700 B Rockledge Drive, Room 5136, MSC 7628, Bethesda, MD 20892, USA.

PERERA Liyanage, National Institutes of Health, 9000 Rockville Pike, Building 10, Room 4B40, Bethesda, MD 20892-1374, USA.

PERVIN Anklesaria, Targeted Genetics Corporation, 1100 Olive Way, Suite 100, Seattle, WA 98101, USA.

PETRICCIANI John, CancerVax Inc., 2110 Rutherford Road, Carlsbad, CA 92008, USA.

PHOGAT Sanjay, Vaccine Research Center, National Institute of Allergy and Infectious Diseases, 40 Convent Drive, Bethesda, MD 20892, USA.

PIERARD Isabelle, GlaxosmithKline Biologicals, 83 Rue de Institut, BE-1330 Rixensart, Belgium.

PRICE Duane B., National Institutes of Health/NIAID, 6700 B, Rockledge Drive, MSC 7616 Room 3147, Bethesda, MD 20892, USA.

PRIOLA Suzette, Rocky Mountain Laboratories, 903 South 4th Street, Hamilton, MT 59840, USA.

RAEBER Alex, Prionics AG, Wagistrasse 27a, CH-8952 Schlieren, Switzerland.

RANDOLPH Valerie, Wyeth Vaccines Research, 401 N. Middletown Road, Pearl River, NY 10965, USA.

RAO Srinivas, National Institutes of Health, NAID/VRC/LAM, 40 Convent Drive, Bethesda, MD 20892, USA.

RAYCHAUDHRUI Gopa, Division of Vaccines and Related Products Application, 1401 Rockville Pike, Rockville, MD 20852, USA.

REHKOPF Carol, Food and Drug Administration, 1401 Rockville Pike, Suite 200 North, Rockville, MD 20852, USA.

REINEKE Elizabeth, Merck & Company, Inc., P.O. Box 4 WP19-161, West Point, PA 19486, USA.

REPIK-BYRNE Patricia, National Institutes of Allergy and Infectious Diseases, 6610 Rockledge Drive, MSC 4048, Bethesda, MD 20892, USA.

ROBERTSON James, NIBSC, Blanche Lane, South Mimms, Potters Bar, Herts. EN6 3QG, United Kingdom.

ROBINSON David, Merck Research Laboratories, 126 East Lincoln Avenue, Rahway, NJ 07065, USA.

ROBINSON Robin, Health and Human Services, Hubert Humphrey Building, Room 636G, Washington, DC 20102, USA.

ROHWER Robert, Baltimore Research and Education Foundation, Inc., VA Maryland Health Care System 10 North Greene Street, MS151, Baltimore, MD 21201, USA

RUBIN Fran, National Institutes of Allergy and Infectious Diseases, 6610 Rockledge Drive, MSC 6604, Bethesda, MD 20892, USA.

RUIZ Maria Sol, Biotechnology Section, Spanish Medicines Agency, Majadahonda, ES-28220 Madrid, Spain.

SABOURAUD Alain, Aventis Pasteur, 1541 Av. Marcel Mérieux, FR-69280 Marcy l'Etoile, France.

SALUZZO Jean-François, Aventis Pasteur, 1541 Av. Marcel Mérieux, Marcy l'Etoile, FR-69280, France.

SCAPPATICCI Kelly, Acambis, 90 Shawmut Road, Canton, MA 01210, USA.

SCHARTON-KERSTEN Tanya, The Biologics Consulting Group, 5001 Weston Parkway, Suite 103, Cary, NC 27513, USA.

SCHIFF Leonard, Regulatory Affairs, Charles River Laboratories, 10091 Dudley Drive, Ijamsville, MD 21754, USA.

SCHOFIELD Timothy, Merck Research Laboratories, Post Office Box 4, WP37C-305, West Point, PA 19486, USA.

SCHREIBER Martha, Aventis Pasteur, 1755 Steeles West, B93, Toronto, M2R 3T4, Canada.

SCHULTZ Alan, International AIDS Vaccine Initiative, 1600 East Gude Drive, Rockville, MD 20886, USA.

SHAPIRO Stuart, NIAID/NIH, 6700-B Rockledge Drive, Bethesda, MD 20892-7628, USA.

SHAW Alan, Merck & Co., Inc., 770 Sumneytown Pike, West Point, PA 19486, USA.

SHEETS Rebecca, NIH/NIAID, 6700 B Rockledge Drive, Room 5145, Bethesda, MD 20892, USA.

SHEN Angela K., Food and Drug Administration/CBER, 1401 Rockville Pike, HFM-676, Rockville, MD 20852, USA.

SHENG Li, Food and Drug Administration/CBER, 29 Lincoln Drive, Building 29A, Room 3D08, HFM-454, Bethesda, MD 20892, USA.

SIERRA-HONIGMANN Ana, Food and Drug Administration/CBER, 29 Lincoln Drive, HFM-457, Bethesda, MD 20892, USA.

SKIADOPOULOS Mario, National Institute of Alergy and Infectious Diseases/NIH, 50 South Drive, MSC 8007, Building 50/ Room 6511, Bethesda, MD 20854, USA.

SNOY Philip, Food and Drug Administration/CBER, 29 Lincoln Drive, NIH HEM-22, Bethesda, MD 20892, USA.

SOWERS Michael, GenVec, Inc., 65 West Watkins Mill Road, Gaithersburg, MD 20878, USA.

STALDER Anna Barbara, Swissmedic, Swiss Agency for Therapeutic Products, Erlachstrasse 8, CH-3000 Berne, Switzerland.

STOKES Anne, GlaxoSmithKline, South Eden Park Road, Beckenham, Kent, BR3 3BS, United Kingdom.

STRAIN Alastair, School of Biosciences, University of Birmingham, Edgbaston, Birmingham, B15 2TT, United Kingdom.

SUBRAMANIAN Ajit, Medimmune Vaccines, 3055 Patrick Henry Drive, Santa Clara, CA 95054, USA.

SYIN Chiang, Food and Drug Administration/CBER, 1401 Rockville Pike, HFM-675, Rockville, MD 20852, USA.

TAFFS Rolf, Food and Drug Administration, 1401 Rockville Pike, HFM-313, Rockville, MD 20852, USA.

TELLIER Marinka, McKesson BioServices, 9210 Corporate Boulevard, Suite 210, Rockville, MD 20850, USA.

TENTE Bill, Product Development, Neurotech USA, Inc., Blackstone Valley Place, Lincoln, RI 02865, USA.

TOWNS Kim, Food and Drug Administration/CBER, 1401 Rockville Pike, Rockville, MD 20852, USA.

TRUBEY Mac, SAIC-Frederick, NCI-Frederick, Post Office Box B, Building 535, 5th Floor, Frederick, MD 21702, USA.

TSAI P.K., Merck Research Laboratories, Post Office Box 4 WP17-101, West Point, PA 19486, USA.

VANNIER Philippe, AFSSA French Food Safety Agency, BP 53, FR-22440 Ploufragan, France

VICARI Giuseppe, Istituto Superiore Di Sanita (retired), Via Principessa Clotilde, 5, IT-00916 Rome, Italy.

VIOLAY Jean-Nichel, Aventis Pasteur, 1541 Av. Marcel Mérieux, FR-69280 Marcy l'Etoile, France.

VUJCIC Luba, Food and Drug Administration, 1401 Rockville Pike, HFM-475, Rockville, MD 20852, USA.

VYAS Girish, UCSF School of Medicine, 513 Parnassus Avenue, Room S-561 UCSF Box 0134, San Francisco, CA 94143-0134, USA.

WARREN Jon, Division of AIDS/NIAID/NIH/DHHS, 6700 B Rockledge Drive, Room 5124, Bethesda, MD 20892, USA.

WATSON Felicia, Immune Response Corporation, Inc., 680 Allendale Road, King of Prussia, PA 19406, USA.

WEISSMAN Charles, National Hospital, Inst. of Neurology, Dept of Neurodegenerative Diseases, MRC Prion Unit, Queen Square, London, WC 1N 3BG, U.K.

WESSMAN Stephen, USDA/APHIS VS CVB PEL ABRM, 1800 Dayton Avenue, Ames, IA 50010, USA.

WHITEMAN Mary, BioReliance Corporation, 14920 Broschart, Rockville, MD 20850, USA.

WISHER Martin, BioReliance, Innovation Park, Hillsfoot Road, Stirling FK9 4NF, Scotland, U.K.

WOLFF Ruth, Biologics Consulting Group, 19809 Meredith Drive, Derwood, MD 20855, USA.

WONG May, National Cancer Institute, Executive Plaza North, Suite 5000, Room 5012, Bethesda, MD 20892, USA.

WOOD David, World Health Organization, Avenue Appia 20, CH-1211 Geneva, Switzerland.

WRIGHT Peter, M.D. Vanderbilt University Medical Center, 1161 21st Ave. S, D-7235 MCN, Nashville, TN 37232, USA.

YE Zhiping, Food and Drug Administration, 8800 Rockville Pike, Bethesda, MD 20892, USA.

YIN Jin, Shire Biologics, Inc., 30 Bearfoot Road, Northborough, MA 01532, USA.

ZOON Kathryn, National Cancer Institute, 31 Center Drive, Bethesda, MD 20892-2440, USA.

Index of Authors

Developments in Biological Standardization

Vol. 68: Geneva, 1986	Cells – Products – Safety
Vol. 69: London, 1987	Cytokines Laboratory and Clinical Evaluation
Vol. 70: Arlington, VA, 1988	Continuous Cell Lines as Substrates for Biologicals
Vol. 71: Utrecht, 1989	Monoclonal Antibodies for Therapy, Prevention and In vivo Diagnosis of Human Disease
Vol. 72: Annecy, 1989	Progress in Animal Retroviruses
Vol. 73: Shizuoka, 1990	Pertussis: Evaluation and Research on Acellular Pertussis Vaccines
Vol. 74: Bethesda, MD, 1990	Biological Product Freeze-Drying and Formulation
Vol. 75: London, 1990	Virological Aspects of the Safety of Biological Products
Vol. 76: Bethesda, MD, 1991	Continuous Cell Lines - An International Workshop on Current Issues
Vol. 77: Annecy, 1991	Standardization of the Immunopharmacology of Natural and Synthetic Immunomodulators
Vol. 78: Bethesda, MD, 1991	Poliovirus Attenuation: Molecular Mechanisms and Practical Aspects
Vol. 79: Ploufragan, 1992	The First Steps Towards an International Harmonization of Veterinary Biologicals: 1993 and Free Circulation of Vaccines within the EEC
Vol. 80: Heidelberg, 1992	Transmissible Spongiform Encephalopathies - Impact on Animal and Human Health
Vol. 81: Cannes, 1992	Virological Safety Aspects of Plasma Derivatives
Vol. 82: Albany NY, 1993	Recombinant Vectors in Vaccine Development
Vol. 83: Annecy, 1993	Genetic Stability and Recombinant Product Consistency
Vol. 84: Langen, 1993	Non-Target Effects of Live Vaccines
Vol. 85: Santa Fé, NM, 1994	Genetics of Streptococci, Enterococci and Lactococci
Vol. 86: Langen, 1994	Replacement, Reduction and Refinement of Animal Experiments in the Development and Control of Biological Products
Vol. 87: Geneva, 1995	New Approaches to Stabilisation of Vaccines Potency
Vol. 88: Bethesda, MD, 1995	Viral Safety and Evaluation of Viral Clearance from Biopharmaceutical Products
Vol. 89: Rome, 1995	Pertussis Vaccine Trials
Vol. 90: Oslo, 1996	Fish Vaccinology
Vol. 91: San Francisco, CA, 1996	Development of Specifications for Biotechnology Pharmaceuticals Products
Vol. 92: Bergen, 1996	Modulation of the Immune Response to Vaccine Antigens
Vol. 93: Annecy, 1996	Safety of Biological Products Prepared from Mammalian Cell Culture
Vol. 94: Bethesda, MD, 1997	Simian Virus 40 (SV40): A Possible Human Polyomavirus
Vol. 95: Paris, 1997	Preclinical and Clinical Development of New Vaccines
Vol. 96: Washington, DC, 1995	Characterization of Biotechnology Pharmaceutical Products
Vol. 97: London, 1997	Biological Characterization and Assay of Cytokines and Growth Factors
Vol. 98: London, 1997	Inactivated Influenza Vaccines Prepared in Cell Culture
Vol. 99: Strasbourg, 1998	Animal Sera, Animal Sera Derivatives and Substitutes Used in the Manufacture of Pharmaceuticals: Viral Safety and Regulatory Aspects
Vol. 100: Geneva, 1998	A Celebration of 50 Years of Progress in Biological Standardization and Control at WHO
Vol. 101: London, 1998	Alternatives to Animals in the Development and Control of Biological Products for Human and Veterinary Use

Developments in Biologicals

Vol. 102: San Francisco, CA, 1999	Advances in Transfusion Safety
Vol. 103: Annecy, 1999	Physico-Chemical Procedures for the Characterization of Vaccines
Vol. 104: Langen, 1999	Development and Clinical Progress of DNA Vaccines
Vol. 105: Paris, 2000	Progress in Polio Eradication: Vaccine Strategies for the End Game
Vol. 106: Rockville, MD, 2000	Evolving Scientific and Regulatory Perspectives on Cell Substrates for Vaccine Development
Vol. 107: London, 2000	The Design and Analysis of Potency Assays for Biotechnology Products
Vol. 108: Langen, 2001	Advances in Transfusion Safety
Vol. 109: Washington, DC, 2000	Biologics 2000 – Comparability of Biotechnology Products
Vol. 110: Palm Cove, 2001	Orphan Vaccines – Bridging the Gap
Vol. 111: Utrecht, 2001	Advancing Science and Elimination of the Use of Laboratory Animals for Development and Control of Vaccines and Hormones
Vol. 112: Bethesda, MD, 2001	Immunogenicity of Therapeutic Biological Products
Vol. 113: Berlin, 2001	Process Validation for Manufacturing of Biologics and Biotechnology Products
Vol. 114: Ames, IA, 2002	Vaccines for OIE List A and Emerging Animal Diseases
Vol. 115: Bergen, 2002	Laboratory Correlates of Immunity to Influenza – A Reassessment
Vol. 116: Los Angeles, CA, 2003	Development of Therapeutic Cancer Vaccines
Vol. 117: Langen, 2003	Consideration of Alternative Licensing Procedures for Vaccines for Minor Species, Minor Indications and Autogenous/Autologous Products
Vol. 118: Langen, 2003	PDA/EMEA European Virus Safety Forum
Vol. 119: Buenos Aires, 2004	Control of Infectious Animal Diseases by Vaccination
Vol. 120: Bethesda, MD, 2003	Advances in Transfusion Safety
Vol. 121: Bergen, 2003	Progress in Fish Vaccinology
Vol. 122: Bethesda, MD, 2003	State of the Art Analytical Methods for Characterization of Biological Products and Assessment of Comparability

IMPRIMERIE MEDECINE ET HYGIENE GENEVE SUISSE

décembre-2005